Chemistry and our world

CHEMISTRY AND OUR WORLD

MUNDIYATH VENUGOPALAN

Western Illinois University

Harper & Row, Publishers

New York

Evanston

San Francisco

London

Sponsoring Editor: John A. Woods
Project Editor: Ralph Cato
Designer: Gayle Jaeger
Production Supervisor: Will C. Jomarrón

CHEMISTRY AND OUR WORLD

Library of Congress Cataloging in Publication Data

Venugopalan, Mundiyath, Date-
  Chemistry and our world.

  Bibliography : p.
  Includes index.
  1. Chemistry. I. Title.
  QD31.2.V46   540   74-23682
  ISBN 0-06-046819-X

To John T. Bernhard, James R. Connor, and John R. C. Morton
on the occasion of the Diamond Jubilee
of Western Illinois University.

Contents

Preface

. . . this world of ours was made
By natural process, as the atoms came
Together, willy nilly, quite by chance,
Quite casually and quite intentionless
Knocking against each other, massed, or spaced
So as to colander others through, and cause
Such combinations and conglomerates
As form the origin of mighty things,
Earth, sea and sky, and animals and men.

TITUS LUCRETIUS CARUS (96?-55 B.C.**)**
Roman poet and philosopher

Lucretius wrote about the world in the first century before Christ. Now, two thousand years hence, we live in another world—a world of "things" created by man. To understand and to function satisfactorily in this modern world of plastics and pollutants requires that everyone possess some knowledge of the materials that compose our environment and the alterations these materials undergo. The science of chemistry brings us that knowledge. Many of the problems of society, such as the threat of war, overpopulation, famine, air and water pollution, and disease can never be solved without an intelligent application of chemistry (although this alone will not solve them). The intent of *Chemistry and Our World* is both to offer knowledge about chemistry and to delineate some of the beneficial and harmful effects of the uses of chemistry in technology.

This book is written for one-quarter or one-semester courses for nonscience students who may *or may not* want to take chemistry. (This latter group may well be in the majority.) The text covers the very basic principles of chemistry and a variety of topics selected to help students develop an appreciation of the importance of chemistry to mankind. Because most students who take this course have relatively little background in science, the style of presentation is informal and nontechnical. There is enough material provided to permit an instructor to exercise a reasonable amount of choice in course content. Emphasis is placed on helping students realize that chemical processes are constantly going on all around us, and within us. Thus, among the topics included are environmental pollution, the energy problem, and chemical phenomena which are relevant to "life" and "death" as we know it.

I am sure my various prejudices and enthusiasms have colored my presentation, and not all topics presented here may interest all instructors and students. Nevertheless, in their diversity, topics such as famine and pesticides, parasitism and drugs, food additives, illegal drugs, household chemicals, plastics, and environmental pollution should be of interest to most classes.

The questions found within the text are for the student to answer on his own. Some questions require only the usual review of material presented in the text. Others, admittedly more difficult, are of an exploratory nature and provide the student with an opportunity for varying degrees of challenge. Answers to some selected problems are to be found at the end of the book. Many of the suggested projects may be assigned as take-home experiments. These combine the concepts of chemistry presented in this text with the things we see and use in everyday life, and they are designed to help students understand the nature of chemical changes and problems within themselves and their environment. An appreciation of chemistry thus derived should arouse in students an interest and concern that will last long after they finish the course.

An instructor's manual has been written to accompany the text. It includes a set of 50 suggested lecture topics for presenting the material in *Chemistry and Our World*. Objectives and a listing of readings and audio-visual resources for these topics are also provided. There are multiple-choice and true-false questions for each chapter, and, for those courses with a lab, the manual provides a list of pertinent experiments.

The material for a textbook necessarily comes from countless sources and is accumulated from studies and discussions with many fine teachers. The selection of topics and the order of their presentation are, however, influenced by the experience gained from teaching over a period of many years. It is impossible to specifically acknowledge these teachers and students, but I am deeply indebted to them and to their work, from which I have evolved many of the ideas for this book. Acknowledgments to numerous sources which contributed illustrative material are included in the text.

Special acknowledgment and appreciation are due my colleagues and friends at Western Illinois University, whose suggestions were most helpful. Two of my colleagues, Professor David J. Rawlinson and Professor Te-Hsiu Ma, as well as many students, especially Ms. Darlene M. Pedersen, have kindly read the manuscript and made many useful suggestions. Several other colleagues, especially Drs. Donald E. Clark, Lee A. Cross, Thomas C. Dunstan, Fred E. Kohler, and David E. Soule, have contributed helpful suggestions for portions of the subject matter. My friend, Dr. John R. C. Morton, has, through the years, been an inspiration and has contributed immensely to the spirit of this work.

I am also grateful to Dr. Marvin Kientz of California State College, Sonoma; Dr. Armine Paul, University of West Virginia; Dr. Jack Sosinsky of the Loop College, Chicago; Dr. J. J. Spurlock of North Texas State University at Denton; and Dr. Sidney Emerman, Kingsborough Community College of the City University of New York at Brooklyn. They reviewed the manuscript with great care, and their comments and criticisms were most helpful. At Harper & Row, Mr. John Woods, Editor for Chemistry, and Mr. Ralph Cato, the project editor, were solid sources of support and good counsel at all times.

I want especially to thank Ms. Debra M. Smith for her patient and efficient typing of the various forms of the manuscript. Much secretarial help was offered by Ms. Carol J. Amos, Ms. Clara M. James, Ms. Debra J. Riebling, Ms. Ruth M. Vawter, and Mr. Merlin P. Lefler. To them for all their help, my heartfelt thanks and sincere appreciation.

Finally, I wish to express my appreciation of the assistance and encouragement from my wife, Hanneke, and my son, Murali. Without their support and cooperation this book would not have been written.

M. Venugopalan

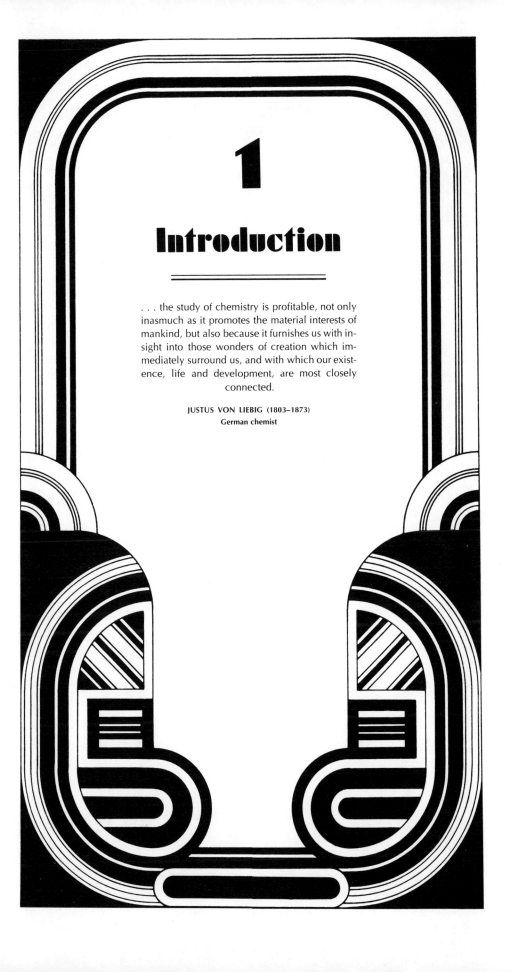

1

Introduction

. . . the study of chemistry is profitable, not only inasmuch as it promotes the material interests of mankind, but also because it furnishes us with insight into those wonders of creation which immediately surround us, and with which our existence, life and development, are most closely connected.

JUSTUS VON LIEBIG (1803–1873)
German chemist

From time immemorial human beings have endeavored to seek and attain knowledge of the universe and to explain natural phenomena. The knowledge systematized from such endeavors is the subject of study known as *science.* Most of what we call scientific knowledge has been accumulated from observations, identifications, descriptions, experimental investigations, and theoretical explanations of natural phenomena in our environment. The materials of our planet earth that surround us make up our *environment,* sometimes called the *terrestrial environment.* It is safe to say that science originated with man's curiosity about his environment. *Technology,* the product of science, developed out of human need and desire to live comfortably and compatibly with the surroundings.

Our attempts to understand nature have been cumulative. By this we mean that in large part we build on knowledge our predecessors have gained; that is to say, we begin where others have left off. The volume of scientific knowledge recorded through human history eventually became so enormous that it had to be divided into somewhat distinct but arbitrary branches. One such branch of scientific knowledge is called *chemistry* (Fig. 1.1).

Chemistry deals with the materials of which our vast universe is composed and the transformations they may undergo. Until recently, our chemical knowledge has been confined mostly to materials on earth and their changes rather than to far away planets of the solar system. Our quest for knowledge has taken an adventurous turn in recent years. Through the technology provided by science, we have been able to explore the material nature of the moon.

CHEMISTRY — ITS DEVELOPMENT

History tells us that the knowledge of chemistry grew slowly over the centuries. Perhaps chemistry took its first steps when primitive people learned to use *fire* to warm themselves and to cook their food. In learning to make fire they discovered charcoal and subsequently found that certain rocks when heated with charcoal could liberate metals such as copper and tin. Following the exploitation of fire and metals, they learned to use fruit juices to color their bodies and garments, bright-colored clays to make up their faces and bodies, bark juices to tan and preserve the skins of their animals, and brews from the leaves, roots, or barks of trees to cure their illnesses. Many of their discoveries were perhaps the result of accidental combination of the ingredients.

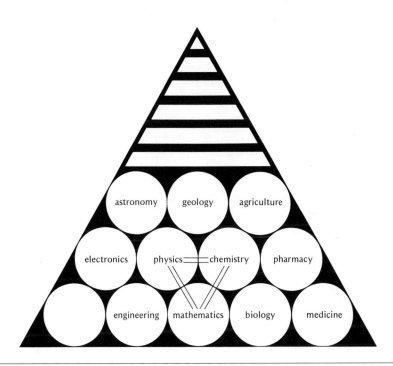

FIGURE 1.1

Some branches of science. Mathematics, physics, and chemistry are considered to be the three basic sciences. All other pure or applied sciences draw upon them in varying degrees.

Various articles made of metals, glass, pottery, and ceramic have been found in excavations and tombs, indicating that civilizations dating back to 4000 B.C. possessed a practical knowledge of chemistry. To these civilizations chemistry was an art to be applied to the manufacture of desired articles.

At one time or another various civilizations have attempted to speculate on the basic qualities of their environment. In direct contrast to the practical aspects of the chemical arts were the speculations of the early Greek and Roman philosophers on the nature of matter — the material things of the earth. About 400 B.C. the Greek philosophers Leucippus and Democritus presented the view that matter is discontinuous; that is, it is composed of particles that are incapable of further subdivision

3

FIGURE 1.2

Democritus, Greek philosopher who speculated on the atomic nature of matter in the fifth century B.C.

and too small to be discernible with the eye. They expounded the "atomic" (derived from the Greek *atomos,* meaning "not cut") nature of matter.

About five centuries later, the Roman poet Lucretius tried to explain the coming into being of matter in well-defined forms. The behavior of solids was attributed to particles hooked together, of liquids to particles with few hooks, and of gases to particles with no hooks (Fig. 1.3).

The Greek philosopher Aristotle, on the other hand, believed that matter is made up of one primordial substance that is continuous and capable of infinite subdivision. The endless kinds of substances known to man were explained as due to different proportions of four "elements"—earth, water, air, and fire—and four "principles"—hot, cold, dry, and moist (Fig. 1.4). From this view of matter came the conviction that some forms of matter could be turned into other forms merely by changing the proportions of these four elements. Thus arose the science of *alchemy,* the practice of which spread from Greece into every part of the civilized world and persisted for nearly 2000 years. The major goals of the alchemists were the transmutation of lead into gold, the discovery of the panacea, and the preparation of the elixir of longevity. Although these goals were not achieved, the alchemists, nevertheless, laid the foundations for the chemical laboratory and the experimental approach to chemistry.

The transition from alchemy to chemistry began in the seventeenth century with the publication of the book *The Sceptical Chymist* by the English scientist Robert Boyle. In this book, Boyle rejected the Aristotelian concept of four elements and arrived at his own idea that elements are simple bodies of matter that cannot be resolved into other bodies of matter, and of which all other bodies of matter are composed. By that time the alchemists were already aware of the need for new concepts. In the following hundred years, probably the most important development was the formulation of the now defunct "phlogiston theory," according to which the principle of fire (phlogiston) is regarded as a material substance. (see Chapter 2). However, the guidelines of chemistry, as it is now known were not set forth until 1789. In that year, in his book *Elements of Chemistry* the French chemist Antoine Lavoisier, who is considered the "father of chemistry," wrote:

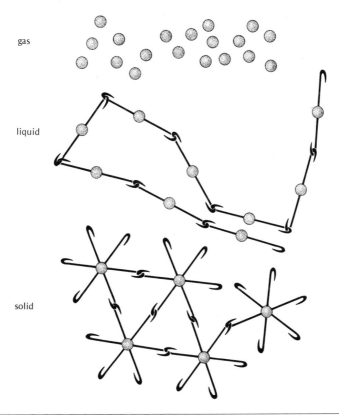

gas

liquid

solid

FIGURE 1.3

The ancient Roman view of matter.

We must lay it down as an incontestable axiom, that in all the operations of art and nature, nothing is created; an equal quantity of matter exists both before and after the experiment . . . upon this principle, the whole art of performing chemical experiments depends.

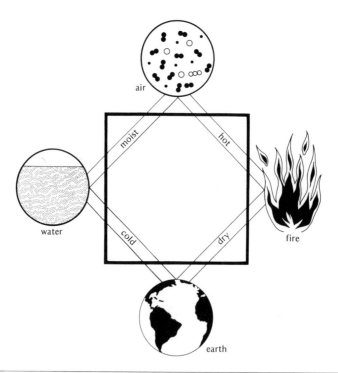

FIGURE 1.4

The ancient Greek view of matter.

From then on chemistry was gradually accepted as a newly developing science whose objectives were quite different from those of alchemy.

At the beginning of the nineteenth century, largely through the efforts of the English mathematician John Dalton, the age-old idea of atoms as the ultimate particles of matter was elaborated. The great conceptual advance represented by Dalton's atoms provided a basis for understanding chemical combination. Many new chemical elements were discovered by the English chemist Humphrey Davy; atomic weights of many elements were accurately determined by the Swedish chemist Jakob Berzelius, to whom we owe our present scheme of chemical symbols (see Chapter 3); the periodic classification of elements was made by the Russian chemist Dmitri Mendeleev, a classification that still forms the basis of systematic descriptive chemistry (see Chapter 5).

The first decade of the present century saw startling scientific discoveries, many of which were to have a profound effect on the development of chemistry. Among the most significant were the discovery of radioactivity, the probe into atomic structure, and the identification and separation of isotopes (see Chapter 4). These discoveries led to a broader understanding of the nature and behavior of matter, which, in turn, laid the foundations on which our present age of advanced technology is largely based.

CHEMICAL TECHNOLOGY—ITS TRIUMPHS AND PROBLEMS

We know from experience that technological progress is always obtained at some cost, but the cost may not be obvious at the outset. The growth in large-scale chemical technology has produced a very large number of effects, both beneficial and harmful, on the environment and on mankind. Here we examine a few cases to show just how complex these effects can be and suggest some remedial measures.

FROM COMBUSTION TO AIR POLLUTION AND THE ENERGY PROBLEM

With time our knowledge of fire advanced so much that today we have an entire field of chemical knowledge known as *combustion,* the burning of substances by fire. Besides heating our homes and cooking our food, combustion helps us to transport ourselves in automobiles, airplanes, and rockets, to prepare many of the material things we need for everyday living, and to destroy refuse and waste materials. However, the technology that developed out of man's understanding of combustion is not without harmful effects on the environment and on mankind. For example, the incomplete combustion of coal has contributed to air pollution (Fig. 1.6), sometimes with disastrous consequences. The poisonous fogs of Donora, Pennsylvania, in 1948 and those of London in 1952 when 4000 "excess deaths" were reported showed that

FIGURE 1.5

Antoine Lavoisier, French chemist of the late nineteenth century, is regarded as the founder of modern chemistry.

air contaminants could accumulate to acutely hazardous levels. The products of incomplete combustion of gasoline in a large number of automobiles have contaminated the atmosphere in densely populated areas. The health effects are complicated because of the action of one contaminant possibly enhancing the effect of another contaminant (*synergistic* action) and because of a wide range of effects upon the young and the aged and upon those suffering from chronic diseases.

FIGURE 1.6

An example of air pollution. Besides contaminating the atmosphere it represents an inefficient use of our limited energy resources. (Photograph courtesy of Robert C. Holt, Jr., *St. Louis Post Dispatch.*)

Besides air pollution, the excessive and inefficient use of fuels (combustible materials) such as coal, natural gas, and oil for comfort and luxury has produced an energy problem. Every day the average American uses nearly 4 gallons (gal) of oil, 300 cubic feet (ft^3) of natural gas, 15 pounds (lb) of coal, and smaller amounts of other sources of energy. This is substantially more than the average energy use of other industrial countries and eight times as much as the world average. In our desire to achieve rapid economic growth and higher standards of living, we have been prodigal with fuel resources that once seemed limitless. Recent developments have made us all more conscious that energy resources, as well as air and water, are finite. This new understanding has generated the need for achieving a balanced, more satisfying way of life—without having to choose between running out of fuel or running out of clean air and water.

FROM CHEMICALS TO WASTES AND DEATHS   Whether we realize it or not, the rise in the standard of living has paralleled the development and growth of chemistry. The myriad of substances we use every day for our existence and comfort are the products from chemical processing of natural resources such as coal, natural gas, petroleum, and minerals which are present on earth. These substances include the textiles, rubber, plastics, metals, and glasses that are needed for our clothing, shoes, utensils, automobiles, housing, weapons, and so forth; the chemical fertilizers and pesticides that have helped to increase food production; the soaps and detergents that have helped to keep us and our clothing clean; medicines and drugs that have kept us alive; and, the many consumer products that have made life easy and convenient for us. The number and quantity of chemical products have increased daily. So have the waste materials originating from the use of these products. It is estimated that in the United States alone garbage collectors pick up nearly 50 billion metal cans, 30 billion plastic containers, 30 billion bottles and jars, and 30 million tons of paper each year. Although much of this waste material is reclaimed and reprocessed, the magnitude of the waste problem is steadily increasing and presents a major source of *esthetic pollution* (Fig. 1.7).

FIGURE 1.7

Esthetic pollution. Simple garbage has become a detriment to the quality of our daily lives. (Photograph courtesy of Robert C. Holt, Jr., *St. Louis Post Dispatch*.)

Chemical contamination of the environment is the inevitable consequence of an expanding industrial society. It has been estimated that more than 500 new and potentially toxic chemicals are produced each year on a scale large enough so that traces of them enter the environment through air, water, and indirectly (or directly) into food. Just consider your immediate environment—the household—alone. Until about 50 years ago, lye, tincture of iodine, rat poison, flypaper, and kerosene comprised the majority of house chemicals that could be fatal if swallowed. Today there are dozens of toxic substances around the house: silver polish, hair preparations, weed killers, pesticides, various cleansers, and a host of drugs. It is no surprise that there are over 10,000 fatal poisonings in the American home each year.

FROM CHEMICAL FERTILIZERS AND PESTICIDES TO FOOD ADDITIVES AND WATER QUALITY With an ever-increasing growth of population, chemical fertilizers and pesticides have an almost ubiquitous and quantitatively expanding use in the agricultural lands of the world. The unequivocal benefit to mankind has been the increased food production. The principal deleterious aspect of the increasing use of fertilizers is their leaching into drainage water. Some of the chemicals present in fertilizers, which find their way into drinking water, have caused harmful effects in infants and livestock. They have also caused the ecological aging of natural waters such as Lake Erie, whereby fish and other aquatic animals may find it impossible to exist. Widespread use of chemical pesticides has prevented insects from taking over the earth and increased farming efficiency. However, public criticism, triggered by the appearance in 1962 of Rachel Carson's book *Silent Spring,* has gradually brought the use of pesticides under control. The persistence and movement of many pesticide chemicals in plants, animals, and soil have led to widespread contamination of our environment and serious ecological problems. In fact, during the past 50 years, public health emphasis on food quality has almost completely reversed from concern about inadequate supply of vitamins and minerals to preoccupation with the presence of trace amounts of pesticides and fertilizers.

One of the major commodities in our daily lives is food, of course, and intentional food additives comprise an important part of the chemical industry. These include preservatives, flavorings, colorings, artificial sweeteners, emulsifiers, mold inhibitors, and a host of others. The Food and Drug Administration checks the toxicity of substances added to foods prior to their approval for human consumption, but there are instances when illnesses have been attributed to the chemicals used as food additives. Fortunately, the human organism is equipped with an efficient array of metabolizing enzymes which have arisen through countless generations of exposure to unwanted chemicals in natural foods. Indeed, food quality will continue to reflect population pressures that give rise to new technologies of crop production, food processing, preservation, and storage, all of which introduce extraneous chemicals in food substances. It will also be modified by the expanding industrialization with its host of chemical effluents that inevitably introduce into food, traces of a vast variety of environmental contaminants.

Another major commodity in our daily life is drinking water. The spread of epidemics of typhoid and enteric diseases through contaminated drinking water has been repeatedly demonstrated. In many parts of the world chlorination (a chemical treatment) of drinking water has become a standard public health practice. Fluoridation of drinking water for the prevention of dental cavities provides an interesting example of a violently controversial problem in environmental contamination in contrast to the public acceptance of water chlorination. Almost nothing is known of the public health effects of the trace contamination of drinking water by the thousands of industrial effluents, by pesticides, fertilizers, detergents, and by heavy metals.

FROM CHEMICALS TO MEDICINES AND THE DRUG PROBLEM Many of us are alive because of chemical products used as medicines, drugs, anesthetics, and antiseptics. Modern surgery is dependent on chemicals such as anesthetics and antiseptics. We use chemicals for relief from fever, pain, and nervous tension. We depend on chemicals for the selective destruction of invading organisms — bacteria in our food and germs in our bodies. The sulfa drugs developed in the 1930s and the antibiotics developed in the 1940s proved to be powerful chemicals in the fight against germs. Many widely used antibiotic preparations today are relatively new and could not be obtained a decade ago. Oral contraceptives and the psychochemicals are among the latest drugs to reach wide-scale use. Psychochemicals, which are found in a third of all prescriptions written, include the antidepressants for the thousands of people plagued by a brooding sense of despair, guilt fear, and insomnia, and the tranquilizers for hypertension and various forms of mental anxiety.

Until vaccines were developed, the dreaded diseases were smallpox, cholera, diphtheria, polio, and tetanus. Until sulfa drugs and antibiotics were developed, infections usually led to the loss of life or limb. The great strides being made in modern medicine are largely the result of an increasing knowledge of the chemical processes occurring in the complex chemical system of the human body. But as is so often the case the triumph is not an unmixed one. The mechanism of how drugs act is still largely unknown. Drug abuse can produce harmful effects. For example, the barbiturates used medically for sedation and sleep are dangerous when ingested

FIGURE 1.8

An example of water pollution in the Mississippi river. (Photograph courtesy of Robert C. Holt, Jr., *St. Louis Post Dispatch.*)

along with alcohol due to a synergistic effect. The depressant effect of the barbiturate is enhanced up to 200 times when taken with or after drinking alcoholic beverages. It has probably led to many deaths, usually reported as due to "an overdose of sleeping pills." Our society has many drug addicts and a flourishing illegal market in heroine, morphine, methadone, opium, and LSD. Addiction results from emotional desire, physical need, and tolerance for the drug. Because drugs are chemicals, some knowledge of their chemistry should be helpful.

CHEMICAL PROCESSES IN AND AROUND US

To appreciate chemistry and its technology we need only examine what is constantly going on all around us and within us. All the action we see, feel, or hear, the materials we use in our daily activities, and our own life processes are examples of chemical processes. In fact, if all chemical processes were somehow to cease, life itself would cease, for life processes consist of a marvellously complex and wonderfully integrated and controlled series of chemical changes. It is safe to say that our lives are made longer, healthier, and more convenient and pleasant because of our

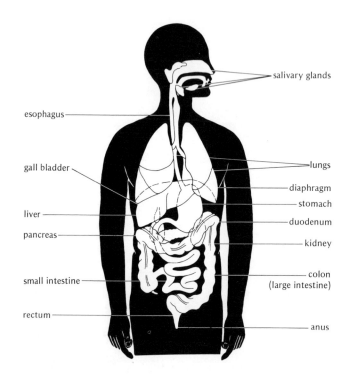

FIGURE 1.9

The human body is a self-propelled chemical factory producing its own energy by the consumption of chemicals. Its reactors are themselves collections of chemicals and operate day and night at a constant temperature of 98.6°F, producing a host of other chemicals—some simple, some extremely complex—from a wide variety of food materials. Unlike man-made chemical factories, the human body can detect its own malfunctions and repair some of its component parts, can detect environmental changes and adapt to them, and can even exert some control over its own environment.

chemical knowledge. In short, chemistry and its technology are part and parcel of us and our environment.

Although we all benefit from chemistry and its technology, we should not ignore the fact that our environment is being altered and can become polluted by some chemical applications. A knowledge of the chemical processes that cause pollution is required to understand environmental changes and to develop the appropriate chemical technology for controlling and preventing pollution. The challenge of our time is the preservation of maximum standards of environmental quality without unduly curtailing the materialistic approach to human welfare. This challenge is immense, for human contributions to environmental contamination are already monumental. To meet this challenge adequately, it is necessary to have some knowledge of the chemical world in which we all live.

THE CHANGING ROLE OF CHEMISTS IN OUR SOCIETY

There was a time when chemists were concerned mainly with the making of materials. We all know that over the years the chemical industry has played an important role in raising the standard of our living. The knowledge and skill of the chemist has increasingly been directed toward the achievement of social goals such as better health, more and better food, housing, and clothing. Pressed by the needs of modern society, many chemists and chemical industries have shifted from the making of materials to the solving of human problems such as air and water pollution, waste treatment, population explosion, depletion of nonrenewable resources, and recycling of materials. For example, some industrial chemists are studying the chemical nature of air pollutants so that the necessary changes can be made in the designs of automobile engines and power plants to eliminate or at least alleviate the air pollution problem. Some others are focusing attention on recycling of materials such as paper, glass, metal, and plastic. Many separation techniques, originally developed to produce pure materials, now find application in solving environmental problems.

Increasingly, the voting population is called upon to make decisions about matters such as prohibition of alcohol, fluoridation of water, building of nuclear power plants, and many others. A knowledge of chemistry is vital to an understanding of the problems involved in such matters. The aim of this book is to provide its readers with that knowledge of chemical principles so that they may make intelligent (and not political) decisions on matters that involve life and death.

QUESTIONS

1.1   What is science?
1.2   How does technology differ from science?
1.3   What is chemistry?
1.4   What is meant by environment?
1.5   Why are chemistry and our environment inseparable?
1.6   Name some discoveries by primitive man that are of a chemical nature.
1.7   What evidence, if any, is there to indicate the existence of chemical industries in the prehistoric period?

1.8 Compare and contrast the views of Democritus and Aristotle on the nature of matter. Supplement the discussion with your own knowledge about material things.

1.9 What is alchemy? What led to its downfall?

1.10 Contrast the Aristotelian concept of the elements with that of Robert Boyle.

1.11 What would you say was Lavoisier's most important contribution to chemistry?

1.12 Give some examples of nineteenth-century chemical investigations.

1.13 What is meant by combustion? Of what use is it?

1.14 List some of the advantages and disadvantages of using chemical fertilizers and pesticides.

1.15 List some dangerous drugs
a. you are familiar with
b. you have heard about

1.16 Outline one application of chemistry with which you are familiar and discuss both advantages and drawbacks for mankind.

1.17 In what way has the role of chemistry in our society changed in recent years?

1.18 How can chemistry be made useful in solving human problems?

2
Methods and measurements

The earth was made so various, that the mind of
desultory man, studious of change and pleased
with novelty, might be indulged.

WILLIAM COWPER (1731–1800)
English poet

The discovery of new phenomena and the making of new materials are sometimes accidental, but these accidents happen mostly to people who are scientists or are being trained in science. The making of fortunate and unexpected discoveries by accident is called *serendipity*. Examples of serendipity in chemistry include the discovery of X-rays and radioactivity (see Chapter 4), of Bakelite, Teflon, celluloid, and artificial silk or rayon (see Chapter 10), and even artificial sweetening agents and some drugs (see Chapter 12), to mention a few. The accidental spilling of a mixture of sulfur and rubber on a hot stove led the American inventor Charles Goodyear to discover the process of *vulcanization* (named after Vulcan, the Roman god of fire and forge). The modern era of psychochemistry may be traced to the accidental ingestion of LSD in 1943 by the Swiss chemist Albert Hofmann.

More often, the making of new materials is the result of logical procedures which include systematic work and objective interpretation by chemists. Following careful planning (and considerable pondering as well) known materials are decomposed or brought together under carefully controlled conditions for the synthesis of new materials. The myriad of substances that we use in everyday life are the products of such logical procedures. All such procedures share the so-called *scientific method* and a *system of measurements.* The development of the scientific method in the late seventeenth century was one of the important factors responsible for the transition from alchemy to chemistry as it is now known.

THE SCIENTIFIC METHOD

In the scientific method, *observation* is the crucial step. Observations about nature or from laboratory experiments that can be reproduced are termed *scientific facts*. Often a small number of scientific facts forms the basis of a *hypothesis*. We are all used to making guesses, and a hypothesis is nothing more than a reasonable guess. It is sometimes used as a basis for further reasoning or for planning experiments to gather more facts. When a large number of related facts is available they can be summarized into concise statements called *laws* or laws of nature. Finally, after various laws have been formulated, it is sometimes possible to propose a *theory*. A theory is an explanation based on facts and reasoning. It is the ultimate in the scientific method of explaining the facts and laws of nature.

The above-described scientific method can be elucidated using a common and ordinary human experience — the movement of things. We see and experience motion in many of our activities, even when we walk or drive to the chemistry building for a chemistry class. Because the movement of things can be observed and reproduced, we can call it a fact. Once we have learned facts about motion our curiosity raises questions as to what causes motion. At this point the scientist departs from observations and begins to make guesses. He hypothesizes that all motion begins with the forces that are at work in nature. He pushes and pulls things to put them in motion or to stop their motion. In the process he gathers a number of related facts which, when summarized concisely, form the laws of motion. Finally, he develops a theory of motion and the forces that cause it.

Theories are subject to revision and rejection if sometime or other they fail to explain new facts and laws that are obviously related. In other words, the scientific method is one that continues but never reaches completion. The development of man's understanding of combustion is a "case history" that demonstrates the interplay among thought, observation, and experiment in scientific endeavor and shows how laws and theories (or models) can be modified or discarded as a result of new experience. At the beginning of the eighteenth century, the formation of ashes from the burning of wood and paper, of soft residues (called *calx*) from the heating of metals in air, of metals from the heating of soft powders with charcoal, and other similar phenomena could not be imagined to take place without the addition or subtraction of some substance. Therefore, the German chemists Johann Becher and Georg Stahl hypothesized the existence of an invisible substance that all combustible material possessed. They named it *phlogiston* and developed a theory that during combustion phlogiston escapes. The theory could explain diverse phenomena such as respiration (giving off phlogiston by a living organism), death (the organism saturates the air with phlogiston from his lungs), the extinguishing of a flame in an enclosed space (the air becoming saturated with phlogiston), the exploding of gunpowder (it is phlogiston-rich), and the small amount of residue on burning charcoal (it loses phlogiston and it is nearly pure phlogiston).

The phlogiston theory dominated scientific thinking for about a hundred years. When chemists began *quantitative* measurements, it became clear that the phlogiston theory could not explain the changes in weight that occur when certain metals are heated or burned in air to form calxes. Following the discovery of free oxygen in air, the French chemist Antoine Lavoisier proved that combustion is the reaction involved when oxygen combines with substances and that combustion and respiration are essentially alike, the difference being essentially in their rates.

17

QUESTIONS

2.1 What is serendipity? Give one example from outside reading.
2.2 Distinguish between the terms hypothesis, law, and theory.
2.3 Is there a single scientific method? Use one of your experiences to prove or disprove it.
2.4 What is the phlogiston theory? What led to its downfall?
2.5 Use your experience to compare and contrast the processes of combustion and respiration.
2.6 Classify each of the following statements as a fact, a law of nature, a hypothesis, or a theory. Explain.
   a. The earth is more or less spherical in shape.
   b. Water tends to run downhill.

SUGGESTED PROJECT 2.1

Observing things (or What can you see for yourself?)

Make a list of as many chemical products in your household as possible. Categorize them as cleansers, cosmetics, drugs, fertilizers, pesticides, and so forth. Exclude food items. Compare your list with the lists of two classmates. See what you missed?

SUGGESTED PROJECT 2.2

Thinking about things (or a shot in the dark)

Ask a friend to fill a shoebox (or similar size box) partially with various materials (rocks, pieces of wood and metal, plastics, etc.), seal off the box, and place it before you. Without any consultation, describe, as best as you can, the contents of the box. Can you now imagine the dilemma of the early scientists when they tried to explain things?

MEASUREMENTS IN CHEMISTRY

Chemistry deals not only with the qualities of materials but also with their quantities. We have already noted the significance of measurements in the overthrow of the phlogiston theory. Precise measurements are a necessary part of much chemical laboratory work and have important applications in industry, medicine, and agriculture. For example, measurements enable us to know the composition (and hence the fertility) of farm soil, the composition (and hence the purity) of a drug synthesized in the laboratory, or the composition (and hence the strength) of an alloy manufactured in a factory.

The units of measurement used in the chemical laboratory are different from those commonly used in everyday life in our country. At one time or the other we all ask questions such as how long, how big, and how heavy? The man on the street in North America would answer such questions in units of feet, cubic feet, and pounds, respectively, which are familiar to us as the units of the *English system*. However, in the chemical laboratory we would answer such questions in units of meters, cubic meters, and kilograms, respectively, which are the units of the *metric* or *International system*. The latter system is widely used in Europe and Asia and gradually is being adopted by nearly every country throughout the world. Unlike the English system, the metric system is a decimal system in which prefixes are used to designate fractions or multiples of the standard unit. These prefixes are italicized in Table 2.1, which lists the commonly used metric units and their abbreviations.

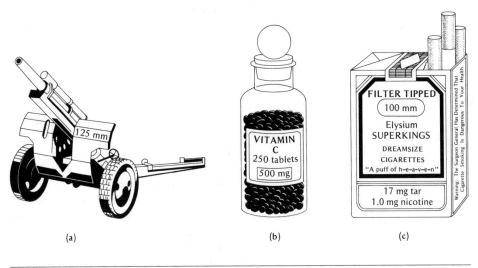

(a) (b) (c)

FIGURE 2.1

Some metric measurements used in the United States. (a) 125-mm Howitzer, (b) a medicine bottle showing both English and metric units on the label, and (c) 100-mm "king-sized" cigarettes.

Even in North America, there already are some metric measurements in general use (Fig. 2.1). These include the 125-millimeter (mm) Howitzer, 35-mm photographic slides, and 100-mm "king-sized" cigarettes. Some prescriptions of drugs are given in milligram dosages, and many food items include both English and metric units on the labels. At this writing two industries, the pharmaceutical and the photographic, are already wholly on metric standards. And some other industries have decided to adopt the metric system.

TABLE 2.1
Metric units and unit prefixes

Quality	Units		
length	*kilo*meter	km =	1000.0 m
	meter	m =	1.0 m
	*centi*meter	cm =	0.01 m
	*milli*meter	mm =	0.001 m
volume	liter	liter =	1.0 liter
	*milli*liter	ml =	0.001 liter
mass	*kilo*gram	kg =	1000.0 g
	gram	g =	1.0 g
	*milli*gram	mg =	0.001 g

What are the standard units in the metric system? The standard unit of length, the *meter* (abbreviated m) was originally defined as one ten-millionth of the distance from the North pole to the equator. However, developments of the space age forced the change to the current definition according to which the meter is 1,650,763.73 times the wavelength of the orange-red light in the spectrum of the gas krypton (see Chapter 4). The meter is equal to 39.37 inches (in.) or a little more than a yard. For measuring the small objects of our environment, usually one-hundredth of a meter, the *centimeter* (abbreviated cm), is used conveniently as the unit of length; for extremely small objects, usually one hundred-millionth of a centimeter, the *Ångström* (abbreviated Å), is used.

Since material objects occupy space, the size or volume of an object can be expressed in terms of cubic length (*cubic centimeter*; abbreviated cc or cm³). We often refer to measurements of volumes in everyday life. For example, we are familiar with units such as pints, quarts, or gallons. In chemistry, the unit of volume is the *liter*. The liter is defined as a volume equal to 1000 cm³. Smaller volumes are measured in terms of one-thousandth of a liter, the *milliliter* (abbreviated ml) and one-millionth of a liter, the *microliter* (abbreviated μl).

The standard unit of mass, the substance of material objects, is the *kilogram* (abbreviated kg). The kilogram was originally defined as the mass of 1000 cm³ of water at 4 degrees centigrade (°C) (see Chapter 3). A piece of platinum metal of precisely this mass, which is kept at the International Bureau of Weights and Measures in France, is the present international standard for the kilogram. Often the *gram* (abbreviated g), which is one-thousandth of a kilogram is used as the basic reference unit, even though it is not the original basic defined unit. The mass of small objects is measured in thousandths of a gram, milligrams (abbreviated mg) or in millionths of a gram, micrograms (abbreviated μg).

Besides the ease of making decimal manipulations, an advantage of the metric system is that the units of mass, volume, and length are related.

Outside of the scientific laboratories, many measurements in the United States are still made in the English system of units. These measurements can be converted to metric units by the use of the relationships given in Table 2.2. A comparison of the metric and English systems of measurements is shown in Fig. 2.2. Some examples of unit conversions are given below.

TABLE 2.2
Some English–metric equivalents

Quality	Equivalents	
length	1 in =	2.54 cm (exact)
	1 ft =	30.0 cm
	1 yd =	0.91 m
	1 mi =	1.61 km
volume	1 teaspoon =	4.0 ml
	1 tablespoon =	15.0 ml
	1 fl. oz =	30.0 ml
	1 cup =	240 ml
	1 pt =	470 ml
	1 qt =	950 ml
	1 gal =	3.8 liters
mass	1 avoir oz =	30.0 g
	1 lb =	454 g

EXAMPLE 2.1

A European athlete wishes to compete in a cross-country footrace of nearly 26 miles (mi) during the Olympic games. How many kilometers should he race during practice sessions?

From Table 2.2, 1 mi = 1.61 km. Therefore, the solution to the problem is

$$26 \text{ mi} \left(\frac{1.61 \text{ km}}{1 \text{ mi}} \right) = 42 \text{ km}$$

Which is the larger unit, the mile or the kilometer?

EXAMPLE 2.2

A Soviet satellite orbits the earth at 1.61×10^4 km per hour (hr). What is its speed in miles per hour?

The answer is

$$1.61 \times 10^4 \text{ km} \left(\frac{1 \text{ mi/hr}}{1.61 \text{ km}} \right) = 1 \times 10^4 \text{ mi/hr}$$

EXAMPLE 2.3

An expensive drug costs $10.00 for 1 g. What is the cost per pound?

From Table 2.2, the conversion relation is 1 lb = 454 g. Thus, the answer is

$$\left(\frac{\$10.00}{1 \text{ g}} \right) \times \left(\frac{454 \text{ g}}{1 \text{ lb}} \right) = \$4540/\text{lb}$$

Note how the units of gram cancel in the calculation.

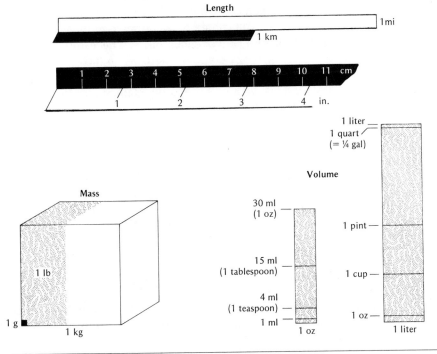

FIGURE 2.2

A comparison of the metric and English systems of measurement.

EXAMPLE 2.4

In Europe gasoline costs about 25 cents a liter. How much would it cost in dollars to fill a 20-gal gasoline tank?

Since the conversion relation is 1 gal = 3.79 liters (see Table 2.2), the answer is

$$20 \text{ gal} \times \left(\frac{3.8 \text{ liters}}{1 \text{ gal}}\right) \times \frac{25\cancel{c}}{1 \text{ liter}} \times \frac{\$1}{100\cancel{c}} = \$19.00$$

Data obtained from measurements cannot be exact because the number of digits expressed in a measurement depends on the limits of the measuring instrument used. Suppose we measure the length of an object and record it as 1.50 m. It is then understood that the measurement is known to the nearest one-hundredth of a meter which is the smallest unit of measure on the ruler. All of the digits in a measurement including the last digit are called *significant digits*. Thus, in the above measurement there are three significant digits. The following rules apply to numbers obtained from measurements:

1. Zeros between nonzero digits and all nonzero digits are significant. For example, the number 5046 has *four* significant digits.
2. Zeros to the left of nonzero digits are not significant, but are used to indicate the position of the decimal point. For example, 0.059 has *two* significant digits.
3. Zeros at the end of a number are significant only when indicated. For example, unless otherwise written, 5540 has only *three* significant digits.

In calculating quantities, the available data may vary in precision; that is, different quantities may have different numbers of significant digits. Therefore, a general rule is that a calculated result should have the same number of significant digits as there are in the least precise item of data. Thus, it is necessary to round off a result, that is, to express fewer digits, by dropping off digits on the right. The following rule applies for rounding off numbers: If the first digit following the digits to be retained is greater than 5, the last digit of those to be retained is increased by one; otherwise, the digits to be retained are not altered.

EXAMPLE 2.5

Calculate the volume of a cube 2.5 cm on edge.

The volume of the cube is 2.5 cm × 2.5 cm × 2.5 cm = 15.625 cm³. This answer is incorrect because it indicates an accuracy of five significant digits, whereas the measurement of the edge of the cube has only an accuracy of two significant digits. Therefore, the correct answer is 16 cm³, obtained by rounding off to two significant digits. This value then indicates the same degree of precision as in the measurement.

Sometimes data obtained from measurements are very small or very large numbers. Large and small numbers can be represented in a form called *scientific notation*. For example, multiples and fractions of 10 can be represented by expressing 10 raised to a *power* or *exponent*. Positive exponents are used to represent multiples of 10:

$$100 = (10)(10) = 10^2$$
$$1{,}000 = (10)(10)(10) = 10^3$$
$$10{,}000 = (10)(10)(10)(10) = 10^4$$

Negative exponents are used to represent fractions of 10:

$$\frac{1}{10} = 0.1 = 10^{-1}$$

$$\frac{1}{100} = 0.01 = 10^{-2}$$

$$\frac{1}{1000} = 0.001 = 10^{-3}$$

To write a number in this form the following procedure is used: The decimal point is moved so that one digit remains to the left of the decimal point. It is then multiplied by ten raised to an exponent corresponding to the number of positions the decimal has been moved. If the decimal is moved to the left, the exponent is positive; if it is moved to the right, the exponent is negative. For example, let us consider the size of objects that are normal size (macroscopic), very large (megascopic) and very small (microscopic). The height of an average man is about 170 cm, the diameter of the earth is about 1,300,000,000 cm, the size of a red blood cell is 0.00076 cm, and the size of a very small virus is 0.000001 cm. By using scientific notation, these numbers can be expressed as 1.7×10^2 cm, 1.3×10^9 cm, 7.6×10^{-4}

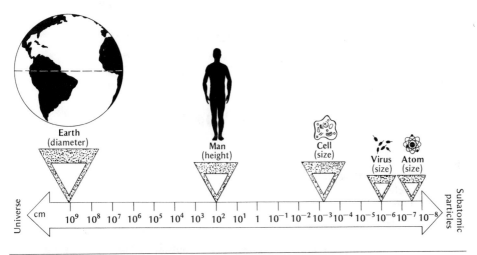

FIGURE 2.3

The use of scientific (exponential) notation to express measurements of large and small objects. The descriptive capabilities of this system are fantastic; for example, the total number of atoms in the *entire known universe* has been estimated at 10^{81}.

cm, and 1×10^{-6} cm, respectively (Fig. 2.3). Clearly, the scientific notation is convenient because it is not necessary to write down numerous zeros. Furthermore, as illustrated by the following example, the scientific notation is also useful in expressing the significant digits in a result.

EXAMPLE 2.6

How many grams are contained in 2 lb?
    The conversion relation is 1 lb = 454 g, therefore

$$2 \text{ lb} \times \left(\frac{454 \text{ g}}{1 \text{ lb}}\right) = 908 \text{ g} = 9 \times 10^2 \text{ g}$$

The correct number of significant digits is obtained by rounding off and using the scientific notation.

Many common measurements required for an understanding of environmental pollution involve large and small numbers. The study of pollution is almost meaningless without some measure of the amount of pollutant. The unit parts per million (ppm) and parts per billion (ppb) are used to express the *concentration* of trace amounts of a substance present in some material. Here concentration means the amount of a specified substance in a fixed amount of a mixture of substances. For example, 1 ppb of gold in the ocean would mean 1 g of gold in 10^9 g of ocean water or 10^{-9} g of gold in 1 g of ocean water. One part per million of carbon monoxide in the air would mean 1 g of carbon monoxide in 10^6 g of air or 10^{-6} g of carbon monoxide in 1 g of air. It could also mean 1 ml of carbon monoxide in 10^6 ml of air or 10^{-6} ml of carbon monoxide in 1 ml of air at the same pressure and temperature.

EXAMPLE 2.7

At any one time 3×10^{15} tons of air is present over the continental United States. Assuming that 300 million tons of pollutants are distributed uniformly in this huge air mass, calculate the concentration of pollutants in parts per million and parts per billion by weight.

The answers are

$$\frac{300 \times 10^6 \text{ tons pollutant}}{3 \times 10^{15} \text{ tons air}} \times 10^6 = 0.1 \text{ ppm}$$

$$\frac{300 \times 10^6 \text{ tons pollutant}}{3 \times 10^{15} \text{ tons air}} \times 10^9 = 100 \text{ ppb}$$

QUESTIONS

2.7 **Why is it necessary to make measurements in chemistry? Give two examples.**

2.8 **Define the following units of measure:**
a. **meter**
b. **liter**
c. **kilogram**
d. **Ångström**

2.9 **Give examples from your own experience or outside reading showing the use of metric measurements in the United States.**

2.10 **In the United States the English units are still in wide use. List the advantages and disadvantages resulting from a complete change to metric system units.**

2.11 **Explain what is meant by**
a. **significant digits**
b. **scientific notation**
Give examples.

2.12 **Suppose you take an American-built car to Europe where speed limits are given in kilometers. What speedometer reading should you not exceed in an 80-km zone to avoid a traffic violation?**

2.13 **A swimming pool used for Olympic competition is 50 m long. What is this distance in yards (yd)?**

2.14 **How many milliliters of water are required to fill an empty fifth (a commonly used American unit equal to one-fifth of a gallon)?**

2.15 **In the United States, first-class postage is 10¢ for each ounce or fraction thereof. How much postage is required for a letter weighing 85 g?**

2.16 **A professional football player signs a contract for an annual salary equivalent to his** weight in gold. If gold sells at $5000 per kilogram and he weighs 220 lb, what is his annual salary?

2.17 A kilogram of ocean water was found to contain 3 mg of salt. Express this concentration of salt in parts per million units.

2.18 A cubic meter sample of city air contains 10 ml of carbon monoxide. What is the level of air pollution in parts per million units?

SUGGESTED PROJECT 2.3
Measuring things (or know yourself)

Make the necessary measurements for the following in the English system of units and, then, convert them into metric units:

a. your dimensions (height, chest, waist, hips, etc.)
b. your weight
c. weight of your favorite candy bar (use the value given on the wrapper)
d. dimensions of your room (length, width, and height)
e. volume of air in your room
f. volume of liquid in a can of your favorite drink (use the value given on the can)

SUGGESTED PROJECT 2.4
Counting your wastes (or what do you throw away?)

Count the number of (1) metal cans, (2) plastic containers, (3) bottles and jars, and (4) boxes (all sizes) discarded each week in the garbage can of your household. Calculate the number of each of these discarded items per person in your household. Assuming your household to be a typical American household determine the number of each of these items discarded by 220 million Americans in a year. Express your result using the scientific notation.

3

Basic qualities of our environment

Atoms move in the void and catching each other up, jostle together, and some recoil . . ., and others become entangled . . ., and thus the coming into being of *composite things* is effected.

SIMPLICIUS (BORN 500 A.D.)
Neo-Platonic philosopher

You probably know that Isaac Newton's curiosity about the falling apple led him to discover the law of gravity. The most obvious consequence is that objects here on earth have *weight*. When we weigh something we are literally measuring the *force of attraction* between that body and the earth. Many substances are produced as the result of changes brought about by forces of attraction that are constantly at work in nature.

When we look around we see changes taking place in our environment. The basis for dynamic changes is one of the qualities of our environment. It is called *energy*. On the earth we derive most of our energy directly from the sun and from the combustion of materials such as coal, petroleum, and natural gas in which energy is presumably stored. In chemistry, we refer to material things as *matter*. It is another basic quality of our environment. Under appropriate conditions, *mass* (the substance of material objects) and energy are interconvertible.

Any study of chemistry concerns itself with the materials of nature and with what these materials do or what can be done to them. Therefore, the concepts of matter, energy, and force are of fundamental importance not only for the study of chemistry but also for understanding the nature of our environment.

MATTER

Matter is defined as anything that takes up *space* and has *mass*. It makes up the material things around us. Earth, air, and water are some familiar examples of matter; they all occupy space and possess mass. Mass is the substance of material objects. The concept of mass is useful because it can be used to compare the different material things in our environment. Sometimes we refer to the weight of an object rather than its mass. Strictly speaking, mass and weight are different: Mass refers to the quantity of matter in an object, whereas weight refers to the gravitational attraction on an object. For example, our astronauts weighed one-sixth of their earth weight on the moon because the gravitational attraction of the moon is about one-sixth that of the earth. However, their masses remained the same.

To make precise descriptions of matter, it is necessary to measure its mass. In the chemical laboratory, we use the gram or fractions of it for the unit of mass; however, the basic defined unit of mass is the kilogram (see Chapter 2). Usually the mass of an object is measured by comparing the mass of the object to standard reference masses called *weights*. These weights are usually pieces of metal that have been calibrated by comparison to the standard unit of mass. Mass comparisons, called weighing, are carried out by use of a *balance*. In principle, a balance consists of an upright with a sharp edge called a knife edge on which is balanced a metal beam such that the mass of the beam on either side of the knife edge is the same (Fig. 3.1a). When an object is placed on one side of the beam, the beam will no longer be in a horizontal, balanced position (Fig. 3.1b). However, the beam can be brought back to its horizontal balanced position by placing a sufficient number of weights (calibrated masses) on the other end of the beam (Fig. 3.1c). The mass of the object is then determined to be the total mass of the weights used in regaining the horizontal balanced position of the beam. The functional principle of most balances is quite similar to this simplified view of weighing even though balances vary quite a bit in design and sensitivity (Fig. 3.2).

Because matter has spatial dimensions, the quality of length is apparent. The idea of length comes from the fact that we can observe distances between points. Although the basic defined unit of length is the meter, in chemical work we often use the centimeter as a basic unit. The size of an object is expressed in terms of cubic centimeter or liter. The liter is defined as a *volume* equal to 1000 cm³ (see Chapter 2). A device used for measuring the volume of liquids is called a *graduated cylinder* (Fig. 3.3). This device also enables one to measure fractions of a liter.

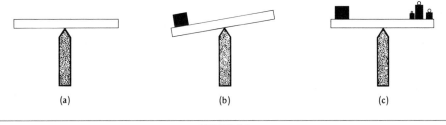

(a)   (b)   (c)

FIGURE 3.1

The balance and its use in the determination of the mass of an object. (a) Balanced position of the beam; (b) unbalanced position of the beam due to an object on one side of the beam; (c) restoration of the beam to the balanced position using weights on the other side of the beam. Note that the mass of the object will equal the total mass of the weights used in regaining the horizontal balanced position of the beam.

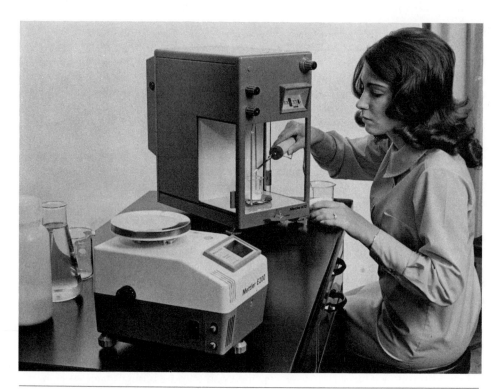

FIGURE 3.2

Two typical balances: a top-loading balance (in front) and an analytical balance. (Photograph courtesy of Mettler Instrument Corporation.)

FIGURE 3.3

A graduated cylinder for measuring an approximate volume of liquid.

Often the *area* occupied by an object is also a measured quantity. It is measured in terms of squared length, for example, square centimeter (abbreviated cm²).

We may distinguish between types of matter by determining either the mass of a specific volume or the volume of a specific mass of each type of matter. However, we shall see later that the volumes of objects can change with temperature so that it is necessary to make the measurements of the amount of mass and volume at a specific temperature. The mass of a substance in unit volume of the substance at a specified temperature is termed the *density* at that particular temperature. The common units used are grams for mass and cubic centimeters or milliliters for volume; the unit of density then is grams per cubic centimeter (g/cm³) or grams per milliliter (g/ml). The density of a substance can be determined by measuring the mass of a sample of the substance and the volume occupied by the sample:

$$\text{density} \left(\frac{g}{cm^3 \text{ or ml}} \right) = \frac{\text{mass (g)}}{\text{volume (cm}^3 \text{ or ml)}}$$

EXAMPLE 3.1

Calculate the density of the liquid if a 68.0 g sample has a volume of 5.00 ml.

$$\text{density} = \frac{\text{mass}}{\text{volume}} = \frac{68.0 \text{ g}}{5.00 \text{ ml}} = 13.6 \text{ g/ml}$$

The liquid with this unusually high density is mercury, which has many uses, as in thermometers, as well as being a serious environmental pollutant, especially its compounds.

EXAMPLE 3.2

Calculate the density of the solid if a 10-g sample displaced 3.6 ml of water in a graduated cylinder.

Because the volume of displaced water is the volume of the solid,

$$\text{density} = \frac{\text{mass}}{\text{volume}} = \frac{10 \text{ g}}{3.6 \text{ ml}} = 2.8 \text{ g/ml}$$

From the definitions of the gram and the milliliter, it follows that 1.00 ml of water at 4°C would have a mass of exactly 1.00 g. The density of water, then, is 1 g/ml at 4°C. Due to the pull of gravity, an object with a density greater than 1 g/ml will sink in water; an object with a density less than 1 g/ml will float in water. The densities of many common substances are known (Table 3.1).

TABLE 3.1
Densities of common substances at 25°C

Substance	Density (g/ml)
salt	2.16
sand	2.32
aluminum	2.70
iron	7.90
copper	8.90
lead	11.30
mercury	13.60
gold	19.30

Besides density there are many other characteristics that can be used to distinguish between different types of matter in our environment. As we shall see in the next section, often the very nature of matter around us serves this purpose.

THE NATURE OF MATTER   If we look around we find that matter in our environment exists in three well-defined forms called *physical states* or *phases:* solid, liquid, and gas (Fig. 3.4). *Solids* are characterized by definite shapes and volumes. *Liquids* have definite volumes but no definite shapes—a given volume of a liquid will assume the shape of the part of a container that it fills. *Gases* have neither definite shapes nor definite volumes—they completely fill any container into which they are introduced. Unlike solids and liquids, gases may expand or be compressed between relatively wide limits.

Under normal earth conditions, matter generally exists in three states. For example, water exists as solid water (ice), liquid water, or gaseous water (water vapor). By proper manipulation of pressure (force per unit area) and temperature, matter in one state can be made to change to another state. Changes from one state to another are called *phase changes*. The conditions under which certain phase changes occur can be used to characterize and identify substances. The temperature at which a solid melts is called the *melting point* of the solid, whereas the temperature at which a liquid boils is called the *boiling point* of the liquid. Measurements of melting and boiling points are often carried out in the laboratory for purposes of identifying substances.

Changes in state such as the melting of a solid and the vaporization of a liquid are examples of *physical changes*. These are changes that do not result in the formation of new kind of matter. Changes that result in the conversion of certain kind of matter into other kinds of matter are called *chemical changes* or *chemical reactions*. The lighting of a match, the burning of a candle, and the digestion of food in our bodies are only a few of the thousands of examples of chemical changes. Chemical reactions are occurring in and around us all the time. Most matter can be made to undergo chemical reactions; the initial matter is called the *reactant* and the new matter the *product*. For example, charcoal (a black solid) is the reactant and carbon dioxide (a colorless gas) is the product in a barbecue fire. Neither in a physical change nor in a chemical change is there any loss or gain of mass. In other words, *mass is conserved during changes in matter.*

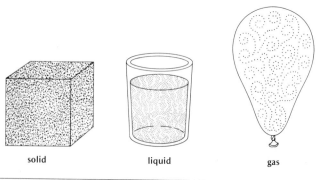

solid   liquid   gas

FIGURE 3.4

The three well-defined forms of matter: solid, liquid, and gas.

Every kind of matter has a set of characteristics or properties that distinguishes it from all other substances. Such properties are called *intrinsic properties*. These characteristics can be observed in any sample of one kind of matter regardless of the shape or size of the sample. Properties such as density, color, physical state, melting and boiling points, electrical conductivity, and so forth are commonly referred to as *physical properties* because they can be observed without causing any change in the chemical composition of the specimen. The *chemical properties* of matter are those that describe the chemical changes or reactions that the substance can undergo. The burning of paper, rusting of iron, decay of wood, and digestion of food are examples of chemical properties. These are also intrinsic properties. Nonintrinsic properties, such as size, shape, length, weight, and temperature, that are not characteristic of any particular type of matter, are known as *extrinsic properties*.

The properties of the various forms of matter in our environment serve to distinguish one type of matter from another. Matter that is uniform throughout in terms of composition and intrinsic properties is referred to as *homogeneous*. Water, sugar, and salt are examples of homogeneous matter. Each of them consists of one kind of matter only and is called a *pure substance*. However, most of the substances around us are made up of *mixtures* of materials. For example, consider food, soil, and concrete in which the different pieces of matter which make them up can be seen. They do not have uniform composition or properties throughout and are, therefore, called *heterogeneous mixtures*. Because there are no restrictions on the proportions in which the various components may be present, an infinite number of heterogeneous mixtures is found in our environment.

Mixtures may consist of solids or liquids or gases only or conglomerations of solids, liquids, and gases. Sometimes when a solid and liquid are mixed a dissolving process occurs as in the case of salt and water (in seawater). Dissolving refers to a process of intimate mixing that produces a mixture called a *solution*. Solutions are homogeneous mixtures and the single phase in which a solution occurs may be gaseous, liquid, or solid. All gases mix with one another in all proportions (air, for example), but for solid and liquid solutions there are usually limits on the solubility of one substance in another. A solution has but a single set of intrinsic properties, but each of these properties depends on the composition of the solution which is variable.

Pure substances are homogeneous forms of matter that have fixed compositions and invariable intrinsic properties. They have been classified into two types: elements and compounds. *Elements* are substances that cannot be separated into simpler substances by ordinary means. *Compounds* are substances that are composed of two or more elements in fixed and definite proportions. For example, oxygen and hydrogen are elements whereas water is a compound made up of 11.2 percent hydrogen by mass and 88.8 percent oxygen by mass.

The above classification of matter is shown schematically in Fig. 3.5.

ELEMENTS AND COMPOUNDS It is now fully recognized that elements are the fundamental forms of matter in our environment. More than three centuries ago, the English chemist Robert Boyle concluded that *elements are pure substances that cannot be decomposed into any simpler pure substances*. Based on this definition, the French chemist Antoine Lavoisier published the first table of elements in 1789 using modern, somewhat systematic names for the elements known at that time (Fig. 3.6). However, his table of elements included some forms of energy, such as light and caloric (heat), and some substances, such as lime and magnesia, which we now know to be compounds. Since then, 105 different elements have been identified and isolated. Each of the elements has a name and symbol. A symbol is a short-hand notation consisting of a one- or two-letter term, the first letter a capital and the second (if any) a lower case letter. For example, H, O, N, and Al are symbols used to denote the elements hydrogen, oxygen, nitrogen, and aluminum, respectively. Many elements have names that are of historical origin; thus it is not unusual to see symbols derived from Latin or Greek names. For example, the symbol Fe originates from *ferrum,* the Latin word for iron. The names and symbols of the elements and their origin are included in Table 3.2. In written work the chemist customarily uses symbols for elements. Therefore, a knowledge of the names and symbols of the elements is fundamental to an understanding of chemistry.

Many of the elements occur in our environment in the uncombined form as well as in the combined form, as constituents of compounds. For example, oxygen exists in the elemental form in our air environment and in the combined form in our water and soil environments. Many of the elements can be separated from the compounds in which they occur. For example, elements such as aluminum and iron are obtained from compounds that are found in nature (see Chapter 8). However, not all the elements occur in nature! Elements such as technetium (Tc), promethium (Pm), astatine (At), francium (Fr), and all the elements heavier than uranium (U) do not occur in nature but were created in laboratories using high-energy machines (see Chapter 6).

Under normal conditions on earth, the uncombined elements occur in the following physical states: 11 elements (argon, chlorine, fluorine, hydrogen, helium, krypton, nitrogen, neon, oxygen, radon, xenon) occur as gases, 2 as liquids (bromine, mercury), and the rest as solids. Elements can be classified according to their properties into three groups: (1) A large number (about 80) of the elements have metallic properties (good electrical conductors, flexible enough to be deformed,

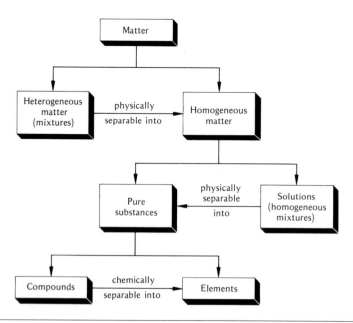

FIGURE 3.5

The classification of matter.

TABLE OF SIMPLE SUBSTANCES

Simple fubftances belonging to all the kingdoms of nature, which may be confidered as the elements of bodies.

New Names.	Correfpondent Old Names.
Light - - -	Light.
Caloric - - -	Heat.
	Principle or element of heat.
	Fire. Igneous fluid.
	Matter of fire and heat.
Oxygen - -	Dephlogifticated air.
	Empyreal air.
	Vital air, or bafe of vital air.
Azote - - -	Phlogifticated air or gas.
	Mephitis, or its bafe.
Hydrogen - -	Inflammable air or gas, or the bafe of inflammable air.

Oxydable and Acidifiable fimple Subftances not Metallic.

New Names.	Correfpondent Old Names.
Sulphur - -	
Phofphorus - -	The fame names.
Charcoal - -	
Muriatic radical -	
Fluoric radical - -	Still unknown.
Boracic radical - -	

Oxydable and Acidifiable fimple Metallic Bodies

New Names.		Correfpondent Old Names.
Antimony -		Antimony.
Arfenic - -		Arfenic.
Bifmuth - -		Bifmuth.
Cobalt - -		Cobalt.
Copper - -		Copper.
Gold - -		Gold.
Iron - - -		Iron.
Lead - -	Regulus of	Lead.
Manganefe -		Manganefe.
Mercury - -		Mercury.
Molybdena -		Molybdena.
Nickel - -		Nickel.
Platina - -		Platina.
Silver - -		Silver.
Tin - - -		Tin.
Tungftein -		Tungftein.
Zinc - - -		Zinc.

Salifiable fimple Earthy Subftances

New Names.	Correfpondent Old Names.
Lime - - -	Chalk, calcareous earth.
	Quicklime.
Magnefia - -	Magnefia, bafe of Epfom falt.
	Calcined or cauftic magnefia.
Barytes - - -	Barytes, or heavy earth.
Argill - - -	Clay, earth of alum.
Silex - - -	Siliceous or vitrifiable earth.

FIGURE 3.6

Lavoisier's table of simple substances (elements). Adapted from Lavoisier's *Traité élémentaire de chimie*, 1789; English translation, 1790.

TABLE 3.2
The elements[a]

Name	Symbol	Source of name and symbol
actinium	Ac	Greek *aktis* or *aktinos* meaning "beam" or "ray"
aluminum	Al	Latin *alumen* meaning "alum"
americium	Am	the Americas
antimony[b]	*Sb*	Latin *stibium* meaning "mark"
argon	Ar	Greek *argon* meaning "inactive"
arsenic	As	Latin *arsenicum* or Greek *arsenikon* meaning "yellow orpiment" or Arabic *az-zernikh* meaning the "orpiment" from Persian *zerni-zar,* gold
astatine	At	Greek *astatos* meaning "unstable"
barium	Ba	Greek *barys* meaning "heavy"
berkelium	Bk	Berkeley, home of University of California
beryllium	Be	Greek *beryllos* meaning "beryl"
bismuth	Bi	German *weisse Masse* meaning "white mass"; later *Wismuth* and *Bisemutum*
boron	B	Arabic *buraq* or Persian *burah*
bromine	Br	Greek *bromos* meaning "stench"
cadmium	Cd	Latin *cadmia*
calcium	Ca	Latin *calx* meaning "lime"
californium	Cf	California
carbon	C	Latin *carbon* for "charcoal"
cerium	Ce	the asteroid Ceres
cesium	Cs	Latin *caesius* meaning "sky blue"
chlorine	Cl	Greek *chloros* meaning "greenish-yellow"
chromium	Cr	Greek *chroma* meaning "color"
cobalt	Co	German *Kobold* meaning "goblin" or "evil spirit"
copper	*Cu*	Latin *cuprum,* from the island of Cyprus
curium	Cm	Pierre and Marie Curie
dysprosium	Dy	Greek *dysprositos* meaning "hard to get at"
einsteinium	Es	Albert Einstein
erbium	Er	Ytterby, a town in Sweden
europium	Eu	Europe
fermium	Fm	Enrico Fermi
fluorine	F	Latin *fluere* meaning "flow" or "flux"
francium	Fr	France
gadolinium	Gd	gadolinite, a mineral named for the Finnish chemist Gadolin
gallium	Ga	Latin *Gallia* for France
germanium	Ge	Latin *Germania* for Germany
gold	*Au*	Latin *aurum* meaning "shining dawn"
hafnium	Hf	Latin *Hafnia* for Copenhagen
helium	He	Greek *helios* meaning the "sun"
holmium	Ho	Latin *Holmia* for Stockholm
hydrogen	H	Greek *hydro* and *genes* meaning "water-forming"
indium	In	Latin *indicum* meaning "indigo"
iodine	I	Greek *iodes* meaning "violet"
iridium	Ir	Latin *iris* meaning "rainbow"
iron	*Fe*	Latin *ferrum* meaning "iron"
krypton	Kr	Greek *kryptos* meaning "hidden"
lanthanum	La	Greek *lanthanein* meaning "to lie hidden"
lawrencium	Lr	Ernest O. Lawrence, inventor of the cyclotron
lead	*Pb*	Latin *plumbum*
lithium	Li	Greek *lithos*

[a] Not included here are the two most recently discovered elements, the names and symbols of which have not been officially recognized at this writing. Both American and Soviet scientists claim their discovery and consequently have named them. The American names are rutherfordium (after Ernst Rutherford, a British physicist) and Hahnium (after Otto Hahn, a German chemist); whereas the respective Soviet names are kurchatovium (after Igor Kurchatov, a Soviet nuclear scientist) and bohrium (after Niels Bohr, a Danish physicist).

[b] Names in italic are not directly derived from their original sources.

36

TABLE 3.2 (Cont.)
The elements[a]

Name	Symbol	Source of name and symbol
lutetium	Lu	Lutetia, ancient name for Paris
magnesium	Mg	Magnesia, district in Thessaly
manganese	Mn	Latin *magnes* for "magnet"
mendelevium	Md	Dmitri Mendeleev, a Russian chemist
mercury	*Hg*	Latin *hydrargyrum* meaning "liquid silver"
molybdenum	Mo	Greek *molybdos* meaning "lead"
neodymium	Nd	Greek *neos* and *didymos* meaning "new twin"
neon	Ne	Greek *neos* meaning "new"
neptunium	Np	planet Neptune
nickel	Ni	German *Nickel,* Satan or "Old Nick"
niobium	Nb	Niobe, daughter of Tantalus
nitrogen	N	Latin *nitrum* or Greek *nitron* meaning "native soda"
nobelium	No	Alfred Nobel
osmium	Os	Greek *osme* meaning a "smell"
oxygen	O	Greek *oxys* and *genes* meaning "acid former"
palladium	Pd	Greek *Pallas,* goddess of wisdom; the asteroid Pallas
phosphorus	P	Greek *phosphoros* meaning "light-bearing"
platinum	Pt	Spanish *platina* for "silver"
plutonium	Pu	planet Pluto
polonium	Po	Poland, native country of Mme Curie
potassium	*K*	Latin *kalium;* English, potash — pot ashes
praseodymium	Pr	Greek *prasios* and *didymos* meaning "green twin"
promethium	Pm	Prometheus, who, according to mythology, stole fire from heaven
protoactinium	Pa	Greek *protos* meaning "first"
radium	Ra	Latin *radius* meaning "ray"
radon	Rn	from radium
rhenium	Re	Latin *Rhenus* for Rhine
rhodium	Rh	Greek *rhodon* for "rose"
rubidium	Rb	Latin *rubidius* for "deepest red"
ruthenium	Ru	Latin *Ruthenia* for Russia
samarium	Sm	samarskite, a mineral
scandium	Sc	Latin *Scandia* for Scandinavia
selenium	Se	Greek *selene* for "moon"
silicon	Si	Latin *silex, silicis* meaning "flint"
silver	*Ag*	Latin *argentum*
sodium	*Na*	English, soda; Latin *natrium*
strontium	Sr	Strontian, town in Scotland
sulfur	S	Sanskrit *sulvere* or Latin *sulphurium*
tantalum	Ta	Greek *Tantalos,* mythological character — father of Niobe
technetium	Tc	Greek *technetos* meaning "artificial"
tellurium	Te	Latin *tellus* for "earth"
terbium	Tb	Ytterby, village in Sweden
thallium	Tl	Greek *thallos* meaning "green shoot or twig"
thorium	Th	Thor, Scandinavian god of war
thulium	Tm	Thule, the earliest name for Scandinavia
tin	*Sn*	Latin *stannum*
titanium	Ti	Latin *Titans,* the first sons of the earth (mythology)
tungsten	*W*	Swedish *tung sten* meaning "heavy stone"; German *Wolfram*
uranium	U	planet Uranus
vanadium	V	Scandinavian goddess Vanadis
xenon	Xe	Greek *xenon* meaning "stranger"
ytterbium	Yb	Ytterby, village in Sweden
yttrium	Y	Latin *yttria*
zinc	Zn	German *Zink,* of obscure origin
zirconium	Zr	Arabic *zargum* meaning "gold color"

possess metallic luster, etc.) and are called *metals*. (2) Some 18 elements (argon, boron, bromine, carbon, chlorine, fluorine, hydrogen, helium, iodine, krypton, nitrogen, neon, oxygen, phosphorus, radon, sulfur, selenium, xenon) do not possess metallic properties and are called *nonmetals*. (3) Some 6 elements (arsenic, germanium, polonium, antimony, silicon, tellurium) display the properties of both metals and nonmetals and are called the *metalloids*.

Many elements are found on earth only in very small amounts, whereas others occur in much greater amounts. If we consider the whole earth including the core, the four most abundant elements by weight are iron (39.8%), oxygen (27.7%), silicon (14.5%), and magnesium (8.7%). If we consider only the earth's crust, which, on an average, is a 10-mi deep outer zone, the most abundant element by weight is oxygen (48.6%), followed by silicon (26.3%), aluminum (7.7%), and iron (4.8%). A comparison of the abundances is made in Fig. 3.7. However, for the entire solar system to which we belong, hydrogen and helium are by far the most abundant elements, with oxygen, neon, carbon, and nitrogen next in order. These data form the basis of our theories concerning the primordial composition of the solar system and the origin of the earth.

Compounds are pure substances composed of two or more elements combined in fixed and definite proportions. Unlike the elements, compounds can be decomposed by simple chemical change into two or more different substances. Water is an example of a compound. Under normal conditions it is a liquid. However, it can be decomposed into the gases hydrogen (11.2%) and oxygen (88.8%), the properties of which are completely unrelated to the properties of water. To date, over 5 million compounds have been identified and isolated. Each of the compounds has a name and a formula. A *formula* is a shorthand notation that indicates what elements and how much of each are present in a compound. Some examples of formulas are CO_2 for carbon dioxide, NH_3 for ammonia, and H_2O for water. The naming of compounds and their formulas are discussed in Chapter 7.

Many compounds are found in nature. But a great majority of the known compounds has been synthesized in the laboratory. Under normal earth conditions, compounds occur in all three physical states. Some familiar examples of gaseous compounds are ammonia (NH_3) and carbon dioxide (CO_2); of liquid compounds, water (H_2O) and alcohol (C_2H_6O); and, of solid compounds, salt (NaCl), sand (SiO_2), and sugar ($C_{12}H_{22}O_{11}$). One of the methods of classifying compounds is based on

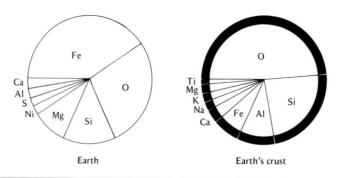

Earth   Earth's crust

FIGURE 3.7

Abundance of the elements on the earth as a whole and on the earth's crust (10-mi deep outer zone only).

whether the compound contains carbon or not. Compounds that contain carbon are called *organic compounds;* those that do not are called *inorganic compounds.*

Another method of classifying compounds is based on whether the molten compound or a solution of it will conduct a current of electricity. If a molten compound or its water solution is a conductor of an electric current, the compound is classified as an *electrolyte;* if not, the compound is classified as a *nonelectrolyte.* There are three common types of electrolytes: acids, bases, and salts. An *acid* is a substance whose water solution has a sour taste, turns blue litmus red, reacts with active metals to form hydrogen, and neutralizes bases. Conversely, a *base* is a substance whose water solution has a bitter taste, turns red litmus blue, feels soapy, and neutralizes acids. *Salts* are the products of the mutual neutralization of acids and bases. The different classes of compounds are discussed at greater length in Chapter 7.

A large number of analyses with a variety of compounds have shown that *a pure compound always contains the same elements in the same relative amounts.* (This observation is often called the *law of definite proportions.*) Water as it occurs in nature or as it is synthesized in the laboratory is always composed of 11.2 percent hydrogen and 88.8 percent oxygen by mass. Another observation concerning compounds is that two elements can sometimes form more than one compound. For example, the elements hydrogen and oxygen form both the compound water and the compound called hydrogen peroxide (5.9% H and 94.1% O). Analyses show that, although the compounds have different compositions by mass, these compositions are related in a simple way. (This observation is called the *law of multiple proportions.*) Further, in the formation of compounds there is no apparent loss or gain of mass. That is, the total mass of the elements that react together is the same as the mass of the compound formed. (This observation is called the *law of conservation of mass.*) Observations such as these give rise to the question: What is the conceivable nature of elements and compounds that could explain such observations?

THE ATOMIC AND MOLECULAR CONCEPTS To explain the nature of matter the atomic concept of matter was developed. The ancient Greeks pondered over such a concept for several centuries, but it was John Dalton, early in the nineteenth century, who postulated the *atomic theory* on the basis of the then existing ideas of elements and compounds and their properties. It took another 150 years to demonstrate that the atomic concept of matter approaches closely the realities of nature, for in May 1970, Albert Crewe of the University of Chicago obtained for the first time the long sought for "photograph" of the atom (Fig. 3.9).

According to Dalton's theory, *atoms* are the smallest particles of elements and *molecules** ("atoms compounded of elemental atoms," according to Dalton) the smallest particles of compounds. To explain the properties of elements, Dalton developed the idea that an element contains only one kind of atom, that the atom is a simple, indestructible particle of matter, and that the atom of each element is different in weight (mass) from the atoms of all elements. Elements, he argued, cannot be changed to simpler substances because their atoms cannot be broken down. The ultimate particle (molecule) of a compound, on the other hand, is made up of one or more atoms of at least two elements, arranged in a definite pattern (Fig. 3.10). Following Dalton's theory, the symbols for the names of elements have come to be used to represent atoms of the elements and the atoms of the elements in com-

---

* Dalton did not advance the molecular concept. It was the Italian chemist Amadeo Avogadro who introduced the term molecule in 1811 to distinguish elemental atoms from the Daltonian compounded atoms.

FIGURE 3.8

John Dalton, English mathematician of the early nineteenth century who formulated the first practical atomic theory and the first table of atomic weights.

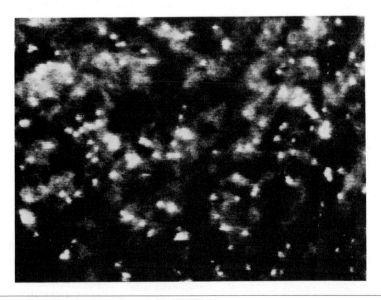

FIGURE 3.9

Chains of thorium atoms as seen through a scanning electron microscope. In each chain the smallest white dots represent single thorium atoms; the larger white dots are aggregates of a few thorium atoms very close together. (Photograph courtesy of Professor Albert V. Crewe, The Enrico Fermi Institute, The University of Chicago.)

pounds. Chemical reactions were conceived to involve the rearrangement of one state of combination of atoms into another (Fig. 3.10).

The significantly new concept of the atom as proposed by Dalton is the definite distinguishing masses attached to atoms of a given element. If atoms are extremely small particles, how is it possible to assign definite distinguishing masses to atoms of different elements? The original attempts were, therefore, to develop a scale of *relative masses* of atoms of elements from a knowledge of the percent composition of the various compounds formed by the elements. In the case of water, for example, the ratio of the mass of an oxygen atom to that of a hydrogen atom is 88.8:11.2 or 8:1. Therefore, Dalton concluded that the oxygen atom is eight times heavier than the hydrogen atom. Of course, Dalton, assumed that one atom of oxygen combines with one atom of hydrogen to form one molecule of water.

In contrast, Avogadro believed that oxygen and hydrogen molecules are composed of two atoms each, O_2 and H_2, and that one molecule of oxygen reacts with two molecules of hydrogen to form two molecules of water, H_2O. On this basis, two hydrogen atoms are 11.2 percent of the mass of the water molecule and one hydrogen atom would weigh one-half of 11.2 or 5.6 percent of the total mass of a molecule. Because the one oxygen atom in the molecule is 88.8 percent of the mass, the ratio of the mass of the oxygen atom to the mass of the hydrogen atom would be 88.8:5.6 or 16:1. In other words, the oxygen atom is 16 times heavier than the hydrogen atom. It thus became clear that if an atomic mass scale was to be firmly established, it would be necessary to know the molecular formulas for compounds in addition to their percent composition.

Using an approach similar to that for hydrogen and oxygen, the Swedish chemist Jakob Berzelius compiled a list of the relative atomic masses of several elements. In order to compile such a list, he had to assign a given value to the mass of

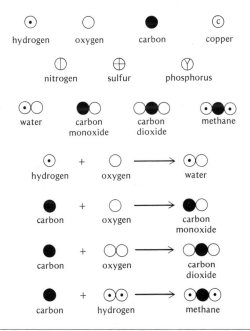

FIGURE 3.10

Dalton's representation of atoms, molecules, and chemical reactions.

one of the elements and then record the masses of the other elements relative to the assigned standard. Berzelius chose the element oxygen as the standard and assigned to it the atomic mass of 100. Several years later oxygen atoms were *arbitrarily assigned* the relative mass of 16. Table 3.3 lists a representative set of Berzelius's values along with a translation of these to the standard of 16 for oxygen and the presently accepted values according to which oxygen has a mass of very nearly 16 (15.9994 exactly). Note that a carbon atom weighs 12 times as much as a hydrogen atom, nitrogen 14 times as much, oxygen 16 times as much, sulfur 32 times as much, and chlorine 35.5 times as much. However, remember that these are relative masses and *not* the actual masses of the atoms.

Modern research, however, has shown that all of the atoms of a given element do not have the same mass, but definite percentages of the atoms of a given element have specific masses. For example, 75.5 percent of the atoms of naturally occurring chlorine have one mass, whereas 24.5 percent have another mass (Fig. 3.11). Atoms of the same element that have different masses are called *isotopes* of that element. Most elements that are found in nature have two or more isotopes, but the percentages of the isotopes for a given element are generally constant irrespective of the source of that element on earth. Only 22 of the elements have no isotopes and, thus, exist in the form of one type of atom. The perfection of a complicated and sensitive apparatus, called the mass spectrometer, enabled the determination of the percentages and relative masses of the isotopes of each naturally occurring element (see Chapter 4).

Atomic masses are, indeed, very small when expressed in units of grams. The heaviest known atom weighs about 10^{-23} g and the lightest known about 10^{-24} g. Therefore, a unit of mass that is more convenient for expressing masses on an atomic scale was devised. This unit is called the *atomic mass unit* (abbreviated amu). It is arbitrary and based on the following definition: *An atom of the most common isotope of carbon (referred to as carbon-12) has a mass of exactly 12 amu.* Therefore, it follows that *1 amu is 1/12 the mass of an atom of carbon-12*. The figure 12 amu was chosen so that no isotopes of an element would have a relative mass of less than 1 amu. The masses of the isotopes of the elements can be expressed in terms of atomic mass units. Table 3.4 gives a list of the masses and percentage abundances of the isotopes of common elements.

When comparing the relative masses of elements, it is more convenient to refer to the masses of the hypothetical average atoms of the elements than to compare the masses of all of the isotopes of each element. Thus, the mass of an average atom of an element came to be known as the *atomic mass* (also referred to as *atomic weight*) of the element. It is the weighted average mass of the isotopes of the element expressed in atomic mass units. If we know the percentage abundances of the iso-

FIGURE 3.11

The two isotopes of chlorine and their distribution. Note that the lighter isotope chlorine-35 is three times more abundant than the heavier isotope chlorine-37.

topes of an element, the atomic mass can be calculated from the masses of the isotopes. The weighted average mass can be calculated by multiplying each isotopic mass by the corresponding fractional abundance (percentage abundance/100). For example, in the case of chlorine 75.4 percent of the atoms weigh 35.0 amu and 24.6 percent of the atoms weigh 37.0 amu. The atomic mass of chlorine is then given by

$$\frac{75.4}{100} \times 35.0 \text{ amu} + \frac{24.6}{100} \times 37.0 \text{ amu} = (0.754)(35.0 \text{ amu}) + (0.246)(37.0 \text{ amu}) = 35.5 \text{ amu}$$

The atomic masses of all the elements are listed in Appendix A.

THE MOLE CONCEPT When working with substances we never deal with individual atoms or molecules, but with amounts of elements and compounds which are measured in grams. To relate the relative masses of elements and compounds in

TABLE 3.3
Relative atomic masses

	RELATIVE ATOMIC MASS		
	Berzelius's value		
Element	O = 100	O = 16	Modern value
aluminum	171	27.4	27.0
calcium	256	41.0	40.1
carbon	76	12.2	12.0
chlorine	221	35.4	35.5
hydrogen	6.2	1.0	1.0
iron	339	54.2	55.8
nitrogen	89	14.0	14.0
oxygen	100	16.0	16.0
silicon	278	44.5	28.1
sulfur	201	32.2	32.0

TABLE 3.4
Percentage abundances (by weight) and masses of the isotopes of common elements

Isotopes	Abundance (wt %)	Mass (amu)
hydrogen-1	99.98	1.008
hydrogen-2	0.02	2.014
carbon-12	98.89	12.000
carbon-13	1.11	13.003
nitrogen-14	99.63	14.003
nitrogen-15	0.37	15.000
oxygen-16	99.76	15.995
oxygen-17	0.04	16.999
oxygen-18	0.20	17.999
iron-54	5.82	53.940
iron-56	91.66	55.935
iron-57	2.19	56.935
iron-58	0.33	57.933

units of grams the chemists have devised a way in which a standard number of atoms or molecules can be referred to as a unit. This unit is called the *mole*. One mole of any substance contains 6.02×10^{23} atoms or molecules, a number that is frequently referred to as *Avogadro's number*.

One mole of a substance is the amount (in grams) that contains the same number of atoms or molecules as the number of atoms that are contained in exactly 12 g of carbon-12. Recall that the same standard, carbon-12, was used for the definition of the atomic mass unit. By using the same standard, the atomic mass of an element is numerically the same as the mass in grams of a mole of the element. For example, if oxygen has an atomic mass of 16.00 amu, then the mass of 1 mole of oxygen atoms is 16.00 g. Thus, the number of grams per mole for each element can be determined from the atomic masses (see Appendix A). Note that the symbol of an element is also used to represent a mole of atoms of the element.

Based on the atomic concept, we can state that each pure compound contains a definite proportion of combined atoms of each element present in the compound. Therefore, compounds can be represented by a formula which indicates the kinds of elements present and the relative number of combined atoms of each element. For example, water can be represented by the formula H_2O which shows that water molecules are composed of two hydrogen atoms and one oxygen atom. In a formula, the symbols of elements indicate the atoms of elements and the subscripts show the number of atoms involved. The formula of a compound can be interpreted as indicating the relative number of moles of each element in the compound. For example, the formula H_2O indicates that 2 moles of hydrogen atoms are combined with 1 mole of oxygen atoms to form 1 mole of water (H_2O) molecules. Thus, a mole of a compound is the amount of the compound (in grams) that contains the number of moles of each element given by the subscripts in the formula. The number of grams per mole of a compound is obtained by multiplying the number of grams per mole of each element in the compound by the corresponding subscript in the formula and then summing the products. For example, the number of grams per mole of water is found as follows:

$$2\left(\frac{1.008 \text{ g}}{1 \text{ mole H}}\right) + 1\left(\frac{16.00 \text{ g}}{1 \text{ mole O}}\right) = \left(\frac{18.02 \text{ g}}{1 \text{ mole H}_2\text{O}}\right)$$

The mole concept is very useful in chemistry because it enables us to convert the number of grams of a substance to the number of moles and then to the number of atoms or molecules. Conversely, if the number of atoms or molecules are known, we can find the number of moles using Avogadro's number and then determine the number of grams of a substance.

EXAMPLE 3.3

How much will 6.02×10^{23} atoms of gold weigh?

There are 6.02×10^{23} atoms of gold in 1 mole of gold. Because gold has an atomic mass of 197 amu, the mass of 6.02×10^{23} atoms is 197 g.

EXAMPLE 3.4

What is the mass of 5.00 moles of oxygen atoms?

The number of moles of oxygen atoms can be converted to the mass by multiplying by the number of grams per mole of oxygen:

$$5.00 \text{ moles O} \left(\frac{16.00 \text{ g}}{1 \text{ mole O}}\right) = 80.00 \text{ g}$$

EXAMPLE 3.5

How many moles of water, H_2O, are contained in a 90.00-g sample of water?

Because 1 mole of water contains 18.02 g, the number of moles of water can be found from

$$90.00 \text{ g} \left(\frac{1 \text{ mole } H_2O}{18.02 \text{ g}} \right) = 5 \text{ moles } H_2O$$

EXAMPLE 3.6

An aspirin molecule weighs 3×10^{-22} g. What is the weight of 1 mole of aspirin?

The weight of 1 mole can be determined by multiplying the weight of 1 molecule by 6.02×10^{23} (Avogadro's number):

$$3 \times 10^{-22} \frac{\text{g}}{\text{molecule}} \times 6.02 \times 10^{23} \frac{\text{molecules}}{\text{mole}} = 180.6 \frac{\text{g}}{\text{mole}}$$

EXAMPLE 3.7

The hallucinogenic drug LSD has the formula $C_{24}H_{30}N_3O$. What will be the weight of (a) 1 mole of LSD and (b) 1 molecule of LSD?

a. The weight of 1 mole of LSD is obtained by multiplying the number of grams per mole of each element in LSD by the corresponding subscript in the formula and then summing the products:

$$24\left(\frac{12.00 \text{ g}}{1 \text{ mole C}} \right) + 30\left(\frac{1.00 \text{ g}}{1 \text{ mole H}} \right) + 3\left(\frac{14.00 \text{ g}}{1 \text{ mole N}} \right) + 1\left(\frac{16.00 \text{ g}}{1 \text{ mole O}} \right) = \left(\frac{376 \text{ g}}{1 \text{ mole LSD}} \right)$$

b. The weight of 1 molecule of LSD is obtained by multiplying the weight of 1 mole by the inverse of Avogadro's number.

$$\frac{376 \text{ g}}{1 \text{ mole LSD}} \times \frac{1 \text{ mole LSD}}{6.02 \times 10^{23} \text{ molecules}} = 6.25 \times 10^{-22} \text{ g/molecule}$$

ULTIMATE PARTICLES IN MATTER   Earlier in this chapter we pointed out that there are two major classes of matter: elements and compounds. In elements, the ultimate particles that retain the properties of the element, are usually the atoms of the element. However, in a few very common elements, the ultimate particle that is normally present is not the single atom, but a molecule composed of two or more atoms. For example, oxygen gas consists of molecules made up of pairs of oxygen atoms. The symbol for oxygen gas is written O_2, indicating that each molecule contains two atoms. Other examples are hydrogen (H_2), nitrogen (N_2), fluorine (F_2), chlorine (Cl_2), bromine (Br_2), and iodine (I_2). All these elements consist of *diatomic* (two-atom) *molecules*. In this case the molecule is the ultimate particle that retains the properties of the element. Some elements such as phosphorus (P_4) and sulfur (S_8) consist of molecules that are made up of more than two atoms and are therefore referred to as *polyatomic molecules*.

In addition to atoms and molecules, an ultimate particle found in nature is the *ion*. An ion is an atom or a bonded group of atoms that carries an electrical charge (Fig. 3.12). The nature of these charged particles is discussed in Chapter 4. They are mentioned here to emphasize the fact that compounds consist of aggregates of atoms (molecules or ions) that exist as stable units. That is, compounds are made up of molecules or ions. Water is an example of a compound in which the ultimate particles are molecules composed of two hydrogen atoms and one oxygen atom. On the other hand, table salt (sodium chloride) is a compound in which the ultimate particles are positive sodium ions (Na^+) and negative chlorine ions (Cl^-). Clearly, a

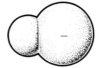

Positive ion   Negative ion

FIGURE 3.12

Idealized representation of ions: a positively charged ion consisting of an atom carrying a positive charge; and a negatively charged ion consisting of two bonded atoms, the entire unit of which carries the negative charge.

---

mixture such as soil might contain all three of the ultimate particles that characterize different substances—atoms, ions, and molecules.

QUESTIONS

3.1   What is meant by the term matter?
3.2   Distinguish between the mass and weight of an object.
3.3   How would you compare the masses of two given objects?
3.4   Give the definition of a liter.
3.5   What is meant by density? Describe how it can be determined.
3.6   Distinguish among
    a. solids, liquids, and gases
    b. physical and chemical changes
    c. physical and chemical properties
    d. homogeneous and heterogeneous matter
    e. elements and compounds
3.7   Give the symbol for each of the following elements: iron, lead, mercury, silver.
3.8   Give the name for each of the following elements for which the symbol is Sb, Au, Na, Sn, W.
3.9   Name four elements that do not occur in nature.
3.10 Which elements are found under normal earth conditions as
    a. gases?
    b. liquids?
    c. diatomic molecules?
3.11 How are the elements and compounds classified?
3.12 What are the most abundant elements in the earth?
3.13 State the postulates of Dalton's atomic theory.
3.14 What were the general observations on compounds prior to Dalton's atomic theory?
3.15 Define the terms:
    a. isotopes
    b. atomic mass unit
    c. atomic weight
    d. mole
    e. Avogadro's number
3.16 If we know that the weight of a carbon atom is 2.0×10^{-23} g, why do we retain the concept of the relative weight of 12 for a carbon atom?
3.17 Using the isotopic masses and the percentage abundances given in Table 3.4, calculate the atomic weights of the elements carbon, nitrogen, and oxygen.

3.18 Calculate the number of grams per mole of nitrogen, methane (CH_4), table salt (NaCl), benzene (C_6H_6), and sugar ($C_{12}H_{22}O_{11}$).

3.19 Why is the mole concept important in chemistry?

3.20 Distinguish between an atom, an ion, and a molecule.

SUGGESTED PROJECT 3.1

Weighing your wastes (or how much do you throw away?)

Collect the newspapers and magazines discarded each week in your household. Obtain their weight in pounds by using a bathroom scale. Convert the weight into kilograms and then calculate the number of kilograms per person in your household. Assuming your household to be a typical American household calculate the tonnage of paper discarded by 220 million Americans in a year (1 U.S. short ton = 2000 lb).

SUGGESTED PROJECT 3.2

Relative densities of some objects (or the sinkers and floaters around you)

Pour about a cup of water into a quart-size paper cup. Drop small objects (such as an ice cube, small pieces of stone, glass, plastic, plywood, Styrofoam and rubber, pennies, nails, broken finger nails, hair, small balloons filled with air and without air) into the water one at a time. Observe which objects sink in water and which objects float on water. Based on your findings determine which objects are less dense than water. Which objects are more dense than water?

SUGGESTED PROJECT 3.3

Counting particles by weighing (or how to get around the problem of counting particles individually)

Weigh an empty bucket using a bathroom scale. Then fill it with sand (SiO_2) and reweigh it. From the weight of sand determine the number of molecules of SiO_2 present in the bucket of sand.

Repeat the experiment using water (H_2O) and determine the number of molecules of H_2O present in the bucket of water.

If you have a balance capable of weighing fractions of an ounce, determine

a. the number of carbon atoms in a piece of charcoal

b. the number of Na^+ and Cl^- ions in a cup of salt

c. the number of $C_{12}H_{22}O_{11}$ molecules in a tablespoon of sugar

ENERGY

Energy is the basis for dynamic changes in our environment. It is not an object like matter, which occupies space, but it is needed to perform work. Therefore, the energy of a body or system is sometimes simply considered to be that body's or system's ability to do work. Although we do not see, taste, smell, hear, or feel energy all the time, we constantly experience its effects. Every action in nature involves energy. Diverse phenomena such as the melting of ice, the burning of wood, or the functioning of living organisms all require the emission or absorption of energy. All physical and chemical changes are accompanied by changes in energy. Thus, the concept of energy is of great importance in chemistry.

There are a number of interrelated forms of energy. The principal forms of energy include potential energy, kinetic energy, electrical energy, radiant energy, heat energy, chemical energy, and nuclear or atomic energy.

Potential energy is the energy that a body possesses because of its position or composition. It can be transformed into *kinetic energy,* the energy a body possesses by virtue of its motion. Our everyday experiences offer ample evidence of this energy transformation. For example, a swinging object shows continual interconversion of kinetic and potential energy. At the bottom of the swing its speed is greatest and so its kinetic energy is a maximum; at the ends of the swing the speed is zero and all the energy has been converted to potential. The sum of kinetic and potential energies is known as the total *mechanical energy* of the object.

Mechanical energy can be converted into *electrical energy,* the energy arising in connection with the flow of electric current. For example, water held in a reservoir behind a dam turns a turbine that at the same time is turning a generator and in the process transforms mechanical energy into electrical energy. Electrical energy is converted in a light bulb to light *(radiant energy)* and some heat energy.

HEAT AND TEMPERATURE   Generally speaking, the total energy an object possesses is called its *internal energy.* The *heat energy* of a body is the energy that flows from a hotter to a cooler body. It is a measure of the internal energy of a sub-

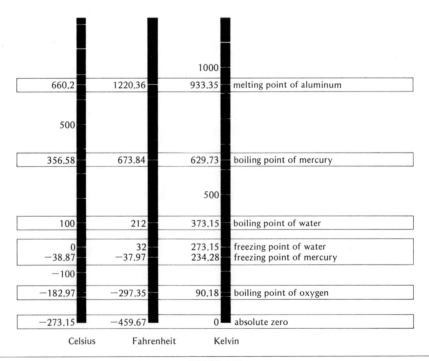

Celsius	Fahrenheit	Kelvin	
		1000	
660.2	1220.36	933.35	melting point of aluminum
500			
356.58	673.84	629.73	boiling point of mercury
		500	
100	212	373.15	boiling point of water
0	32	273.15	freezing point of water
−38.87	−37.97	234.28	freezing point of mercury
−100			
−182.97	−297.35	90.18	boiling point of oxygen
−273.15	−459.67	0	absolute zero

FIGURE 3.13

The relationships among the Celsius, Fahrenheit, and Kelvin temperature scales.

stance due to its *temperature*. Temperature is a measure of the intensity of heat. It is important to realize that temperature and heat are two entirely different concepts.

In our country, temperatures are usually measured in *degrees Fahrenheit* (abbreviated °F). On this scale, water freezes at 32°F and boils at 212°F. In chemical laboratories, however, temperatures are measured in *degrees Celsius* or *centigrade* (abbreviated °C). On the centigrade scale, water freezes at 0°C and boils at 100°C. The relationship between the Fahrenheit scale and the centigrade scale is

$$\frac{°C}{°F - 32} = \frac{5}{9}$$

The degree centigrade is not the basic defined unit of temperature in the metric system, but it is often used to express temperature measurements. The basic defined unit of temperature in the metric system is *degrees Kelvin* (abbreviated °K). The size of the Kelvin degree is the same as the size of the Celsius degree, but 0°C corresponds to 273°K. The relationships among the Kelvin, Celsius, and Fahrenheit temperature scales are illustrated in Fig. 3.13.

EXAMPLE 3.8

Convert 98.6°F (normal body temperature) to degrees centigrade.

$$\frac{°C}{°F - 32} = \frac{5}{9}$$

Therefore,

$$\frac{°C}{98.6 - 32} = \frac{5}{9}$$

or

$$°C = (98.6 - 32)\frac{5}{9}$$
$$= \frac{66.6 \times 5}{9}$$
$$= 37$$

EXAMPLE 3.9

Convert −40°C to degrees Fahrenheit.

$$\frac{°C}{°F - 32} = \frac{5}{9}$$

Therefore,

$$\frac{-40}{°F - 32} = \frac{5}{9}$$

or

$$9(-40) = 5(°F - 32)$$

Thus,

$$°F = \frac{9(-40)}{5} + 32$$
$$= -40$$

Note that −40° is the one temperature that has the same number in degrees centigrade and degrees Fahrenheit.

EXAMPLE 3.10

Convert 212°F to the Kelvin scale.

First the 212°F must be converted to the Celsius scale; then 273°K must be added to the Celsius temperature after it has been converted to Kelvin degrees (°K = °C + 273).

$$\frac{°C}{°F - 32} = \frac{5}{9}$$

Therefore,

$$\frac{°C}{212 - 32} = \frac{5}{9}$$

or

$$°C = (212 - 32)\frac{5}{9}$$

$$= 100$$

$$°K = \left(\frac{1°K}{1°\cancel{C}}\right)(100°\cancel{C}) + 273°K$$

$$= 373°$$

Any device used to measure temperature is called a *thermometer*. It makes use of some property of a material that varies in a regular fashion as the temperature of the material changes. The most commonly available thermometer is based on the

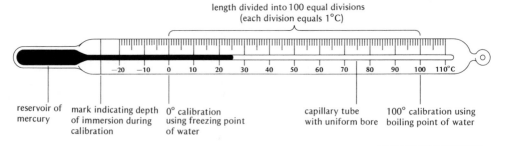

length divided into 100 equal divisions
(each division equals 1°C)

reservoir of mercury | mark indicating depth of immersion during calibration | 0° calibration using freezing point of water | capillary tube with uniform bore | 100° calibration using boiling point of water

FIGURE 3.14

The mercury thermometer.

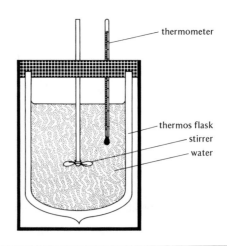

thermometer

thermos flask

stirrer

water

FIGURE 3.15

The calorimeter.

expansion and contraction of the volume of a certain amount of mercury as the temperature rises and falls. A typical mercury thermometer is shown in Fig. 3.14. It consists of a glass bulb containing pure mercury connected to a capillary tube of uniform bore. Normally, the 0° ice–water and the 100° water–steam points are used for two temperature marks, and the interval is divided into 100 equal degrees. The height of the mercury will give an accurate reading of the temperature in degrees centigrade depending on the uniformity of the capillary bore and the expansions of glass and mercury.

When heat is added to an object the temperature usually increases. Thus, temperature measurements can be used to determine heat energy. For example, the amount of heat absorbed by a given amount of water is directly related to the number of degrees increase in temperature of the water. In fact, this is how the commonly used unit of heat, the *calorie* (abbreviated cal), was first established. A calorie is the amount of heat required to raise the temperature of 1 g of water by 1°C (from 14.5 to 15.5°C). However, the amount of heat in a body is determined not only by its temperature but by the amount and kind of matter present as well. For example, 1 g of iron at 100°C contains less heat than 1 kg of iron at 100°C; also, 1 g of iron at 100°C contains less heat than 1 g of aluminum at 100°C, but more heat than 1 g of gold at 100°C.

The amount of heat needed to raise the temperature of 1 g of a substance by 1°C is called its *specific heat:*

$$\text{specific heat} = \frac{\text{number of calories}}{(\text{number of grams})(°C)}$$

The specific heat of water is high as compared with most substances. The specific heats of a few common substances are given in Table 3.5. A device used to measure heat is called a *calorimeter* (Fig. 3.15). It is an insulated vessel, for example, a Dewar (thermos) flask, equipped with thermometer and stirrer. Often a known amount of water is used so that the heat gained or lost by a substance is found from the heat lost or gained by the water and the calorimeter. To calculate the heat in calories gained or lost by a substance, a useful relationship is

heat in calories = (specific heat of substance) × (temp. change) × (mass of substance).

TABLE 3.5
Specific heats of common substances

Substance	Specific heat (cal/g °C)
water	1.000
aluminum	0.212
iron	0.108
gold	0.031
alcohol	0.581
salt	0.185
sand	0.188
carbon	0.127

EXAMPLE 3.11

Calculate the heat required to raise the temperature of 1 kg of water from 25 to 35°C.

Because 1 cal is needed to raise the temperature of 1 g of water by 1°C, the heat required to raise the temperature of 1 kg of water by 10°C (from 25 to 35°C) is

$$\frac{1 \text{ cal}}{(1 \text{ g})(1°C)} \times 10°C \times 1 \text{ kg} \times \frac{1000 \text{ g}}{1 \text{ kg}} = 10{,}000 \text{ cal} = 10 \text{ kcal}$$

EXAMPLE 3.12

The addition of 200 g of a metal at 20°C to 100 g of water heated to 86°C caused the temperature of water to drop to 79°C. What is the specific heat of the metal?

$$\text{heat gained by metal} = \text{heat lost by water}$$
$$= 100 \text{ g } (86{-}79°C)$$
$$= 700 \text{ cal}$$
$$\text{heat gained by 1 g metal} = 700 \text{ cal} \times \frac{1}{200 \text{ g}}$$
$$= 3.5 \text{ cal/g}$$
$$\text{temperature rise of the metal} = (79{-}20°C)$$
$$= 59°$$
$$\text{specific heat of the metal} = \text{calories of heat per gram of metal per °C}$$
$$= 3.5 \frac{\text{cal}}{\text{g}} \times \frac{1}{59°C}$$
$$= 0.059 \frac{\text{cal}}{\text{g °C}}$$

Because heat energy is measured so simply, it is common practice to convert other forms of energy into the equivalent amount of heat energy and then measure the heat.

SUGGESTED PROJECT 3.4
Energy changes from lowering and raising water temperatures

Take a cup of water in a quart-size Styrofoam cup. Measure the water temperature using an outdoor thermometer (in degrees Fahrenheit). Heat another cup of water in a pan to 110°F and pour it into the water in the Styrofoam cup. Stir the water gently with the thermometer and measure its new temperature. Calculate the amount of heat energy needed to raise the temperature of 1 g water by 1°C.

Repeat the experiment using a cup of cold water at 40°F (obtained by melting ice) instead of the hot water at 110°F. Calculate the amount of heat energy needed to lower the temperature of 1 g water by 1°C. Compare the results.

RADIANT ENERGY   The form of energy associated with ordinary light, X-rays, radio waves, or infrared rays is radiant energy. Collectively, all such radiations are known as *electromagnetic radiations*. They travel through space with a speed of about 3.00×10^{10} cm/sec (= 186,000 mi/sec), but without the apparent transmission of matter. However, each kind of radiation has a different energy (see Fig. 3.16). For example, radio waves have very low energy, X-rays have high energy.

All forms of radiation exhibit a dualistic nature; that is, they can be thought of either as *waves* or as *particles*. If we consider radiation as waves, such as tidal waves, we can speak of it in terms of *wavelength* (Fig. 3.16). For example, radio waves have

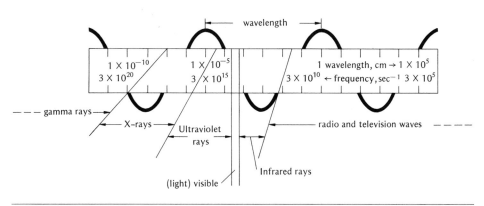

wavelength

| 1×10^{-10} | 1×10^{-5} | 1 wavelength, cm → 1×10^5 |
| 3×10^{20} | 3×10^{15} | 3×10^{10} ← frequency, sec^{-1} 3×10^5 |

- - - gamma rays

X-rays

Ultraviolet rays

radio and television waves - - - -

Infrared rays

(light) visible

FIGURE 3.16

The different types of radiant energy and the electromagnetic spectrum.

a wavelength of approximately 1×10^6 cm, whereas X-rays have wavelengths of the order of 10^{-7} to 10^{-9} cm. The shorter the wavelength the greater the energy. The velocity of light (3×10^{10} cm/sec) divided by the wavelength is termed the *frequency* (per sec) of the radiation. A common unit of frequency, cycles per second (now termed Hertz), is familiarly applied to radio frequencies by television and radio-broadcasting stations. Frequency is often used to distinguish one type of radiation from another. When radiation is thought of as a stream of discrete particles the particles or units of energy are referred to as *photons* or *quanta*. Because photons have extremely small mass, the larger the number of quanta the greater the total energy.

CHEMICAL ENERGY A substance can have energy by virtue of its ability to undergo chemical reactions. This energy is potential, but not because of any particular position with respect to the earth. For this reason, this type of energy is more appropriately called *chemical energy*. When matter undergoes change, chemical energy is transformed to other kinds of energy. For example, when coal and gasoline burn, chemical energy is converted into heat or light energy or both. In fact all life processes involve the change of chemical energy to other forms of energy. Conversely, other kinds of energy can be transformed to chemical energy by the proper kind of change. For example, radiant energy from the sun is transformed in plants to chemical energy that is stored in the substances constituting the plant. Humans and animals eat parts of the plant and, through chemical processes, gain some of the energy that is used in life processes (metabolism) and for performing useful work.

If substances change in such a way that heat energy is given to the surrounding environment, the change is said to be *exothermic*. Conversely, a change in which substances take up heat energy from the surroundings is said to be *endothermic*. Naturally, a process that is exothermic in one direction is always endothermic in the opposite direction. For example, the freezing of water is exothermic but the melting of ice is endothermic.

ATOMIC ENERGY Earlier in this chapter it was stated that matter is made up of tiny discrete particles called atoms or combinations of atoms which are called molecules. The kind of energy which is called *atomic energy* or *nuclear energy* is associated with the way in which atoms are constructed. It is possible to transform this type of energy to heat, light, and other kinds of energy. We shall discuss more about this in Chapters 6 and 16.

QUESTIONS

3.21 Distinguish between matter and energy.

3.22 Name the principal forms of energy.

3.23 Cite several examples to illustrate the conversion of energy from one form to another.

3.24 Distinguish between heat and temperature.

3.25 What is the relationship between a degree on the Celsius scale and one on the Fahrenheit scale?

3.26 The boiling point of oxygen is −297.35°F. What is the boiling point in
   a. degrees centigrade
   b. degrees Kelvin

3.27 The freezing point of gold is 1063.0°C. What is the freezing point in °F?

3.28 A European nurse reported that a patient's temperature was 40°C. Was the patient sick or healthy?

3.29 Describe the procedure that you would use for the calibration of a mercury thermometer.

3.30 Give the definitions of
   a. calorie
   b. specific heat

3.31 How much heat energy is required to change the temperature of 10 g water from 20 to 30°C?

3.32 A sample of matter weighing 100 g is heated to 96.5°C and then dropped into a calorimeter containing 1 kg water at 20.0°C. If the final temperature is 21.5°C, what is the specific heat of the substance?

3.33 List the various parts of the electromagnetic spectrum in the order of increasing energy.

3.34 What are photons?

3.35 Discuss the dual behavior of radiations.

3.36 What is meant by chemical energy?

3.37 Give several examples of the conversion of chemical energy to other kinds of energy.

3.38 What is the difference between exothermic and endothermic processes?

EQUIVALENCE OF MASS AND ENERGY

Observations of energy changes have shown that energy can be stored and transformed from one form to another, but cannot be created or destroyed in any transformation of matter. This observation is often called the *law of conservation of energy*. However, in 1905, Albert Einstein theorized that energy and mass are actually related in a special manner and that under the proper circumstances mass may be converted to energy and energy to mass according to the relationship $E = mc^2$. Here E is the energy in *ergs* (4.184×10^7 ergs = 1 cal), m is the mass equivalent in grams, and c is the velocity of light in centimeters per second ($= 3.00 \times 10^{10}$ cm/sec). Einstein's calculations showed that over 21 trillion (1 trillion $= 10^{12}$) cal of energy are obtained in the conversion of 1 g of matter into energy.

Because energy changes in physical and chemical processes are several orders of magnitude smaller than in mass-to-energy conversions, it is believed that only minute changes in mass occur and that these changes are too small to measure by weighing. Thus, for practical purposes, it can be said that mass is conserved during transformations of matter (the *law of conservation of mass*). However, in nuclear reactions, where the energy change is much greater than in chemical reactions, there is an observable difference in mass between reactants and products. Thus, some matter is converted to energy in a nuclear reactor (see Chapter 6). Note that all exothermic processes are accompanied by mass losses to the surroundings and all endothermic processes by mass gains from the surroundings.

According to Einstein's law of the equivalence of mass and energy, mass must be recognized as another form of energy. A particle at rest possesses energy that is proportional to what Einstein termed its *rest mass,* energy that can be thought of as energy of matter. A particle in motion has a *relativistic mass* which includes the rest mass and an added mass due to the kinetic energy (or other forms of energy) that the body possesses. Calculations have shown that at the speed of light the relativistic mass of an object is about 50 times its rest mass. However, atoms and molecules move about with velocities of only a few kilometers per second and, therefore, their relativistic masses are very close to their rest masses. In chemistry, the treatment of mass in a nonrelativistic manner introduces no serious errors and is thus justified.

QUESTIONS

3.39 **What is meant by the equivalence of mass and energy?**
3.40 **Distinguish between rest mass and relativistic mass.**
3.41 **Calculate the calories of energy evolved when 10 g of a substance is completely converted into energy.**
3.42 **Justify the validity of the law of conservation of mass in the light of Einstein's theory.**

FORCE

What is force? We know intuitively what a force is. We might simply call it "push" or "pull." More formally, *force is an action that may cause some effect.* For example, the action of pushing may cause the effect of motion—the movement of an object. There are various types of forces. *Gravitational force* is the attraction between the earth and objects present in the gravity field the earth produces in the space around it (Fig. 3.17). Indeed, it is a force that arises when any object is near another (Fig. 3.18).

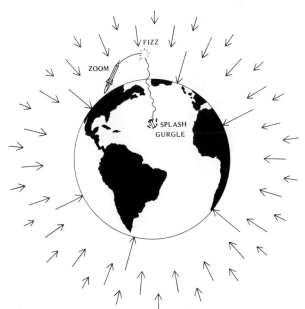

FIGURE 3.17

The earth's gravity field pulls objects toward the earth.

Some other examples of forces that are familiar to us are *magnetic* and *electrical* forces. We know that in the vicinity of a magnet there is a magnetic force field that can attract certain types of objects such as pieces of iron. Electrical forces result from the existence of positive and negative electric charges. Objects with electric charges have electrical force fields about them. Objects of unlike charge tend to attract one another, whereas objects of like charge tend to repel one another. These *electrostatic forces* of attraction and repulsion are easily demonstrated by the *pith-ball experiment* (Fig. 3.19). Electric charges constantly leak off charged bodies. The *lightning rod,* which affords protection against lightning strokes, is based on this principle and functions by leading the electric charges in clouds harmlessly into the ground (Fig. 3.20). It is important to realize that when objects interact an attractive or repulsive force may arise between them. The forces of interaction due to electric charges and magnetic fields are very important in understanding the nature and chemistry of our environment (see Chapter 4).

Just as in the case of macroscopic objects, there are forces of attraction between the ultimate microscopic particles (atoms, molecules, or ions) in matter. These forces are responsible for holding the particles together in liquids and in solids, but are much weaker than the forces that hold atoms together within a molecule (see

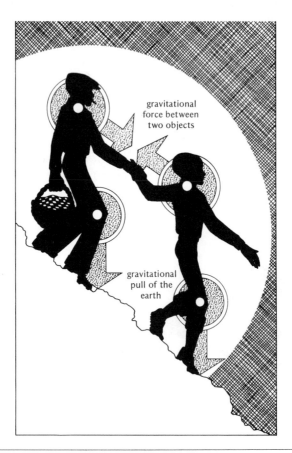

gravitational
force between
two objects

gravitational
pull of the
earth

FIGURE 3.18

Gravitational force between two objects.

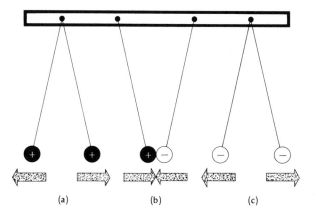

(a)   (b)   (c)

FIGURE 3.19

Pith-ball experiment showing electrical forces of attraction and repulsion. In (a) two pith balls repel each other because they have been positively charged by touching with silk-rubbed glass; and in (c) two pith balls repel each other because they have been negatively charged by touching with wool-rubbed amber. When a pith ball charged as in (c) is brought near another pith ball charged as in (a), an electrical force of attraction results as shown in (b).

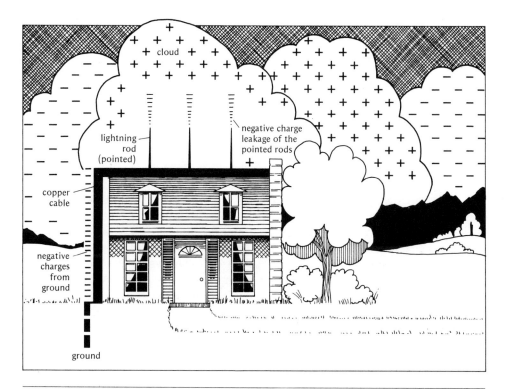

FIGURE 3.20

Electrical forces of attraction between clouds and lightning rods.

Chapter 4). In fact, for some substances, these forces are so weak that the particles fly apart entirely to form a gas. Consider the forces of attraction between water molecules. These *intermolecular forces* keep the solid ice or liquid water intact, except for a few molecules that manage to break away as gas or vapor (Fig. 3.21). If a piece of ice is heated it melts to form liquid water. The heat energy increases the kinetic energy (energy of motion) of the water molecules so that they begin to slide easily over one another rather than just stick together. This sliding motion results in the liquid state of water. Continued heating increases the motion of the molecules and the molecules at the liquid surface gradually tear free from the forces of inter-molecular attraction to form the gaseous state. Note the terms that are used in Fig. 3.21 for the processes of changing from one state to another.

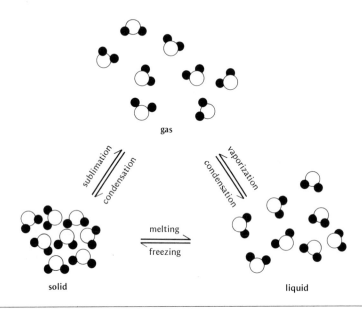

FIGURE 3.21

A molecular view of change of state.

The process of mixing or dissolving results from the preferential attraction of one particle for another. For example, when table salt (sodium chloride) is added to water it dissolves because the sodium and chlorine ions are attracted more to the water molecules than to themselves (Fig. 3.22a). On the other hand, when oil is added to water it does not dissolve because the forces between water molecules are so great that oil molecules cannot penetrate between them. Therefore, they *spread* over water (Fig. 3.22c). Gases always mix uniformly because there are no forces sufficient to restrain them from penetrating into one another (Fig. 3.22b). Nearly every form of environmental pollution is the result of the mixing, dissolving, or the spreading process. The reverse process of separation is therefore important in pollution and environmental work. Figure 3.23 illustrates some of the separation methods commonly used in the chemical laboratory and now finding application in the solution of many environmental problems.

When a mixture of oxygen and hydrogen gases is ignited with a flame a *chemical change* occurs and liquid water is produced (Fig. 3.24). Even at room temperature, the reaction proceeds with explosive violence due to the strong attraction between oxygen and hydrogen molecules. The change can be represented by the chemical equation

$$2H_2 + O_2 \longrightarrow 2H_2O$$

which gives the following information:

1. Both hydrogen and oxygen are diatomic (two-atom) molecules.
2. One molecule of water contains two atoms of hydrogen and one atom of oxygen.
3. The elements hydrogen and oxygen combine to form the compound water.
4. Two diatomic molecules of hydrogen and one diatomic molecule of oxygen yield two molecules of water.
5. Two moles of diatomic hydrogen molecules and 1 mole of diatomic oxygen molecules yield 2 moles of the compound water.

Usually the arrow is read as "forms" or "yields." The number written before a formula is called the *coefficient*. It indicates both the number of molecules and the number of moles of the substance involved.

You have just learned how much information a chemical equation supplies to the reader who understands the language of chemistry. However, the equation as written does not specify the state of the reactants and products and the energy changes involved in the reaction. This additional information may be included by modifying the equation as follows:

$$2H_2(gas) + O_2(gas) \xrightarrow{\text{flame}} 2H_2O(liq) + 137 \text{ kcal}$$

It is interesting to note that it takes energy to initiate the combination and that much energy is liberated in a clean way, making the reaction potentially important for reducing air pollution.

Sometimes it is necessary to heat a reaction mixture to produce the desired chemical change. For example, when we use a flashbulb in a camera, we pass a small electric current from a battery through some magnesium (Mg) metal and thus heat the metal in the presence of oxygen gas that is contained in the bulb to produce magnesium oxide (MgO):

$$2Mg + O_2 \longrightarrow 2MgO$$

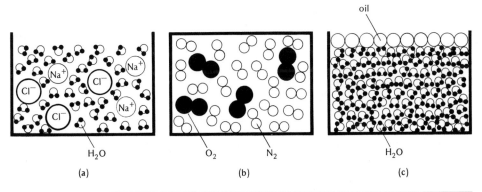

FIGURE 3.22

A molecular view of (a) the process of dissolving table salt in water, (b) the uniform mixing of oxygen and nitrogen in air, and (c) the spreading of oil over water.

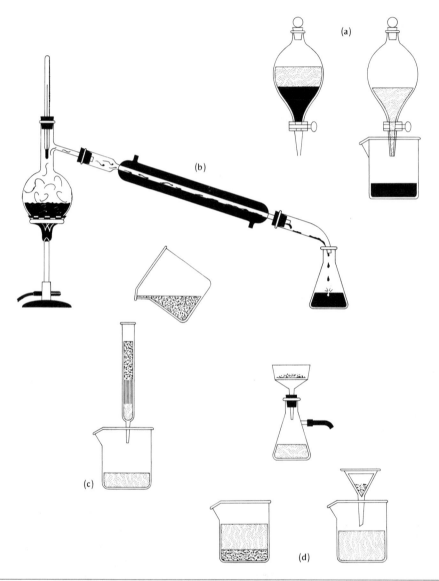

FIGURE 3.23

Separation methods: (a) *extraction* in which a separating funnel and two nonmixing liquids are used to separate substances on the basis of the preferential attraction of one substance for a liquid; (b) *distillation* in which selective vaporization of a substance from a mixture of substances is followed by condensation to form a pure or relatively pure liquid; (c) *chromatography* in which a mixture of substances in solution is at first poured through a material (which adsorbs one substance more strongly than another substance) to separate them and is subsequently collected by washing the material using a liquid; (d) *filtration* in which a filter with many small holes is used to separate solids from liquids that are forced through the filter.

Why does the magnesium have to be heated before it will burn? At room temperature the interatomic forces between magnesium atoms are strong enough to hold the atoms together tightly. When the electric current is passed, the heat energy causes the magnesium atoms to move about rapidly, so that they are separated from one another and the interatomic forces between them are weakened. At the same time, the oxygen atoms of the oxygen gas also move apart. Because magnesium atoms are attracted strongly to oxygen atoms, therefore magnesium oxide is formed. When this happens, both heat and light energy (the flash) are released.

Similarly, when a barbecue fire is started, the heat energy causes the carbon atoms of the charcoal to move about rapidly, so that they become more easily separated from one another and combine with oxygen atoms of the air to form the new molecules of carbon dioxide (CO_2):

$$C + O_2 \longrightarrow CO_2$$

When the new molecule is formed, energy is given up — heat energy that keeps the fire burning and cooks the meat. However, if there is not sufficient oxygen, then the poisonous carbon monoxide (CO) gas is formed:

$$2C + O_2 \longrightarrow 2CO$$

This reaction has contributed considerably to increasing air pollution, particularly because of the widespread use of coal for heating purposes in many industrial and domestic operations.

Within our own body cells, the carbon and hydrogen atoms from the food we eat are combined with the oxygen from the air taken in by the lungs to produce molecules of carbon dioxide and water. When these new molecules are formed, energy is released — heat energy that keeps us warm and acts as the force for our movements.

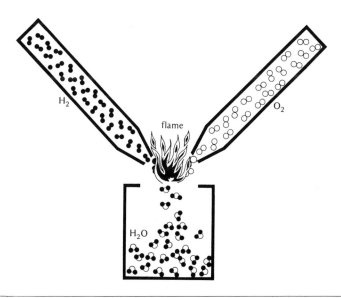

FIGURE 3.24

A molecular view of a chemical change. A mixture of oxygen and hydrogen gases is ignited with a flame to produce liquid water.

QUESTIONS

3.43 List from your personal experience as many consequences of the law of gravitation as you can.

3.44 Do you think jumping off a table demonstrates the law of gravity?

3.45 What is meant by the term force? Is it a basic quality of our environment? If so, why?

3.46 Explain the functioning of a lightning rod.

3.47 Why is it advisable to ground television and radio sets with outside aerials?

3.48 What are intermolecular forces? Outline the nature of these forces in the three well-defined forms of matter.

3.49 From the standpoint of intermolecular forces, explain the processes of mixing and dissolving.

3.50 What do you think are the dangerous consequences of spreading oil over our oceans?

3.51 What happens when a candle (a compound of carbon and hydrogen) burns in air? Explain the phenomenon by taking into account the intermolecular forces involved.

3.52 Summarize the information that you derive from the equation:

$$2CO + O_2 \longrightarrow 2CO_2$$

SUGGESTED PROJECT 3.5

Charging and discharging (or how to shock yourself and test your patience)

Make two pith balls from pieces of thoroughly dried raw potato and coat them with aluminum paint. Suspend the balls near each other from a support. Note that the thread by which they are suspended insulates them from their surroundings (see Fig. 3.19).

Rub a glass rod with a piece of silk and touch one of the pith balls with it. Observe what happens to the pith ball. Rub a rubber rod with wool or fur and touch the second pith ball with it. Observe what happens to the pith balls. Explain your findings by taking account of the electrical charges.

Caution! Do not perform this experiment on a really humid day. (Why?)

4

Fundamental particles of our environment

All the changes we can produce consist in separating particles that are in a state of cohesion or combination, and joining those that were previously at a distance.

JOHN DALTON (1766–1844)
English mathematician

Even before Dalton set forth his theory of hard and indivisible atoms, there was some evidence that matter is electrical in nature. In 1800 the two British chemists William Nicholson and Anthony Carlisle had passed an electric current through acidified water and decomposed it into hydrogen and oxygen (Fig. 4.1); and, before 1810, the English chemist Humphrey Davy had liberated elements such as potassium, sodium, magnesium, and calcium from their molten salts by using an electric current. Soon after, Michael Faraday theorized that the electric current was carried through the melt or solution by charged Daltonian atoms which he called *ions*. In fact, Faraday distinguished between two types of ions: positively charged ions, called *cations*, and negatively charged ions, called *anions*. Further work with solutions and melts established beyond doubt that the structure of the atom is electrical in nature, but the atom as such is electrically neutral.

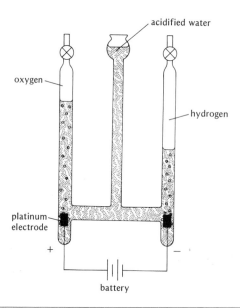

FIGURE 4.1

Apparatus for the electrolysis of water.

However, it took at least another 100 years of systematic work for the composition and structure of atoms and molecules to be elucidated. Many phenomena discovered during the latter part of the nineteenth century and the early twentieth century contributed to these researches. Except for two accidental discoveries, the advances made resulted from extensive investigations bearing on the relationship between electrical charge and structure. In this chapter we shall examine these developments which are of great interest to us today.

Living in the **atomic age**, we are well aware that the tremendous energy of the atom arising out of its complicated structure can be harnessed for both constructive and destructive purposes. In everyday life most of us use substances, such as plastics, rubber, drugs, and medicines, that were made by combining different atoms in a laboratory. How can we combine atoms to form desirable products? What must be the conceivable structure of atoms that will explain the formation of these products? The answers to these questions are best understood from an exposition of the structure of atoms and the developments that led to the understanding of this structure.

It is important to note that the structure of the atom, as we know it today, was developed by slowly fitting together one discovery after another. Often one discovery provided the means by which a later one could be made.

FIGURE 4.2

Sir Humphry Davy, English chemist of the early nineteenth century who discovered 12 elements.

NATURE AND COMPOSITION OF ATOMS

Following up on his work with melts and solutions, Faraday attempted to pass electricity through a tube that had part of the air pumped out. However, it was the English physicist William Crookes who, in 1876, succeeded in passing an electric current through a tube containing metal wires or plates connected to electric terminals (Fig. 4.3). With almost all of the air from the tube pumped out, Crookes observed a brilliant glow on the tube opposite the negative terminal (now referred to as the *cathode*). At that time the glow was attributed to some sort of radiation emanating from the cathode and was called *cathode rays*. Crookes' apparatus has since been known as *Crookes' tube* or simply the *gas discharge tube*; it is the forerunner of TV picture tubes so familiar to us today.

Soon after Crookes' discovery that highly rarified gases are good conductors of electricity, a considerable amount of systematic research in this field of work was begun by a number of investigators in Europe. However, before the results of the systematic work were correctly interpreted, two accidental discoveries occurred. One was the discovery of *X-rays* in 1895 by the German physicist Wilhelm Konrad Roentgen; the other the discovery of *radioactivity* in 1896 by the French physicist Antoine Henri Becquerel. Interestingly, both discoveries, accidentally made by scientists who were looking for something different, provided the tools by which the

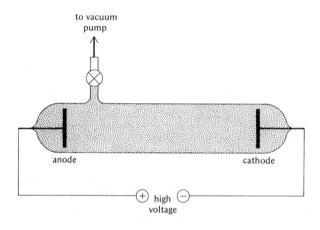

FIGURE 4.3

Crookes' cathode ray tube.

structure of the atom would finally be explored. In addition, the discovery of radioactivity gave immediate proof for the existence of particles of matter smaller than the atom. Therefore, we shall briefly discuss the nature of these accidental discoveries.

DISCOVERY OF X-RAYS AND RADIOACTIVITY   In 1895, while studying the glow produced in certain substances by cathode rays, Roentgen noted that whenever the cathode ray tube was operating, an invisible radiation was given off from the positive plate (now referred to as the *anode*). It could pass through solid objects and darken photographic plates on the other side of these objects. Surprised, Roentgen called the mysterious, penetrating radiation X-rays (high-energy light of short wave-lengths). As we shall see later in this chapter, Roentgen's new radiation found immediate application in the study of atomic structure. Today, we all know that X-rays are very valuable in the diagnosis of disorders within human bodies.

At about the same time that Roentgen discovered X-rays, Becquerel was experimenting with the visible light emission (fluorescence) of some natural compounds of the element uranium. In 1896, he noted that, besides fluorescent light, these compounds emitted a radiation that passed through black paper, which cannot be penetrated by ultraviolet light. Further experimentation proved that the new phenomenon was not related to X-rays but that it was a property of the element uranium. Later, the element thorium was found to exhibit the same property, and Pierre and Marie Curie followed with the famed discovery of the element radium. Elements that spontaneously give off radiation were termed *radioactive elements,* and the phenomenon of the release of radiation was called *natural radioactivity.* The discovery of natural radioactivity left little doubt that atoms were divisible; some naturally occurring atoms spontaneously decompose, producing fragments that are either atoms of lighter elements or subatomic particles.

DISCOVERY OF SUBATOMIC PARTICLES   In 1886, the German physicist Eugen Goldstein experimented with gas discharge tubes that had perforated cathodes (Fig. 4.4). To his expectation he found that while cathode rays were formed and sped off toward the anode, *positive rays* were formed and shot through the holes in the

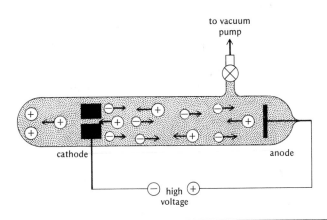

to vacuum
pump

cathode

anode

high
voltage

FIGURE 4.4

Goldstein's positive-ray apparatus.

cathode in the opposite direction. However, it was not until 1897 that the precise nature of both these rays was determined, largely due to the work of the English physicist Joseph Thomson. He first noted that the cathode rays were deflected in electric and magnetic fields. Thomson's apparatus showing deflection of the cathode rays in an electric field is shown in Fig. 4.5. Because the cathode rays were deflected toward the positive plate (anode) Thomson concluded that these rays must be composed of negatively charged particles and that each particle carried an identical negative charge. Subsequently, the name *electron* (abbreviated e) was given to these units of negative charge.

Thomson could only measure the mass-to-charge ratio of the electron by determining the degree of deflection in a magnetic field of known strength. However, he found that this ratio was the same regardless of the kind of rarified gas in the tube or the type of material comprising the cathode. In the waning years of the nineteenth century, the significant conclusion was made that *all matter contains electrons; all electrons are the same irrespective of their source.*

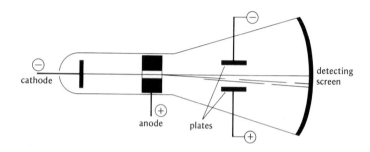

FIGURE 4.5

Thomson's apparatus for determining mass-to-charge ratio of the electron.

The beginning of the twentieth century was marked by the accurate measurement of the charge of the electron. The American physicist Robert Millikan sprayed microscopic oil drops into a chamber in which some of the air had previously been ionized by X-rays so that the oil drops would acquire electric charges. He then applied an electric field to make the charged oil droplets move upward or downward (Fig. 4.6). With only a few oil drops expended, Millikan was able to calculate the charges on droplets from the rate of upward motion of the droplets and the known electric field strength. It was found that *the charges on the droplets were always an exact multiple of some fundamental unit of electric charge which happened to be the charge on a single electron.* Further, this celebrated *oil drop experi-*

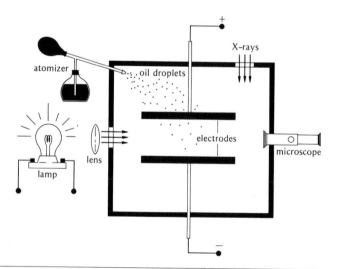

FIGURE 4.6

Millikan's apparatus for measuring the charge on the electron.

ment paved the way for the calculation of the mass of the electron, by coupling Millikan's value for the charge and Thomson's value for the mass-to-charge ratio. It was thus found that the electron is an extremely light particle, having a mass of only 9.1×10^{-28} g when it is at rest. We now know that this mass is 1/1838 the weight of a hydrogen atom; in atomic mass units this is equal to 5.5×10^{-4} amu.

Subsequent detailed studies of gas discharge tubes placed in magnetic fields revealed that many different types of positive particles could exist, depending on the gas used to flush out the tube prior to evacuation. The lightest positive particle was formed when there was a little hydrogen gas in the tube. It had a charge equal in size but opposite in sign to that of the electron and has a mass 1837 times that of the electron, that is, 1.67×10^{-24} g. This subatomic positive particle was subsequently named the *proton* (abbreviated p). In atomic mass units its mass is 1.007 amu.

Following the discovery and characterization of the electron and the proton, many scientists were looking for an answer to the question: Could atoms be thought of as being constituted solely of protons and electrons? It soon became clear to them that it was necessary to determine the masses of specific atoms and compare these masses with the total masses of protons and electrons per atom. Seeking to compare the masses of atoms directly, the English physicists Joseph Thomson and Francis Aston began their experiments on the deflection of streams of charged gaseous particles by means of known electrical and magnetic forces. Their experiments climaxed with the perfection of an apparatus called the *mass spectrometer* (Fig. 4.7) with which they could compare the masses of atoms precisely. It became

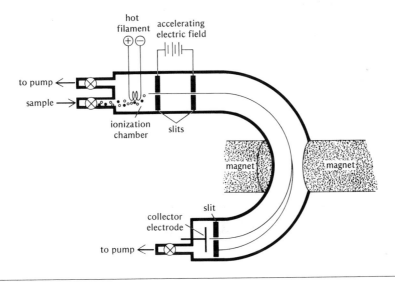

FIGURE 4.7

The mass spectrometer. In this instrument, positive ions are produced from atoms or molecules exposed to a heated filament. The positive ions are then accelerated by means of an electric field and passed through a slit into a magnetic field. The slit serves to select a beam of ions. The charged particles follow a curved path in the magnetic field which is determined by the mass-to-charge ratio of the ion. The heavier ions are more difficult to deflect. If the chamber is highly evacuated, there will be no air to interfere with the ion paths. When two ions with the same charge travel through the tube at the same velocity, the one with the greater mass will tend to follow the wider circle. By increasing the strength of the magnetic field, it becomes possible to bring first the less massive ions, and then the more massive ions, through the second slit, producing a signal at the collector plate.

clear that atoms of the same element differ in mass; such atoms were called *isotopes* (see Chapter 3). It also became clear that the masses of the protons and electrons do not account for the total masses of atoms.

To Thomson and the New Zealand physicist Ernest Rutherford, the different masses for atoms of the same element meant the existence of a different number of heavy, uncharged particles. They introduced the concept of a neutral particle called the *neutron* (abbreviated n) whose presence would explain the additional mass of an atom but not upset the balance of charges between protons and electrons. Thus, two isotopes of the same atom would contain the same number of protons but a different number of neutrons. However, the neutron turned out to be a rather elusive particle which escaped detection for years because it had no charge. Not until 1932 did the English physicist James Chadwick identify the neutron as a neutral particle with a mass of 1.0087 amu, which is about the same as that of a proton. The experiments that led to the identification of the neutron are discussed in Chapter 6. With the discovery of the neutron, scientists were able to discard completely the solid, indestructible Daltonian atom in favor of structured atoms composed of the subatomic particles listed in Table 4.1.

DISCOVERY OF THE NUCLEAR ATOM   Soon after the discovery of natural radioactivity, a simple experiment demonstrated that the radiation emanating from uranium and thorium was of three types. When passed through a strong electric field, a narrow beam of radiation was separated into three parts, as shown on a photographic plate (Fig. 4.8). One part was slightly deflected toward the negatively charged plate, indicating that there were heavy positive charges in the radiation; another part was substantially deflected toward the positively charged plate, indicating that there were light negative charges in the radiation; and, the remaining third part of the beam passed through the electric field undeflected, indicating that

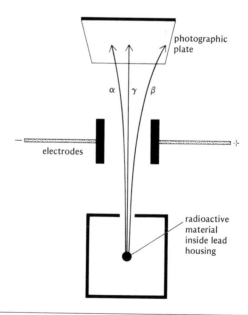

FIGURE 4.8

Behavior of radioactive rays in an electric field.

part of the radiation was uncharged. In 1899, Ernest Rutherford named these three types of radiation *alpha* (α), *beta* (β), and *gamma* (γ) *rays,* respectively. Subsequently, α-rays were identified as streams of positive helium ions, He^{2+} (helium atoms that have lost two electrons); β-rays were identified as streams of electrons; and γ-rays were shown to be very much like radiant energy, only more penetrating than X-rays. Due to the particular nature of α- and β-rays, they are also referred to as α- and β-particles. Table 4.2 summarizes pertinent information about these three types of radiations.

TABLE 4.1
Subatomic particles

Particle	Symbol	Mass (amu)	Relative charge
electron	e	0.00055	−1
proton	p	1.00728	+1
neutron	n	1.00867	0

TABLE 4.2
Characteristics of α-, β-, and γ-rays

Ray	Particle	Mass (amu)	Relative charge	Energy
α	helium ion	4	+2	low
β	electron	5.5×10^{-4}	−1	medium
γ	radiant energy	none	none	high

With the discovery of α-rays, Rutherford and his students began to study the interactions between α-particles and various kinds of matter. One of their experiments, which subsequently became known as the *gold foil experiment,* provided a clue to the location of protons and electrons within atoms. In this experiment, conducted in 1911, a narrow beam of alpha particles were allowed to bombard a very thin piece of gold foil (about 4×10^{-5} cm thick). By a fluorescent method, Rutherford and his students were able to show that most of the α-particles went straight through the foil; however, a small percentage of the α-particles was deflected and a very few were bounced back from the gold foil (Fig. 4.9). From these observations Rutherford reasoned that the atom is mainly empty space and that a massive but tiny particle with a large positive charge must be somewhere within the atom. He then proceeded to picture a relationship between the bombarding α-particles and the gold atoms (Fig. 4.10). As a result the concept of the existence of a center or *nucleus* within the atom emerged for the first time.

The fact that the proton was already known to be about 1837 times as heavy as the electron supported Rutherford's concept of the nuclear atom. Additional support was given by the English physicist Henry Moseley who, in 1913, conducted experiments to measure the *nuclear charge*. By using an X-ray tube similar to that shown in Fig. 4.11, Moseley found that various elements emitted characteristic X-rays when bombarded with cathode rays and that the wavelengths of the emitted X-rays (in the vicinity of 10^{-8} cm) decreased regularly as the atomic weights of the elements increased. A close examination of the experimental data enabled Moseley

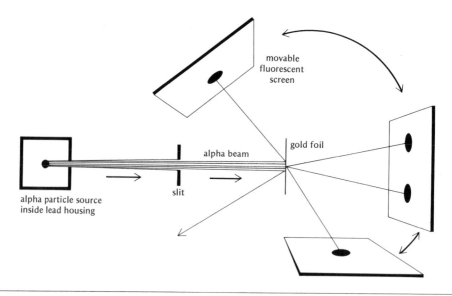

FIGURE 4.9

Rutherford's celebrated gold foil experiment.

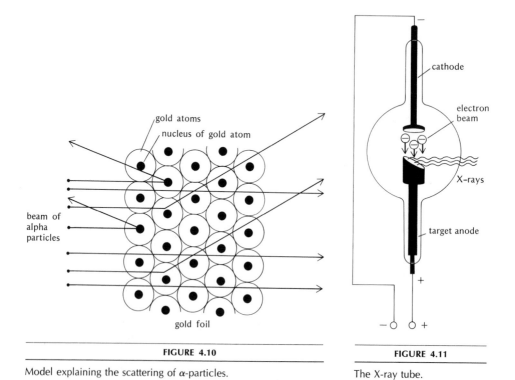

FIGURE 4.10

Model explaining the scattering of α-particles.

FIGURE 4.11

The X-ray tube.

to conclude that *the number of positive charges on the nucleus increases from atom to atom by a single electronic unit. This number of positive charges was termed the atomic number.* As we shall see in Chapter 5, the atomic number is the basis for a classification of the elements. It was later concluded that nuclear charge was caused by protons within the nucleus. Recall that each proton bears a positive charge equal in magnitude to the negative charges carried by electrons. On this basis, *the atomic number is equal to the number of protons in the nucleus of an atom, or the number of electrons in an uncharged atom.* Atomic numbers of all the elements are listed in Appendix A. The number of protons plus neutrons in the nucleus has been termed the *mass number* of the atom, and the term *nucleon* is used to refer to a nuclear particle, irrespective of whether it is a neutron or a proton.

The discoveries cited in the preceding paragraphs of this chapter formed the basis for developing structural models for the atom. Before we proceed with the discussion of these models let us summarize the main features of atoms.

MAIN FEATURES OF ATOMS An atom is an extremely small, electrically neutral particle that has a tiny but massive positive core or nucleus and one or more electrons relatively far outside its nucleus. The total space occupied by the orbiting electrons in an atom is quite large relative to the volume of the nucleus. The composition of individual atoms may be predicted using the following facts:

1. The atomic number is the number of protons or electrons.
2. The number of protons equals the number of electrons.
3. The mass number is the number of nucleons or the total number of neutrons and protons.
4. The mass number minus the number of protons equals the number of neutrons.
5. The isotopes of an element differ from each other only in the numbers of neutrons in their nuclei.

Note that the above rules allow for atoms losing or gaining charge to form ions and for two atoms of the same element having different masses. However, the question remains: How are the electrons placed around the nucleus? This question will be dealt with in the section on Atomic Structure (p. 74).

Just how big are the masses and sizes of atoms? We have stated earlier in this chapter that a proton weighs 1.67×10^{-24} g and that an electron weighs 9.11×10^{-28} g. Clearly, the weight of a hydrogen atom, which is composed of a single proton and an electron, is only slightly greater than 1.67×10^{-24} g. One of the heaviest atoms known, namely the uranium atom, weighs about 240 times that of the weight of a hydrogen atom. Even this mass is so minute that we usually compare atoms and subatomic particles with one another using their masses in atomic mass units. A simple calculation will show that it requires about 6.02×10^{23} amu to equal 1 g of hydrogen atoms.

Atomic diameters have been calculated from measurements of the volume occupied by a known number of atoms and by assuming the atom to be spherical. These calculated diameters are of the order of 10^{-8} cm. A common unit for describing the size of atoms is the Ångström which is equal to 10^{-8} cm. The calculations for various atoms show that atoms range in size from the hydrogen atom with a diameter of less than 1 Å to the cesium atom with a diameter of about 5 Å. It has also been estimated that an atomic nucleus has a diameter of only about 1/50,000 that of the atom, that is, about 2×10^{-5} Å. One can easily visualize how difficult it is to compare atomic dimensions with the dimensions of ordinary objects.

QUESTIONS

4.1 Which discoveries caused scientists to start thinking that atoms are destructible?

4.2 Distinguish between X-rays and cathode rays.

4.3 Suggest evidence favoring electrons as the fundamental particles of all matter.

4.4 Suggest an experiment to prove that a beam of fast moving particles consists of protons and not electrons.

4.5 Characterize the three types of emissions from naturally radioactive substances as to charge, relative mass, and the relative penetrating power.

4.6 Distinguish between the mass number of an atom and the mass of an atom.

4.7 How do the following discoveries indicate that Dalton's concept of the atom is inadequate?
   a. cathode rays
   b. positive rays
   c. natural radioactivity
   d. nucleus
   e. isotopes

4.8 If all the elements known at present are arranged in order of increasing atomic number, how do the successive elements differ in
   a. the number of protons?
   b. the number of electrons?
   c. the number of neutrons?
   d. atomic weight?

4.9 Name some practical applications of cathode ray tubes and X-ray tubes.

4.10 Briefly describe atomic dimensions.

ATOMIC STRUCTURE

Following the discovery of the electron and the proton, in 1898, Thomson proposed the earliest model for the atom. He suggested that the atom is a positively charged sphere on whose surface the electrons remain just like "raisins in a pudding" (Fig. 4.12a). With the completion of Rutherford's gold foil experiments in 1911, there emerged a new model called the *nuclear atom*. It envisaged a positively charged, tiny but massive nucleus surrounded by electrons some of which formed the outside surface of the atom (Fig. 4.12b). An important new concept then introduced was that some of the electrons are so far away from the nucleus that the space between the nucleus and the outer electrons is empty except for other floating electrons. Two years later, the Danish physicist Niels Bohr described the atom as "a miniature solar system" in which the electrons are in motion around a tiny, positively charged nucleus in definite elliptical orbits just as the planets around the sun (Fig. 4.12c). It was argued that the electron possesses some energy because of its motion and position and that the distance between the electron and the nucleus depends on the *electron energy*. Further, he made the revolutionary suggestion that electrons cannot have just any energy, but only certain allowed values, that is, the total energy of an electron is *quantized*. In other words, electrons can only be found at certain specific distances from the nucleus in specific orbits. The possible quantized positions of the electron were subsequently called the *energy states* or *energy levels* of the electron. Bohr's model did provide a good model for the simpler atoms but could not be used to explain atoms with many electrons.

The modern picture of the atom emerged as a result of the work primarily by the German physicists Paul Dirac, Werner Heisenberg, and Erwin Schrödinger in the third decade of this century. Although Thomson had proved that electrons were particles, the French physicist Louis de Broglie had noted that a beam of electrons can also behave very much like a beam of light; that is to say, the electrons also have wavelike properties. The new model of the atom was based on this wave nature of the electron. It abandoned the definite, planetary electron orbits of the Bohr atoms in favor of what are now known as *electron clouds*. The actual theory involves a complicated form of mathematics, often referred to as *quantum mechanics* or *wave mechanics*. However, a knowledge of this mathematics is not necessary in order to understand the concept of the atom finally developed. Actually, the present-day picture of the atom is somewhat similar to Bohr's model with the exception that the energy states or levels are more complicated than those in Bohr's model. The basis for such detailed description originated from the results of detailed analyses of the interaction of matter and radiant energy.

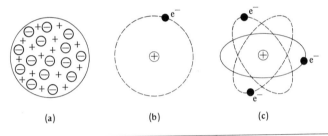

(a)   (b)   (c)

FIGURE 4.12

Atomic models: (a) Thomson's sphere model; (b) Rutherford's nuclear model; (c) Bohr's solar system model.

MATTER AND RADIANT ENERGY   Most of us are familiar with the rainbow phenomenon. One of the earliest discoveries of the nineteenth century was that when ordinary light is passed through a prism it is separated into a continuous spectrum of colors as in a rainbow. On the other hand, when the light from an electric discharge tube (such as those used by Thomson and Goldstein in their discoveries of the electron and the proton, respectively) is passed through a prism, only a few narrow lines are obtained. The setup of the simple prism (Fig. 4.13) is the forerunner

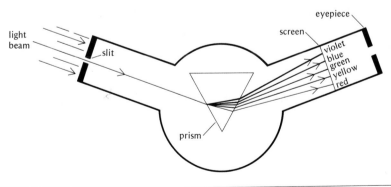

FIGURE 4.13

A simple prism spectroscope.

of the modern *spectroscope,* a device that made possible many studies of the features of radiant energy (light) emitted by matter as well as absorbed by matter.

As a result of several spectroscopic studies, it was concluded that the electrons outside the nucleus are normally in places of relatively low energy known as *ground states* or *levels.* However, when the atoms are subjected to the high temperatures of flames or to electric fields in discharge tubes, the electrons, especially those that are far away from the nucleus, absorb energy and are forced out to places of higher energy or *excited states* (Fig. 4.14). When these excited electrons fall back to lower-energy levels, a certain amount of energy is given off, sometimes as visible light. More intensive studies of the radiations emitted by excited atoms revealed that there are several possible energy states and that the energies of electrons within a given energy state differ from one another. It was thus necessary to postulate the existence of several *main energy levels* and *energy sublevels* within a given main energy level. Further, electron energies were found to increase with increasing distance from the nucleus.

The recognition of the energy sublevels was the result of discoveries of different series of lines in the spectra of excited atoms. As each new series of lines in the spectra was discovered the sublevels were given names such as *sharp, principal, diffuse,* and *fundamental* (abbreviated as *s, p, d,* and *f,* respectively). These names

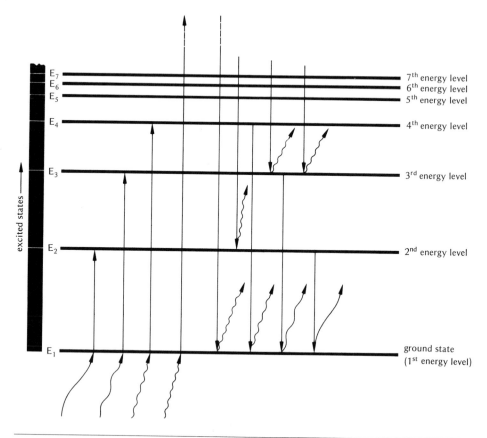

FIGURE 4.14

Excitation of electrons from low-energy levels (E_1) to higher-energy levels (E_7) and their de-excitations.

were based on the nature and importance of the lines in the spectrum. The energy levels were numbered, starting with the lowest (ground level) as 1, the next lowest as 2, and so on. The sublevels within each main energy level were identified by prefixing the number of the main energy level to the sublevel symbols or names. For example, 1s is the s sublevel in the first main energy level, 2p the p sublevel in the second main energy level, and so on. Detailed calculations showed that the number of sublevels in any main energy level is equal to the number of the main energy level. That is to say, the first main level has only one energy level, the second main level has two energy sublevels, the third main level has three energy sublevels, and so on. Furthermore, it was found that in the first main level there is only the s sublevel, in the second there can be s and p sublevels, in the third there can be s, p, and d sublevels, and so on. Finally, it was shown that just as the energy of the main level increases with increasing distance from the nucleus, so does the energy of the sublevels within a main level. In other words, within a main level the f sublevel has higher energy than the d sublevel, the d sublevel has higher energy than the p sublevel, and the p sublevel has higher energy than the s sublevel, and so on.

A schematic representation of the relative energies of the main energy levels and the energy sublevels for an atom is shown in Fig. 4.15. Such a representation is often referred to as an energy-level diagram. Note that the energies of the lowest

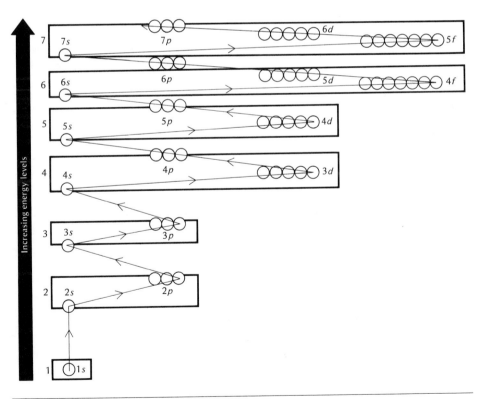

FIGURE 4.15

An energy-level diagram showing the relative energies of the main levels and sublevels.

levels differ considerably from one another, but at higher levels the differences become smaller. In some cases, a sublevel of a lower main level may be of higher energy than a sublevel of a higher main level, as for example in the case of the $3d$ and $4s$ sublevels.

ROLE OF WAVE MECHANICS Spectroscopic studies of the interaction of matter and radiant energy provided considerable information on electrons and their energy levels in atoms. However, these studies did not provide any information about the movement of electrons in atoms. As a matter of fact Heisenberg had already stated that it is not possible to measure precisely the position and the velocity of an electron at the same time. Our present knowledge of the position of electrons in atoms is the result of the use of mathematical methods known as wave mechanics. According to the wave mechanical model, electrons are not considered as particles with specific positions in the atom, but are viewed with respect to the likelihood or probability of being located in certain regions of space about a nucleus. These regions of space are referred to as *orbitals*. An electron may be in any place within an orbital at a given time, but there can be no more than two electrons in an orbital at the same time. Further, the two electrons in the same orbital must be spinning in opposite directions. That is to say, if one electron is spinning clockwise about an axis, the other electron in the same orbital must be spinning counterclockwise.

An important question at this point is, how many electron orbitals and electrons are there in a given energy level or sublevel? An s sublevel has *one* s orbital, a p sublevel has *three* p orbitals, a d sublevel has *five* d orbitals, and an f sublevel has *seven* f orbitals. Because an orbital can contain a maximum of two electrons, an s sublevel can accommodate only 2 electrons, a p sublevel a maximum of 6 electrons, a d sublevel a maximum of 10 electrons, and an f sublevel a maximum of 14 electrons. Note that the orbitals making up a sublevel correspond to the same electron energies and are sometimes called *degenerate orbitals*. Table 4.3 summarizes the energy-level structures in the first four energy levels.

The next question is, what is the order of filling orbitals? Because the electrons in an unexcited atom tend to occupy the lowest available energy positions, the orbitals are filled in the order of increasing energy of the orbitals or sublevels. This

TABLE 4.3
Energy-level structures in the first four energy levels

Main level	Sublevel Number	Sublevel Type	Orbitals Number	Orbitals Type	Electron state notation	Maximum number of electrons Sublevel	Maximum number of electrons main level
1	1	s	1	s	$1s$	2	2
2	2	s	1	s	$2s$	2	8
		p	3	p	$2p$	6	
3	3	s	1	s	$3s$	2	18
		p	3	p	$3p$	6	
		d	5	d	$3d$	10	
4	4	s	1	s	$4s$	2	32
		p	3	p	$4p$	6	
		d	5	d	$4d$	10	
		f	7	f	$4f$	14	

order is generally according to the simple scheme shown in Fig. 4.16. It is important to recall that although the first two main energy levels are widely separated in energy value, for the third, fourth, and higher main levels there can be overlapping of energies (see Fig. 4.15). As the number of electrons in an atom increases, the order in which sublevels are filled usually follows the energy ranking, but, within a given sublevel, each orbital is generally first occupied by a single electron before any orbital may have two electrons. Although there are exceptions to the order in which orbitals are filled, the scheme outlined above applies in enough cases to make it a reliable guide in arriving at the electronic structures of different atoms.

ELECTRONIC STRUCTURES AND CONFIGURATIONS OF ATOMS   We can now write the electronic structure for the atoms of any element in terms of the sublevel distribution of the electrons if we know the total number of electrons. The total number of electrons is readily known from the atomic number of the element. For example, hydrogen, which has an atomic number of 1, has its one electron located in the s sublevel of the first energy level. A convention used to indicate this is called the *orbital notation*:

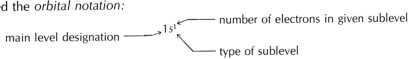

Similarly, the electronic configurations of helium (atomic number 2), lithium (atomic number 3), beryllium (atomic number 4), boron (atomic number 5), and carbon (atomic number 6) are $1s^2$, $1s^2 2s^1$, $1s^2 2s^2$, $1s^2 2s^2 2p^1$, and $1s^2 2s^2 2p^2$, respectively. To illustrate a rather complicated case, let us consider the electronic configuration of the atom mercury with atomic number 80. The 80 electrons in a mercury atom are distributed according to the following configuration:

$$1s^2 2s^2 2p^6 3s^2 3p^6 4s^2 3d^{10} 4p^6 5s^2 4d^{10} 5p^6 6s^2 4f^{14} 5d^{10}$$

Sometimes the same configuration is written abbreviated as [xenon core] $6s^2 4f^{14} 5d^{10}$, where the term xenon core stands for the electron arrangement of the atom xenon (atomic number 54) and indicates the filling of all sublevels through $5p^6$. Summing up, the number of electrons per main energy level for mercury is 2, 8, 18, 32, 18, 2.

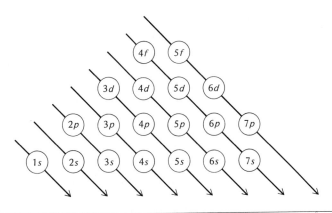

FIGURE 4.16

Order of increasing energy for sublevels and orbitals.

Thus, by using the order of energies of the sublevels and the electron capacities of the sublevels, the electronic configuration of an element can be deduced given the atomic number. The electronic configurations of all the elements are given in Appendix B of this text.

It is customary in deducing the orbital distribution of electrons to use a box and arrow representation for each orbital and electron: Electrons in filled orbitals are denoted by two arrows pointing in opposite directions to indicate opposite spins. This type of representation is particularly useful in delineating the distribution of electrons in partly filled sublevels. For example, let us consider the distribution of electrons in carbon (atomic number 6) and nitrogen (atomic number 7) atoms:

$$2p\ \boxed{\downarrow}\ \boxed{\downarrow}\ \boxed{\ }\qquad\qquad 2p\ \boxed{\downarrow}\ \boxed{\downarrow}\ \boxed{\downarrow}$$

$$2s\ \boxed{\uparrow\downarrow}\qquad\qquad\qquad\qquad 2s\ \boxed{\uparrow\downarrow}$$

$$1s\ \boxed{\uparrow\downarrow}\qquad\qquad\qquad\qquad 1s\ \boxed{\uparrow\downarrow}$$

carbon ($1s^2 2s^2 2p^2$)   nitrogen ($1s^2 2s^2 2p^3$)

Notice that the boxes are arranged so that the increasing energies of the orbitals are indicated. Also, within a given sublevel (p sublevel in the above cases), each orbital is occupied by a single electron before any orbital has two electrons. Such single electrons are referred to as *unpaired* electrons.

The electrons in the outer energy level are called *valence electrons*. When atoms combine to form molecules the outer energy electrons are usually the only electrons involved. Therefore, it is useful to indicate the valence electrons as dots around the symbol of the atom or element:

Li·   ·Be·   ·B·   ·C·   ·N·   :O·   :F·   :Ne:

Such symbols are called *electron-dot symbols* and provide a convenient means of representing the valence-level electron distribution in many elements. Notice that the elements Ne, Ar, Kr, Xe, and Rn all have completely filled s and p sublevels and, therefore, eight electrons in their outer energy or valence level. This is called the *noble gas* or *octet configuration,* and in the section on Bonding and Molecular Structure (p. 82), we shall see that there is a certain stability associated with such a configuration.

SHAPES OF ATOMS   The shapes of atoms are developed from the shapes of their electron orbitals, and it is therefore necessary to determine the general shapes of the electron orbitals. Because electrons are negatively charged and are in rapid motion within the atoms, they can be visualized as clouds of negative charge, that is, electron clouds. The shapes and densities of these clouds correspond to the likelihood or probability of finding the electrons in specific portions of the atom. Wave mechanical determinations allow us to approximate the shapes of these clouds.

In our previous discussion of the electronic structures of many-electron atoms, we considered four different types of orbitals, namely, the s, p, d, and f orbitals. An s orbital has been pictured as a spherically shaped cloud, and a p orbital as a dumbbell-shaped cloud (Fig. 4.17). The p orbitals comprising a given p sublevel are oriented perpendicular to one another as shown in Fig. 4.18. The d and f orbitals are thought to have more complicated shapes than the s and p orbitals. Further, the sizes of orbitals increase as the electron energy level increases; that is, at higher-energy levels the orbitals are larger in size than at lower-energy levels.

With these shapes of the orbitals in mind, it is possible to develop a three-dimensional picture of the atom. To do this we must consider the atom as consisting of a tiny nucleus surrounded by the large electron orbital clouds in which the electrons are located. We must also remember that the electron energy levels are made

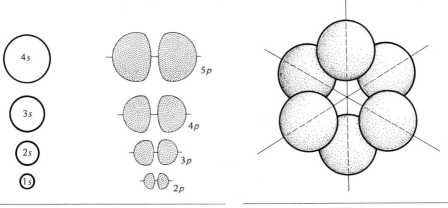

FIGURE 4.17

Shapes of s and p orbitals.

FIGURE 4.18

Orientation of p orbitals shown as a composite of the three orbitals in a given sublevel.

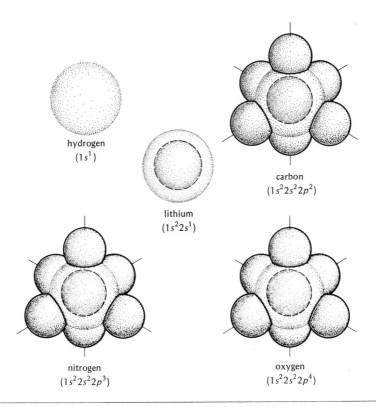

FIGURE 4.19

Shapes of some simple atoms.

up of energy sublevels and that these sublevels contain specific numbers of electron orbitals. Consider, for example, the hydrogen atom. It can be pictured in the ground state as consisting of the nucleus (1 proton) surrounded by a spherical cloud of negative charge ($1s^1$, one electron in $1s$ orbital), which is the electron located in the lowest possible energy state (Fig. 4.19). Excited hydrogen atoms may be diagrammed in a similar manner if we know the energy state of the electron on which the shape and size of the electron cloud depend. The lithium atom is somewhat more complicated than the H atom, for it has a $1s$ orbital with two electrons and a $2s$ orbital with 1 electron (Fig. 4.19). Notice that the $2s$ orbital is larger in size than the $1s$ orbital. In atoms such as those of carbon, nitrogen, and oxygen the three $2p$ orbitals are also involved, as shown in Fig. 4.19. We can extend this picture to atoms of other elements. However, for many-electron atoms the pictorial representation becomes complicated so that often idealized diagrams are used.

QUESTIONS

4.11 Compare and contrast the various models of the atom.
4.12 What is an excited atom according to the Bohr concept?
4.13 Distinguish between the terms orbit and orbital.
4.14 Name the various energy sublevels.
4.15 How many orbitals are there in the second, third, and fourth main energy levels?
4.16 What are degenerate orbitals?
4.17 What is meant by energy overlap? Illustrate with examples.
4.18 Write in orbital notation the electronic configurations of oxygen, iron, and aluminum atoms.
4.19 Distinguish between unpaired electrons and valence electrons.
4.20 What are the rules governing the filling of electron orbitals?
4.21 Draw the shape of the fluorine atom.
4.22 Write the electron-dot symbols for sodium, silicon, calcium, sulfur, phosphorus, and argon.

BONDING AND MOLECULAR STRUCTURE

Hitherto in this chapter we have explored the composition, structure, and shape of the atoms. The next logical step is to elucidate the manner in which atoms combine to form stable molecular structures such as those found in air, water, the earth, and our own bodies. Once this is done the likely structures and shapes of molecules can be understood.

A LOOK ALIKE TREND   Because valence electrons form the periphery of atoms, the combination of atoms to form stable molecules must necessarily involve some forces of interaction among the valence electrons of different atoms. From a study of the chemical behavior of various elements, Walther Kossell and Gilbert Lewis proposed in 1916 the *octet rule* which states that *atoms interact to change the number of electrons in their outer energy levels in an effort to achieve an electronic structure similar to that of a noble gas.* We have already noted that, with the exception of helium, all noble gases have eight electrons in their outer energy levels. Compared to the atoms of other elements the noble gas atoms are relatively very inert chemically. The octet must, therefore, represent a condition of great stability. If atoms other than the noble gas atoms must achieve a stable structure, then they must lose, gain, or share valence electrons with atoms in their neighborhood. On this basis two general bonding schemes have been devised and applied to account for the formation of stable molecules of two or more atoms.

Scheme I: transfer of electrons As the atom as such is electrically neutral, a loss or gain of electrons by an atom must result in the formation of a positive or negative ion. Let us illustrate this by considering two simple atoms: a sodium atom that has one electron in its outer energy level and a chlorine atom that has seven electrons in its outer energy level. If sodium loses one electron, then it will achieve the electron configuration of the noble gas neon:

$$\text{Na·} \longrightarrow \text{Na}^+ + e^-$$
$$[\text{Ne}]3s^1 \qquad [\text{Ne}]$$

Here [Ne] stands for the electron configuration of the neon atom. On the other hand, if chlorine gains one electron, it will achieve the electron configuration of the noble gas argon:

$$:\!\ddot{\text{Cl}}\!\cdot \; + e^- \longrightarrow \; :\!\ddot{\text{Cl}}\!:^-$$
$$[\text{Ne}]3s^23p^5 \qquad\qquad [\text{Ar}]$$

Compared to the reactive atoms of sodium and chlorine, their ions are very stable species and have completely different physical and chemical properties. Notice that an ion does not have the same number of protons and electrons; for example, Na^+ ion has 11 protons and 10 electrons. The difference is +1, the charge on the ion.

The reaction between sodium metal and chlorine gas can be considered to involve the transfer of electrons from sodium atoms to chlorine atoms to form sodium ions and chloride ions:

$$\text{Na} \quad + \quad \text{Cl} \quad \longrightarrow \quad \text{Na}^+ \quad + \quad \text{Cl}^-$$
$$1s^22s^22p^63s^1 \quad 1s^22s^22p^63s^23p^5 \quad 1s^22s^22p^6 \quad 1s^22s^22p^63s^23p^6$$
$$[\text{Ne}]3s^1 \qquad\quad [\text{Ne}]3s^23p^5 \qquad\quad [\text{Ne}] \qquad\quad [\text{Ar}]$$

The electrostatic attraction that one charged ionic species has for another results in the type of chemical bond called an *electrovalent* or *ionic bond*. When ions achieve

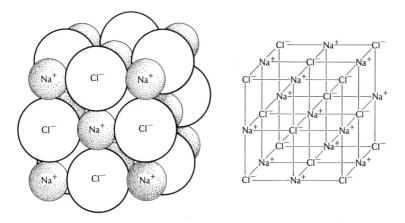

FIGURE 4.20

Crystal structure of the ionic solid sodium chloride (table salt). On left is the orderly aggregation of ions; on right is the relationship of ions in sodium chloride.

ionic bonding they arrange in the solid state in a three-dimensional pattern that satisfies the electrostatic attractive forces between the ions. This geometrical pattern involves a *crystal lattice* in which the positive and negative ions occupy specific lattice sites, as shown in Fig. 4.20. Thus the ionic compound Na^+Cl^- is actually a large molecule comprising many positive sodium ions bound to many negative chloride ions.

Generally speaking, elements with few (one to three) valence electrons (mostly metals) tend to lose electrons, whereas elements with many (four to seven) valence electrons (mostly nonmetals) tend to gain electrons to achieve the nearest possible noble gas electron configuration.

Scheme II: sharing of electrons   Another way in which atoms can acquire a noble gas electron configuration is by the *sharing* of electrons. For example, the formation of chlorine gas from chlorine atoms can be represented as

$$:\ddot{C}l\cdot + :\ddot{C}l\cdot \longrightarrow :\ddot{C}l:\ddot{C}l: \quad \text{or} \quad :\ddot{C}l—\ddot{C}l:$$

Notice that the chlorine atoms share a pair of electrons (indicated by the line) to achieve an octet of electrons around each chlorine atom. Sharing of electron pairs by atoms results in the type of chemical bond known as a *covalent bond* and the type of species known as a *molecule*.

A simple way to represent molecules is to use the electron-dot symbols of the atoms involved, so that the shared pairs of electrons are indicated. Such an arrangement of electron-dot symbols is called a *Lewis (electron-dot) structure*. Compounds in which the atoms are combined in molecules are referred to as molecular or covalent compounds. In general, nonmetals tend to form covalent compounds readily. As a matter of fact, several of the nonmetals exist in the form of diatomic molecules. However, there are two special cases of covalent bonding which we must consider.

Case 1: multiple covalent bonding. In the case of some diatomic molecules such as oxygen and nitrogen, it is observed that more than one electron pair must be shared by their atoms in order to obey the octet rule:

$$:\ddot{O}: + :\ddot{O}: \longrightarrow :\ddot{O}::\ddot{O}: \quad \text{or} \quad :\ddot{O}=\ddot{O}:$$

$$:\dot{N}\cdot + :\dot{N}\cdot \longrightarrow :N:::N: \quad \text{or} \quad :N\equiv N:$$

Notice that each of these atoms has access to eight electrons. The sharing of two pairs of electrons between two atoms (such as oxygen) is called a *double bond;* the sharing of three pairs of electrons between two atoms (such as nitrogen) is called a *triple bond.* The sharing of more than three pairs of electrons between two atoms does not occur.

Case 2: coordinate covalent bonding. In the case of some molecules and ions, the pair of electrons shared by two atoms is contributed by only one of the atoms. For example, the molecule BCl_3 combines with gaseous NH_3 to give a new crystalline compound, $BCl_3{:}NH_3$.

$$
\begin{array}{ccc}
:\ddot{C}l: & H & :\ddot{C}l{:}H \\
:\ddot{C}l{:}B \ \ + :N{:}H \longrightarrow & :\ddot{C}l{:}B{:}N{:}H \\
:\ddot{C}l: & H & :\ddot{C}l{:}H
\end{array}
$$

Notice that the BCl_3 molecule does not obey the octet rule (see Exceptions to the

Octet Rule, below). Here, the pair of electrons shared by the boron and the nitrogen atoms is contributed by the nitrogen atom. The sharing of an electron pair between atoms when both of the electrons of the electron pair come from one of the atoms constitutes a special kind of covalence and is referred to as *coordinate covalence*. The formation of a *coordinate bond* thus depends on the existence of a pair of un-used valence electrons. Not only molecules but also ions may add to unused electron pairs. For example, ammonia may add H^+ to form the polyatomic ammonium ion, NH_4^+, and water may add H^+ to form the polyatomic hydronium ion, H_3O^+:

$$
\begin{array}{c}
H \\
H:\!\overset{\cdot\cdot}{N}\!: + H^+ \longrightarrow \left[H:\!\overset{\cdot\cdot}{N}\!:\!H \right]^+ \\
H \qquad\qquad\quad H
\end{array}
$$

$$
H:\!\overset{\cdot\cdot}{\underset{\cdot\cdot}{O}}\!:\!H + H^+ \longrightarrow \left[H:\!\overset{\cdot\cdot}{O}\!:\!H \right]^+
$$

Actually, there is no difference once the bonds are formed, for each bond consists of a pair of electrons.

EXCEPTIONS TO THE OCTET RULE Generally, when atoms combine to form molecules, each atom obeys the octet rule. The reason behind the formation of octets is that in most instances only the outer *s* and *p* orbitals, which can accommodate a total of eight electrons, are involved in bonding. However, there are some exceptions, as exemplified by the electron-dot structures of the BCl_3 and PCl_5 molecules:

$$
\begin{array}{ccc}
& & Cl \\
:\!\overset{\cdot\cdot}{\underset{}{Cl}}\!: & Cl & \\
:\!\overset{\cdot\cdot}{Cl}\!:\!\overset{\cdot\cdot}{B} & & P\ Cl \\
:\!\overset{\cdot\cdot}{\underset{\cdot\cdot}{Cl}}\!: & Cl & \\
& & Cl
\end{array}
$$

Boron has 6 bonding electrons and phosphorus has 10 bonding electrons. Just as in the case of hydrogen

$$
H\!\cdot + H\!\cdot \longrightarrow H\!:\!H
$$

where a duo (two) of electrons produces a stable molecule, so in the case of boron and phosphorus, it must be assumed that there is a certain stability associated with 6 and 10 electrons, respectively.

RESONANCE CONFIGURATIONS For some molecules no single electron configuration can be drawn that will show the arrangement of the bonds between the atoms of the molecule and will be in agreement with its observed physical and chemical properties. In such cases it has been assumed that the actual configuration lies between two or more limiting or *resonance* configurations. For example, we may write the electron-dot structure of ozone (O_3) as

$$
:\!\overset{\cdot\cdot}{O}\!:\!:\!\overset{\cdot\cdot}{O}\!:\!\overset{\cdot\cdot}{O}\!: \quad \text{or} \quad :\!\overset{\cdot\cdot}{O}\!:\!\overset{\cdot\cdot}{O}\!:\!:\!\overset{\cdot\cdot}{O}\!:
$$

$$
O\!=\!O\!-\!O \quad \text{or} \quad O\!-\!O\!=\!O
$$

Both structures show two kinds of bonds between the oxygen atoms, a double bond and a coordinate bond (see Case 2, p. 84). Actually, the properties of ozone indicate no difference in the bonds between oxygen atoms and, hence, the bonds must have equal strengths. The ozone molecule is, therefore, considered to be something intermediate between the two resonance forms shown above. Such an intermediate structure is termed a *resonance hybrid*.

COEXISTENCE OF IONIC AND COVALENT BONDS   Numerous molecules exist in which more than one kind of bonding is found. In general, salts that contain atoms of more than two elements exhibit both ionic and covalent types of bonding. Consider, for example, the salts sodium sulfate (Na_2SO_4) and sodium phosphate (Na_3PO_4). These salts are ionic and exist as sodium and sulfate or phosphate ions in the crystalline form. Within the sulfate (SO_4^{2-}) and phosphate (PO_4^{3-}) ions, the atoms are held together by covalent coordinate bonds, that is, by sharing of electron pairs:

$$2Na^+ \left[\begin{array}{c} \ddot{O} \\ \ddot{O} \ddot{S} \ddot{O} \\ \ddot{O} \end{array} \right]^{2-} \qquad 3Na^+ \left[\begin{array}{c} \ddot{O} \\ \ddot{O} P \ddot{O} \\ \ddot{O} \end{array} \right]^{3-}$$

SHAPES OF MOLECULES   When atoms combine by sharing electrons, the bond formed results from an overlap of two atomic orbitals, each of which is occupied by a single electron. Naturally, the bond will be oriented in the direction of the orbitals that provide the electrons and will thus influence the shape and structure of the resulting molecule. Let us illustrate this by overlapping the orbitals of some atoms such as hydrogen, oxygen, and nitrogen to form the simple molecules H_2, H_2O, and NH_3. Figure 4.21 illustrates the formation of a hydrogen molecule as a result of the overlap of the 1s atomic orbitals of the hydrogen atoms. Recall that the 1s orbital of the hydrogen atom is half-filled and, as a result of the overlap of two half-filled orbitals, an electron pair is shared to form a covalent bond. Similarly, if we consider the overlap of the 1s orbitals of the hydrogen atoms with two perpendicular 2p orbitals of the oxygen atom or three perpendicular 2p orbitals of the nitrogen atom, we have the idealized shapes of the water and ammonia molecules as shown in Fig. 4.22. Note that as a result of the overlap the electron density is increased between the combining atoms; it is this concentration of electron density that constitutes the covalent bonds and produces the molecule.

In general there is a very pronounced tendency for single electrons in an atomic orbital to pair up with another single electron to form a filled orbital. However, an electron pair will repel another electron pair so that the various electron pairs exist as far apart as possible. Let us illustrate this by considering the example of the methane (CH_4) molecule which has the following electron-dot structure:

$$\begin{array}{ccc} & & H \\ H & & | \\ H \ddot{:} C \ddot{:} H & \text{or} & H—C—H \\ H & & | \\ & & H \end{array}$$

Clearly, the carbon atom occupies a somewhat central position because it has four electrons available for bonding, whereas each of the hydrogen atoms has only single bonding electrons. Thus, in the methane molecule, carbon is the central atom, surrounded by four pairs of valence electrons. To satisfy the criterion of repulsion among electron pairs, the four pairs of valence electrons are best arranged in a *tetrahedral* fashion about the central carbon atom. That is to say, the electron pairs will be directed toward the apices of a tetrahedron with an angle of about 109.5° between a given pair of electrons (Fig. 4.23). Such a structure indicates an equivalence of the bonds within the molecule. The same approach can be used to predict

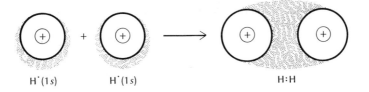

FIGURE 4.21

Interpenetration of the 1s atomic orbitals of two hydrogen atoms to form the hydrogen molecule.

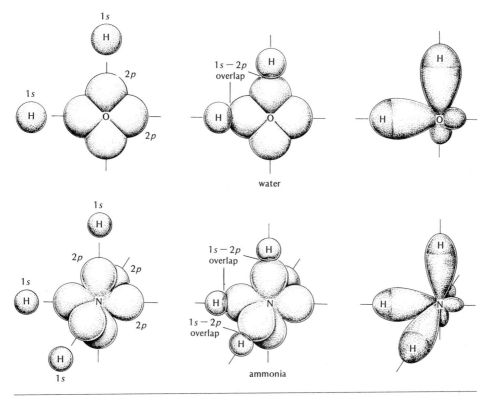

FIGURE 4.22

Idealized shapes of water and ammonia molecules. Atomic orbitals which are not used in bond formation are not shown here. (Reproduced, by permission, from T. R. Dickson, *Introduction to Chemistry*, p. 120. Copyright © 1971 by John Wiley & Sons, Inc.)

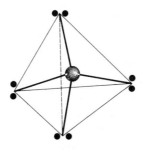

FIGURE 4.23

Tetrahedral distribution of four electron pairs about a central atom.

the geometric shapes of other molecules, even in cases in which the central atom has less than or more than four pairs of electrons. The predicted shapes of various molecules are shown in Fig. 4.24.

Molecule	Shape	Type	Electron–dot structure
carbon dioxide (CO_2)	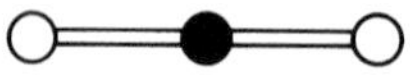	linear	:Ö: :C: :Ö: or O=C=O
water (H_2O)	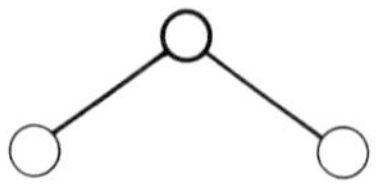	angular	:Ö: H ˙˙ H or H ·Ö· H
ammonia (NH_3)	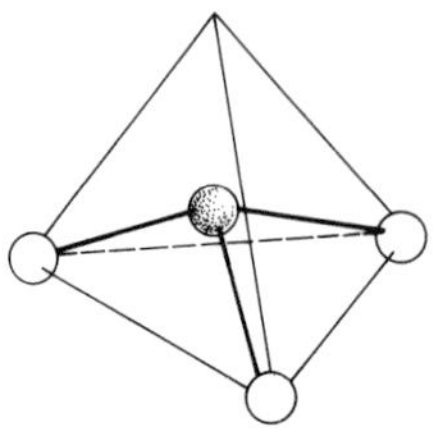	triangular pyramid	·N· H ˙˙˙ H H or :N: H N H H
methane (CH_4)	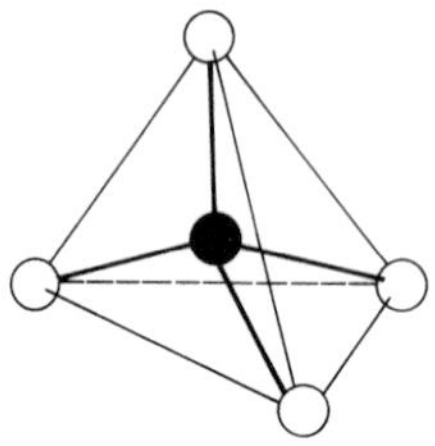	tetrahedral	H H:C:H H or H H—C—H H

FIGURE 4.24

Geometric shapes of various molecules.

The geometric shape of the methane molecule clearly indicates that the centers of the atoms are located at certain distances from one another. This distance is called the *bond length*. The angle formed by the bonds of any two atoms to the central atom is called the *bond angle*. Actually the bonded atoms vibrate in a manner that constantly changes the bond lengths and the bond angles by small amounts. However, it is possible to determine experimentally average bond lengths and angles in many molecules. Table 4.4 lists bond distances and angles for a few simple molecules in our environment. Finally, it is possible to determine the strength of a bond by measuring the energy required to separate and isolate the atoms from their molecular and ionic combinations. These energies, called *bond energies,* have been determined experimentally for many combinations of atoms. Some representative bond energies are given in Table 4.5. Note that ionic and covalent bonds are the strongest bonds between atoms and are roughly of the same order of magnitude. Moreover, the bond strength increases with bond multiplicity: The carbon-to-carbon triple bond (three pairs of shared electrons) has an energy of 199 kcal/mole compared with 146 kcal/mole for the double and 83 kcal/mole for a single carbon-to-carbon bond. We shall use this information in studying the chemistry of carbon compounds (see Chapter 9).

If we examine the electronic configuration of carbon ($1s^2 2s^2 2p^2$), we notice that a carbon atom has only two unpaired electrons available for bonding. The symmetrical shape and the equivalence of the four C—H bonds in the methane molecule (see Fig. 4.23) require four orbitals to be involved in bond formation. Here is a situation that requires the "promotion" of one of the $2s$ electrons to the vacant $2p$ orbital of the carbon atom when the methane molecule is formed:

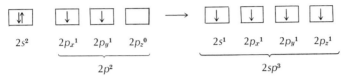

We have to assume that the energy for this promotion comes from the energy of reaction resulting from the combination of atoms. In any case, four orbitals become available for bond formation, with an unpaired electron in each orbital. To account for the equivalence of the four C—H bonds, we assume that four orbitals (one s and three p) are equalized or *hybridized* in such a way that the four become exactly alike and are oriented toward the four corners of a regular tetrahedron, such as is pictured in Fig. 4.23. This rearrangement of the orbitals to give four equivalent orbitals capable of bond formation is referred to as the *hybridization of orbitals,* and

TABLE 4.4
Bond distances and angles of some environmentally significant molecules

Molecule	Bond	Bond distance (Å)	Bond angle (degrees)
H_2O	O—H	0.97	104.5
H_2S	S—H	1.32	92
SO_2	S—O	1.43	118
CO_2	C—O	1.16	180 (linear)
CH_4	C—H	1.09	109 (tetrahedral)
NH_3	N—H	1.01	107

TABLE 4.5
Representative bond energies

Single bonds		Multiple bonds	
Bond	Bond energy (kcal/mole)	Bond	Bond energy (kcal/mole)
H—H	104	O=O	118
H—Cl	103	C=O	178
C—C	83	C=C	146
C—O	83	C≡C	199
C—H	87	N≡N	225
O—H	111		
Si—O	106		

the bonds derived from one s and three p orbitals are referred to as sp^3 hybrid bonds. Several other types of hybrid bonds are possible. For example, for carbon atoms forming a double bond the hybridization is an sp^2 arrangement and for carbon atoms forming a triple bond the hybridization is an sp arrangement:

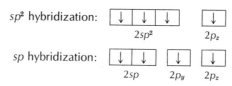

Note that the structure of the BCl_3 molecule can be explained on the basis of sp^2 hybrid bonds.

How the concept of hybridization of orbitals leads to particular molecular formulas and shapes is discussed in Chapter 9.

POLAR BONDS AND POLAR MOLECULES   Like atoms, molecules must have no overall charge. The total number of electrons must equal the total number of protons present in the various nuclei involved in the molecule. Because the electrons surround the nucleus in an atom the average center of negative charge usually coincides with the position of the nucleus. However, the average positions of some of the electrons may change as a result of the sharing of electron pairs in molecules. That is, the centers of positive and negative charge may not coincide in some molecules and as a result a region of positive and negative charge may be associated with certain parts of some molecules.

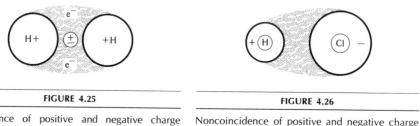

FIGURE 4.25	FIGURE 4.26
Coincidence of positive and negative charge centers in a hydrogen chloride molecule.	Noncoincidence of positive and negative charge centers in a hydrogen chloride molecule.

Usually when a molecule involves similar atoms, such as in the case of the hydrogen molecule, the two nuclei are the same and they equally share the bonding electron pair, the orbital of which occupies a position between the two positive nuclei. In consequence, the average center of positive and negative charges coincide as shown in Fig. 4.25. However, when a molecule involves different types of atoms, as in the case of the hydrogen chloride (HCl) molecule, the nuclei do not generally share electron pairs equally because the chlorine attracts the bonding electron pair more than the hydrogen. In other words, there is a greater probability of locating the electron pair in the vicinity of the chlorine than in the vicinity of hydrogen (Fig. 4.26). In consequence the average centers of positive and negative charge do not coincide and we say that there is an uneven distribution of charge. When this occurs in a covalent bond it is referred to as a *polar bond*. The extent of the polarity depends on the distance between the centers of unlike charge. The greater the separation of charge the greater the polarity of the bond. A molecule, such as hydrogen chloride, in which there is a net separation of positive and negative charge centers is called a *polar molecule* and is said to possess a *dipole moment*. Dipole moment is a measure of the extent of charge separation in a polar molecule. The situation is similar to ionic combinations and a polar molecule may be considered as having some ionic character.

In the case of a polyatomic molecule, whether or not it is a polar molecule depends on whether or not the average centers of positive and negative charge are separated. Thus the structure of the molecule is the deciding factor even if polar bonds are present in the molecule. For example, H_2O molecules that have polar O—H bonds are polar molecules possessing a dipole moment because they are angular molecules and their average centers of positive and negative charge are separated. In contrast, carbon dioxide that has polar C═O bonds is a *nonpolar molecule* because it has a linear structure and the average centers of negative and positive charge of the molecules coincide as there is no net separation of charge.

O
H ⊖ H   O═C═O
⊕   ⊕

water   carbon dioxide

Generally, in molecules with tetrahedral structures, such as CH_4, the centers of positive and negative charge coincide, and they are, therefore, nonpolar molecules with polar bonds. Molecules with angular structure and triangular pyramid structure, such as H_2S and NH_3 respectively (see Fig. 4.24), have their average centers of charge separated so that they are considered to be polar molecules.

An expected consequence of the highly polar nature of H_2O molecules, based on the rule that like dissolves like, is that the totally nonpolar O_2 molecules do not dissolve to an appreciable extent in water. This is borne out by the fact that despite the enormous reservoir of O_2 in the atmosphere and its high relative concentration in air (21% by volume), the O_2 that dissolves from air into water reaches only about 10 ppm at full saturation. On the other hand, the high polarity of the H_2O molecule allows water to dissolve variable amounts of ionic material substance from the earth's crust. For example, table salt (sodium chloride), in which sodium ions (Na^+) and chlorine ions (Cl^-) are held together by ionic bonds, dissolves in water because the negative oxygen atom of H_2O attracts Na^+ with a force that is strong enough to pull Na^+ away from its crystalline Cl^- neighbor (Fig. 4.27). Once pulled loose, the Na^+ ion is surrounded by other H_2O molecules, each tending to bind the Na^+

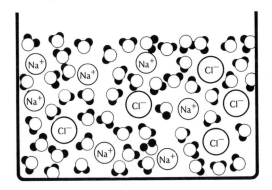

FIGURE 4.27

The dissolving of salt in water.

through its negative oxygen atom.* Sodium chloride dissolves, then, because H_2O binds Na^+ and Cl^- tightly enough to pull them out of the NaCl crystal. This fact, combined with the abundance of NaCl in nature, explains why the oceans are salty and why salt is a frequent contaminant of inland waters as well (see Chapter 14).

We may now ask the question as to how it is possible to predict whether a bond is polar or nonpolar. Clearly, the polarity of a bond depends on whether or not an atom involved in the bond attracts the shared electron pairs to a greater extent than the other atom involved. The American chemist Linus Pauling has devised a scale of numbers, called *electronegativities,* which represent the tendency of an atom of a given element to attract electrons (Table 4.6). These electronegativity values range from 4.0 for fluorine to 0.7 for cesium. The greater the electronegativity of an element, the greater the tendency for the atoms of that element to attract electron pairs in covalent bonds and form polar molecules. Metals generally have low electronegativities and for this reason they are sometimes referred to as *electropositive elements.*

TABLE 4.6
Electronegativities of some elements

Elements	Electro-negativity
F	4.0
O	3.5
N, Cl	3.0
Br	2.8
C, I, S	2.5
Au, Se	2.4
H, Pt	2.2
P, Te	2.1
Ag, Bi, Cu, Hg, Pb, Sb	1.9
Fe, Co, Ni, Mo, Si, Ge, Sn	1.8
Cr, Zn	1.6
Mn	1.5
Mg	1.2
Li, Ca	1.0
Na	0.9
K	0.8
Cs	0.7

Although many substances appear to fall into either the ionic or covalent classification, several others seem to exhibit a bonding intermediate between the two. In fact, sometimes we refer to *partially ionic* or *partially covalent* substances. The scale of electronegativity values is particularly useful in predicting the degree of ionic character, or polarity, of compounds. The greater the difference in electronegativity values between two atoms, the more ionic, or more polar, the bond between these atoms. When the difference is small, the bonding is essentially covalent.

---

* In the same way, Cl^- is bound by the positive part of several H_2O molecules. The binding of ions by the charged ends of the H_2O molecule is called *hydration* and the cluster of H_2O molecules with an ion in the center is called a *hydrated ion* (see Chapter 5).

Let us illustrate this by taking the example of a chlorine atom combining with another chlorine atom, a hydrogen atom, and a sodium atom, respectively:

Cl—Cl	H—Cl	Na$^+$Cl$^-$
$(3.0 - 3.0 = 0)$	$(3.0 - 2.1 = 0.9)$	$(3.0 - 0.9 = 2.1)$
pure covalent	polar covalent	primarily ionic
(nonpolar)	(partially ionic)	

These combinations represent the transition from a purely covalent to a polar covalent and, finally, to an ionic substance. In general, if the difference in electronegativities of the atoms is 2 units or more the bonds are primarily ionic; if the difference is less than 2 the bonds are considered covalent but polar in some degree; and, if the difference is zero the bonds are purely covalent.

THE HYDROGEN BOND Because polar molecules have separate centers of positive and negative charge, polar molecules can be attracted to one another by electrostatic forces; that is, the negative ends of polar molecules are attracted toward the positive ends of others and vice versa. This is called a *dipole-dipole* interaction.

A very important case of dipole-dipole interaction occurs in collections of polar molecules that involve hydrogen bonded to a highly electronegative atom such as oxygen or nitrogen. Because of the high polarity of these bonds and the lack of any inner energy-level electron shielding around the hydrogen, it is possible for a strong intermolecular electrostatic attractive force to arise. This force of attraction between the hydrogen of one molecule and the electronegative atom of another molecule is strong enough to be considered to be a type of chemical bond. It is called a *hydrogen bond*.

The hydrogen bond is important because it gives rise to fairly stable aggregates of water molecules:

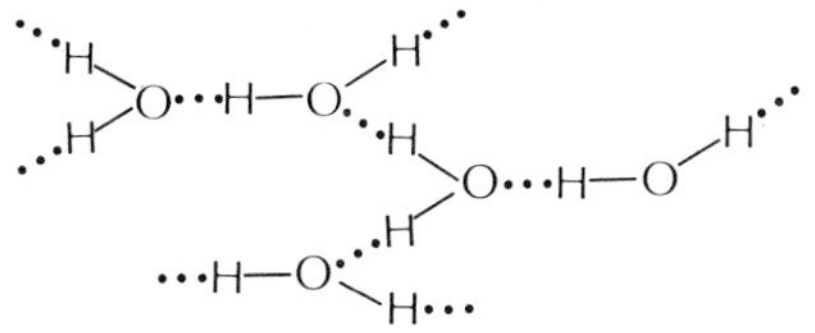

Here the dotted lines represent the hydrogen bonds. In fact, hydrogen bonds are responsible for holding water molecules together as a liquid under normal earth conditions.

The hydrogen bond is an important factor involved in the structures of many molecules of biological importance (see Chapter 11). It is easily broken and re-formed, and is apparently involved in many important biological processes. However, hydrogen bonding in water is so extensive that the large amount of energy needed to break hydrogen bonds makes water an excellent energy absorber. Large bodies of water, such as the oceans, rivers, and lakes moderate extremes of climate by absorbing solar energy. In the same way, water is an excellent medium (coolant) for carrying away industrial heat (see Chapters 8 and 14).

QUESTIONS

4.23 **What is the octet rule?**
4.24 **Distinguish between ionic and covalent bonds.**
4.25 **Name two compounds that do not obey the octet rule.**
4.26 **What is a hydrogen bond?**

4.27 Define electronegativity. Which elements are the most electronegative? The least?

4.28 Distinguish between polar and nonpolar molecules.

4.29 Name two compounds that exhibit both ionic and covalent bonding.

4.30 Sketch the likely shapes of the phosphine (PH_3) molecule and the phosphonium (PH_4^+) ion.

4.31 Define the terms bond length, bond angle, and bond energy.

4.32 Predict whether or not CH_3Cl is a polar molecule.

4.33 What is meant by hybridization of orbitals? Give examples.

4.34 Write the electron-dot structures for chloromethane (CH_3Cl), oxygen difluoride (OF_2), and tetrafluoroethene (C_2F_4).

4.35 Discuss the transition from purely covalent to completely ionic compounds.

4.36 What do you think John Dalton's response would be if he were to read this chapter? Comment with particular reference to the opening quotation from Dalton at the beginning of this chapter.

SUGGESTED PROJECT 4.1

Building models of atoms and molecules (or atomic and molecular architecture)

We have seen that an atom can be visualized as consisting of a nucleus (protons and neutrons) surrounded by the electron orbital clouds in which the electrons are located. If we know the general shapes of the orbital clouds and their spatial distribution around the nucleus we can build models for different atoms. Because molecules are composed of atoms, the spatial distribution of the atoms within a molecule will give rise to the shape of the molecule. A simple way to determine the shape of a molecule is to consider the spatial arrangement of the orbitals that are involved in binding the atoms together to form the molecule.

Another way to deduce the shape of a molecule is to consider the distribution of the valence electron pairs around the atoms in a molecule. The electron pairs involved in covalent bond formation must be arranged about the central atom in a molecule as far apart as possible. If we know the angle between a given pair of electrons, we can develop geometric shapes for the molecules.

In this project the student will build various models using a block of soft Styrofoam and a molecular model-building kit.

Atomic orbital shapes.   Cut out from the Styrofoam block several cubes of 2 and of 3 cm. From each of these cubes carve out spheres of 2 and 3 cm diameters and use these spheres to represent the 1s and 2s orbital clouds, respectively. Also, cut out from the Styrofoam block pieces $2 \times 3 \times 3$ cm in dimensions and carve out of them dumbbell shapes (see Fig. 4.17). Orient two dumbbell shapes perpendicular to one another using toothpicks, as shown in Fig. 4.18. Several such pairs are to be used for representing 2p orbitals.

Shapes of atoms.   Construct the shapes (see Fig. 4.19) of

a. a hydrogen atom

b. a nitrogen atom

c. an oxygen atom

Use a 1s orbital cloud to represent the hydrogen atom; a 2s orbital cloud and three 2p orbitals for nitrogen and oxygen atoms, assuming that the 1s orbital cloud (which is smaller in size than the 2s orbital) is within the 2s sphere. Indicate the presence of one-electron orbital by painting the orbital shapes with a lighter shade and two-electron orbitals using a darker shade. Submit the models thus prepared to your instructor for his comments.

Shapes of ionic solids. Use 2- and 3-cm spheres to represent Na^+ and Cl^- ions, respectively. Glue them together in the fashion shown in Fig. 4.20 to obtain a representation of the ionic solid sodium chloride. Note the orderly aggregation of ions in a geometrical pattern which involves crystal lattice. Submit the model to your instructor.

Shapes of molecules. Construct the shapes of

a. a water molecule
b. an ammonia molecule

by overlapping the appropriate orbitals of the atoms involved in each molecule. For example, overlap the 1s orbitals of the hydrogen atoms with two perpendicular 2p orbitals of the oxygen atom to obtain the shape of a water molecule (see Fig. 4.22). It is not necessary to show the atomic orbitals of oxygen that are not used in the formation of the bonds. Carve out of the Styrofoam block the final shapes of the water and ammonia molecules and submit them to your instructor.

Note that the models you have constructed have angular shapes. Use the appropriate color models and sticks in your molecular model-building kit to construct

a. a water molecule
b. an ammonia molecule
c. a methane molecule

Draw on paper possible geometric shapes for the molecules noting that the central atom in the molecule has around it four pairs of electrons distributed in a tetrahedral fashion. Measure the H—O—H, H—N—H, and H—C—H angles and report them to your instructor.

SUGGESTED PROJECT 4.2
Testing for molecular polarity

Molecules may be of two types—polar and nonpolar. By a polar molecule we mean a molecule in which the centers of positive and negative charge do not coincide. That is to say, there is a net separation of positive and negative charge centers within the molecule. These charge centers are attracted by electrical charges of opposite sign. Naturally a molecule in which there is no charge separation, that is, a nonpolar molecule, cannot be attracted by electrical charges.

In this project the student will test several household liquids for the polar nature of their molecules. Common household liquids are water, alcohol, acetone (nail polish remover), and gasoline. Besides a clean, dry squeeze bottle the only other materials required to perform this experiment are pieces of plastic and wool.

First fill the squeeze bottle with a known liquid. Then rub a piece of plastic several times with some wool so that the plastic acquires a charge of static electricity. Squeeze the bottle so that the liquid runs down into a sink and bring the freshly rubbed plastic near the flowing liquid. Observe whether or not there is a shift in the flow of the liquid. If there is a shift, is it toward the plastic or away from it? Repeat the experiment with other available liquids. Assuming the charge acquired by the plastic to be negative, report your conclusions to the instructor.

SUGGESTED PROJECT 4.3
Growing single crystals (or chemical gardening)

Atoms, ions, or molecules that compose matter in the solid state can be arranged in either a random (amorphous) or in an orderly (crystalline) fashion. When such an orderly arrangement is made to occur in a repeating pattern, beautiful crystals form. Growing beautiful crystals is an art like growing beautiful flowers and shrubs.

In this experiment we shall attempt to grow some large single crystals. Recommended substances are common salt ($NaCl$), copper sulfate ($CuSO_4$), and potassium dichromate ($K_2Cr_2O_7$). Note that the growing of a large, single crystal requires much care and attention over a period of several weeks.

Shake sufficient amounts of each of the above substances in separate bottles with about 50 ml of cold water so that saturated solutions are obtained. Observe the color of the solutions. Then decant the clear, supernatant liquids into clean wide-mouthed glasses. Cover the glasses with paper, label them, and set aside. Examine your solutions every other day or so. As soon as a single well-formed crystal is found, take it out with a spoon and suspend it in a fresh, cold, saturated solution of the same substance. Examine the growing crystal frequently to make sure that smaller crystals do not grow on it. Every three days or so change the old solution for a freshly prepared saturated solution of the substance. You may prepare more than one crystal if facilities allow.

One week before the end of the quarter or semester, take the crystals out and dry them using scraps of filter paper. Submit the crystals thus prepared to your instructor along with a report giving their names, colors, and sketches of their shapes and sizes.

5

Elements and their families

With the periodic and atomic relations now shown to exist between all the atoms and the properties of their elements, we see the possibility not only of noting the absence of some of them but even of determining and with great assurance and certainty the properties of these as yet unknown elements.

DMITRI MENDELEEV (1834–1907)
Russian chemist

To date 105 elements are known. Each of these elements has its own unique set of properties. Some elements are radioactive, some are extremely rare, and others do not occur in nature and can only be made in the laboratory. Some elements occur naturally in impure or combined forms and must be purified or produced industrially. But, nearly all of the naturally occurring elements play an essential role in life. Our bodies contain oxygen, carbon, hydrogen, nitrogen, calcium, phosphorus, potassium, sodium, chlorine, sulfur, magnesium, iron, iodine, and traces of many other elements (Fig. 5.1). Many of these elements are present in extremely small amounts, but they are absolutely essential for normal bodily functions that depend on a fantastically large number of complicated chemical reactions. Although a specific concentration of an element may be beneficial to our well-being, an excess or deficiency could have detrimental effects. The

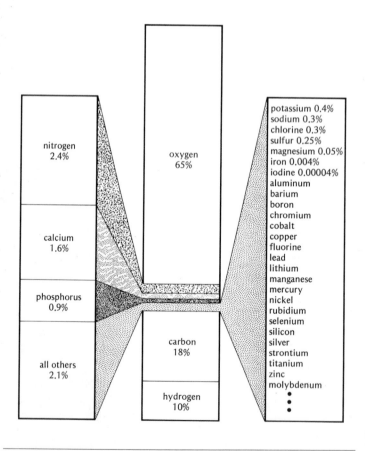

FIGURE 5.1

Elemental composition of man.

purpose of this chapter is to introduce you to the chemistry of the known elements and to their biological and environmental importance.

To study the chemistry of 105 elements individually would be an enormous task. Even at the beginning of the nineteenth century when fewer elements were known, attempts were made to group elements that were alike in many ways. It was not until 1869 that an essentially correct grouping of the elements was made independently by the Russian chemist Dmitri Mendeleev and by the German chemist Lothar Meyer. Their classifications were based on properties, such as melting points, boiling points, and chemical activities of the elements, which were found to vary in a roughly *periodic* manner, that is, rising and falling in a repetitious way, as the atomic weights of the elements increased (Fig. 5.2). With the discovery of the atomic nucleus, the order of increasing atomic weights was replaced by atomic numbers in periodic classifications of the elements. Today, the various elements are studied in groups or families based on similarities in the electronic configurations of atoms of the elements.

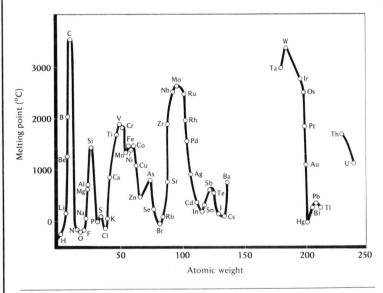

FIGURE 5.2

Periodic variation of melting points of the elements.

DEVELOPMENT OF THE CONCEPT OF PERIODICITY AND THE PERIODIC TABLE

TRIADS AND OCTAVES   Whether the chemists of the nineteenth century were inspired by the music of the times or not is unclear, but the earliest classifications of the elements were called the *triads* and the *octaves*. During the early part of the nineteenth century the German chemist Johann Döbereiner grouped the elements with similar chemical properties in sets of three which he called *triads*. He observed that chlorine, bromine, and iodine were alike in many ways, as were also iron, cobalt, and nickel. Unfortunately, as new elements became known it was found that many elements do not fit into a triad. In 1864 the English chemist John Newlands arranged the elements into groups of eight called *octaves*. Newlands observed that when the elements are listed in order of increasing atomic weights, physical and chemical similarities appear at intervals of eight elements. Today, more than eleven decades later, we know that Newlands was closer to the truth than many of his contemporaries who doubted him.

FIGURE 5.3

Dmitri Mendeleev, Russian chemist of the late nineteenth century who first devised and published the periodic table of the elements.

THE RUSSIAN OCTAVES   It was the Russian chemist Mendeleev and the German chemist Meyer who in 1869 correlated independently the properties of the elements in the form of what is today known as the original *periodic law*. Their periodic law states that *when the elements are arranged in sequence according to increasing atomic weights, there is a periodic repetition of elements with the same properties.* The meaning of this statement is best understood by reference to Mendeleev's periodic table published in 1871 (Fig. 5.4). As in Newlands' classification there are eight vertical columns called *groups* or *families* of elements with similar properties. In each horizontal row called a *period,* the elements are arranged primarily in order of increasing atomic weight. The outstanding aspect of Mendeleev's work was that, unlike Meyer, he left gaps in his table to predict the existence and properties of elements yet to be discovered. It was possible to predict properties of elements unknown at that time by comparing the properties of elements located above, below, and on either side of the gaps in Mendeleev's periodic table. For example, Mendeleev predicted in 1871 the existence and properties of undiscovered elements such as gallium (below Al in group III) and germanium (below Si in group

IV). (In later years, the predicted elements were discovered and found to have the properties Mendeleev had forecast.) The wide acceptance of Mendeleev's periodic table can be attributed largely to its predictive aspects.

THE MODERN PERIODIC TABLE   As our knowledge of the structure of the atoms of various elements advanced the original statement of the periodic law was revised. The modern periodic law states that *the physical and chemical properties of the elements are periodic functions of their atomic numbers.* In the modern periodic table (Fig. 5.5), elements are listed strictly in the order of increasing atomic number. Notice that there are at least four instances (elements with atomic numbers 18 and 19, 27 and 28, 52 and 53, and 90 and 91) in which an increase in atomic number is not accompanied by an increase in atomic weight.

As our knowledge of the electronic configuration of the atoms of elements improved, an explanation for the observed periodic properties of the elements was feasible. It was found that elements that have similar properties have similar electronic configurations. In other words, the properties of elements are directly related to the electronic configuration of the atoms of the elements. Let us illustrate this by considering the electronic configurations of the elements Li, Na, K, Rb, and Cs which are in group I and which have similar properties:

$$
\begin{aligned}
&\text{Li} \quad 1s^2 2s^1 \\
&\text{Na} \quad 1s^2 2s^2 2p^6 3s^1 \\
&\text{K} \quad 1s^2 2s^2 2p^6 3s^2 3p^6 4s^1 \\
&\text{Rb} \quad 1s^2 2s^2 2p^6 3s^2 3p^6 3d^{10} 4s^2 4p^6 5s^1 \\
&\text{Cs} \quad 1s^2 2s^2 2p^6 3s^2 3p^6 3d^{10} 4s^2 4p^6 4d^{10} 5s^2 5p^6 6s^1
\end{aligned}
$$

These configurations indicate that all of these elements have one electron in the outer energy level. Therefore, they have similar physical properties and form similar compounds with other elements.

The modern periodic table is best remembered by interpreting it in terms of the distribution of the electrons in the various orbitals according to the order of energy of the orbitals discussed previously (see Chapter 4). For this reason, the periodic table shown in Fig. 5.5 emphasizes the positions in the table corresponding to the various orbitals. There are, however, some exceptions in that in some instances one electron is placed in the d sublevel before the f sublevel is used. For example,

Row	Group I R_2O	Group II RO	Group III R_2O_3	Group IV RH_4 RO_2	Group V RH_3 R_2O_5	Group VI RH_2 RO_3	Group VII RH R_2O_7	Group VIII RO_4
1	H=1							
2	Li=7	Be=9.4	B=11	C=12	N=14	O=16	F=19	
3	Na=23	Mg=24	Al=27.3	Si=28	P=31	S=32	Cl=35.5	Fe=56, Co=59,
4	K=39	Ca=40	−=44	Ti=48	V=51	Cr=52	Mn=55	Ni=59, Cu=63
5	(Cu=63)	Zn=65	−=68	−=72	As=75	Se=78	Br=80	Ru=104, Rh=104,
6	Rb=85	Sr=87	?Yt=88	Zr=90	Nb=94	Mo=96	−=100	Pd=106, Ag=108
7	(Ag=108)	Cd=112	In=113	Sn=118	Sb=122	Te=125	I=127	
8	Cs=133	Ba=137	?Di=138	?Ce=140	−	−	−	−
9	−	−	−	−	−	−	−	Os=195, Ir=197,
10	−	−	?Er=178	?La=180	Ta=182	W=184	−	Pt=198, Au=199
11	(Au=199)	Hg=200	Tl=204	Pb=207	Bi=208	−	−	
12	−	−	−	Th=231	−	U=240	−	−

FIGURE 5.4

Mendeleev's periodic table of 1871. Adapted from *Annalen der Chemie*, Suppl. 8, 1872.

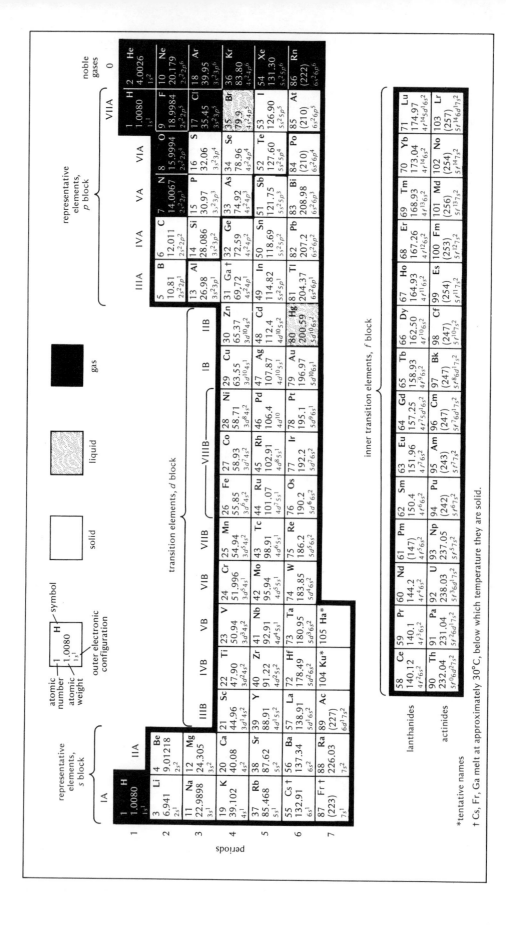

lanthanum has the configuration ending in $5d^1 6s^2$ rather than $4f^1 6s^2$ and actinium has the configuration ending in $6d^1 7s^2$ rather than $5f^1 7s^2$, despite the energy overlap between the $5d$ and $4f$ sublevels and the $6d$ and $5f$ sublevels.

Based on the orbital interpretation of the periodic table the elements have been classified according to the part of the periodic table in which they are located. Figure 5.5 illustrates what are called the s block elements, the p block elements, the d block elements, and the f block elements. Elements that comprise the s and p blocks (with the exception of the noble gases, all of which have completely filled s and p sublevels) are called the *representative elements*. The groups of representative elements have reference designations and family names, as listed in Table 5.1. Elements that comprise the d block elements are called the *transition elements;* they also have group designations (such as IIIB, IVB, and so on) and family names called after the first member of a group. The f block elements are called the *inner transition elements.* Those involving the $4f$ sublevel are called the *lanthanides* or the *rare earths,* and those involving the $5f$ sublevel are called the *actinides.*

THE PERIODIC TABLE AND TRENDS IN BEHAVIOR Before we discuss the families of elements, it is worthwhile to examine the general trends in behavior of the elements in relationship to the periodic table. The trends in behavior are best illustrated by referring to Fig. 5.6 which shows the relative sizes of the atoms of elements in the periodic table and their tendencies to attract electrons, that is, their electronegativities. In general, atoms become smaller from left to right in a period suggesting that the electrons are pulled in closer to the nucleus. The smaller the atom, the closer the valence electrons can come to the nucleus, and the more tightly they are held. Thus the tendency of atoms to gain electrons increases from left to right in a period (i.e., the electronegativity of atoms increases from left to right in a period). On the other hand, the larger the atom, the farther away are the valence electrons from the nucleus, and the more loosely they are held. Moreover, in the larger atoms the valence electrons are shielded from the positive nucleus by the inner core of the lower-energy level electrons; this *shielding effect* tends to decrease the attraction of the nucleus for the valence electrons. Thus, the tendency for atoms to lose electrons increases from top to bottom in a group or family of elements. However, Fig. 5.6 indicates that this trend is much more evident in the groups of representative ele-

---

TABLE 5.1
Group designations and family names of representative elements

Reference or group designation	Family name
IA	alkali metals
IIA	alkaline earth metals
IIIA	boron–aluminum family
IVA	carbon family
VA	nitrogen family
VIA	oxygen family
VIIA	halogen family

FIGURE 5.5 (OPPOSITE)

Modern periodic table of the elements.

ments than in the groups of transition elements. In fact, in some instances these trends do not apply at all (Fig. 5.6).

The chemical behavior of the elements depends on how the elements lose, gain, and share electrons to form chemical bonds. A *loss* of electrons is referred to as *oxidation,* whereas a *gain* of electrons is referred to as *reduction.* In many chemical changes involving two or more elements, one element loses electrons and the other elements gain the electrons so that oxidation and reduction proceed side by side in the very same reaction. For example, when a strip of zinc is immersed in a solution of copper sulfate the following reaction takes place:

$$Zn + CuSO_4 \longrightarrow ZnSO_4 + Cu$$

Actually, the reaction occurs between zinc atoms and copper ions:

$$Zn + Cu^{2+} \longrightarrow Zn^{2+} + Cu$$

Here zinc loses two electrons in going from Zn to Zn^{2+} and is oxidized, whereas copper gains two electrons in being transformed from Cu^{2+} to Cu and is reduced.

Reactions such as this are referred to as oxidation-reduction (or *redox*) reactions and are characterized by a change in the *oxidation state* (or *oxidation number*) of the elements involved. Oxidation numbers are positive or negative numbers assigned to the various elements according to arbitrary rules (see Chapter 7). In the reaction discussed above the oxidation state of zinc changes from 0 to +2 and that of copper goes from +2 to 0. Some of the oxidation numbers can be predicted from the

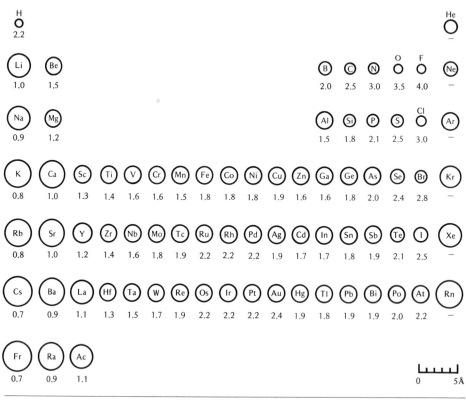

FIGURE 5.6

Relative sizes of the atoms and their electronegativities.

periodic table as they are periodic functions of the atomic numbers (Table 5.2). However, it is important to note that certain elements have a number of oxidation states that are not evident from the position of the element in the periodic table. When there is no true change in the oxidation states of any of the elements involved in a reaction we have a *nonoxidative* or *metathesis* reaction. Such a reaction is characterized by a simple exchange of atoms, groups of atoms, or ions between the reacting substances. The reaction of an acid with a base to form a salt and water, for example,

$$HCl + NaOH \longrightarrow NaCl + H_2O$$

constitutes one of the most important classes of nonoxidative reactions (see Chapter 7).

TABLE 5.2
Common oxidation numbers from the periodic table

group	I	II	III	IV	V	VI	VII
oxidation number	+1	+2	+3	+2, +4	+5	+6	+7
				−4	−3	−2	−1

QUESTIONS

5.1   How was the periodic table developed before modern atomic theory?

5.2   Locate and explain the anomalies that would exist in a periodic table based on increasing atomic weights of the elements.

5.3   Give a modern statement of the periodic law.

5.4   Compare and contrast Mendeleev's periodic table with a modern periodic table (see Figs. 5.4 and 5.5).

5.5   "The properties of elements are directly related to the electronic configuration of the atoms of the elements." How do the electronic configurations of the elements in the halogen family bear out this statement?

5.6   Classify elements according to the electronic features of the sublevels and locate them in a periodic table.

5.7   Comment on the change in electronegativities of elements in the periodic table (see Fig. 5.6).

5.8   Distinguish between oxidation and reduction as applied to elements.

SUGGESTED PROJECT 5.1
Verification of the periodic law

Obtain from a Handbook of Chemistry (available in a science library) the boiling points of elements with atomic numbers from 1 to 30, 32 to 42, and 44 to 55. Illustrate the periodicity of properties by plotting the boiling points of elements against their atomic numbers. Use Fig. 5.2 as a model for constructing the graph. Examine the graph for recurring features. Explain how this graph indicates a periodic relation between atomic number and the property plotted. From your graph and from the Handbook of Chemistry, find the boiling points of elements with atomic numbers 31 and 43, respectively.

FAMILIES OF REPRESENTATIVE ELEMENTS

s BLOCK ELEMENTS   The lone member: hydrogen   Hydrogen is listed by itself because it cannot be satisfactorily fitted in with any other family of elements. For centuries it was called "inflammable air" or "air that bursts forth like wind." Hydrogen has never been found to occur in the uncombined form in nature on our planet, although about 75 percent of our sun and other live stars are composed of elemental hydrogen. The hydrogen found on earth is mostly in the combined form as water, the most abundant and essential of compounds available to man. In fact, the name hydrogen is derived from the Greek *hydro* and *genes* meaning "producer of water." Elemental hydrogen occurs in the form of diatomic molecules, H_2, and is a colorless, tasteless, odorless gas of very low density (0.09 g/liter).

Hydrogen exists as three isotopes: ordinary hydrogen (protium), deuterium (symbol D), and tritium (symbol T). These are shown diagrammatically in Fig. 5.7. Like hydrogen, deuterium and tritium are diatomic gases under normal conditions on earth. Unlike hydrogen and deuterium, however, tritium is radioactive and is produced mostly in high-energy machines (see Chapter 6). In nature there are as many as 7000 ordinary hydrogen atoms for 1 deuterium atom. Because electrolysis of water releases ordinary hydrogen about six times as readily as deuterium (due to difference in masses), an enrichment of deuterium occurs in the residual water. This process is used on the Savannah River in South Carolina to obtain heavy water (D_2O) which serves as a moderator in nuclear reactors (see Chapter 6) and offers promise as a source of heavy hydrogen for fusion power in the future (see Chapters 6 and 16).

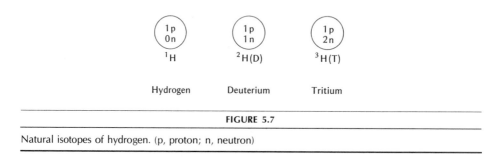

^1H $^2H(D)$ $^3H(T)$

Hydrogen   Deuterium   Tritium

FIGURE 5.7

Natural isotopes of hydrogen. (p, proton; n, neutron)

The isotopes of an element exhibit slight differences in ordinary physical and chemical properties. The differences in chemical properties, usually referred to as *isotope effects,* concern mainly the speed of reaction rather than the kind of reaction. The relative production of hydrogen versus deuterium in the electrolysis of water is an example of isotope effect. It has attracted the attention of many biologists because water is our body fluid as well as one of the raw materials from which our food is synthesized. Enough work has already been done to demonstrate that the deuterium isotope effect may have a far-reaching influence on life processes. Chemical reactions necessary to life, such as the oxidation of blood sugar (glucose), are very much slowed by the exchange of protium and deuterium between body chemicals and heavy water. A lag from several weeks to several months occurs before plants placed in heavy water start to grow. The rate of growth of plants is much slower in heavy water than in ordinary water.

Like the alkali metals, hydrogen atoms have a single outer *s* electron, and, like the halogens, hydrogen atoms require a single electron to achieve a noble gas electronic structure. Thus, the reactions of hydrogen are in some respects similar to those of the alkali metals and in some other respects similar to those of the halogens. For example, the reactions of hydrogen with oxygen, nitrogen, and chlorine are similar to some of the reactions of the alkali metals:

$$2H_2 + O_2 \longrightarrow 2H_2O$$
$$N_2 + 3H_2 \longrightarrow 2NH_3$$
$$H_2 + Cl_2 \longrightarrow 2HCl$$

Hydrogen does not have the same tendency to combine with metallic elements as with nonmetallic ones. However, under proper conditions, very reactive metals do combine with hydrogen to form metal hydrides:

$$2Na + H_2 \xrightarrow{\text{heat}} 2NaH$$
$$\text{sodium hydride}$$

$$Ca + H_2 \xrightarrow{\text{heat}} CaH_2$$
$$\text{calcium hydride}$$

These reactions of hydrogen are similar to some of the reactions of the halogens.

The acids constitute a noteworthy group of hydrogen-containing compounds. In fact, in the laboratory hydrogen is prepared by the reaction of metals such as Zn, Mg, or Al with acids:

$$Zn + 2HCl \longrightarrow ZnCl_2 + H_2$$

Metals such as Na and K are not usually used because they react too vigorously.

Commercially, most hydrogen is manufactured by the reaction of coke (carbon) or natural gas (methane) with steam:

$$C + H_2O \longrightarrow CO + H_2$$
$$CH_4 + H_2O \longrightarrow CO + 3H_2$$

Another method is the electrolysis of acidified water (see Fig. 4.1):

$$2H_2O \xrightarrow{\text{electric current}} 2H_2 + O_2$$

However, the reaction is highly endothermic (68.32 kcal/mole) and the high-energy requirement makes it too expensive for commercial production. Nevertheless, it has been suggested that hydrogen derived from electrolysis of water could supplant all fossil fuels as well as electrical distribution networks.

A hydrogen-based energy system has a simple cycle — water is both the source of hydrogen and the product of its consumption (Fig. 5.8). Because hydrogen is very clean burning and fully recyclable, it should be enthusiastically welcomed by environmentalists as a fuel. Storage problems also may be resolved in the future by packaging hydrogen as a metal hydride (see above). The vast amounts of by-product oxygen could have a big impact on technology and our lifestyle. However, public fear of potential hazards in handling hydrogen, linked mainly to the 1937 explosion and burning of the hydrogen-filled Zeppelin *Hindenburg*, will have to be assuaged before switching to a hydrogen-fueled technology.

On the chemical manufacturing side, large quantities of hydrogen are used to convert liquid fats into solid fats (such as Crisco); these solid fats are, in turn, used for foods and for making soap. Hydrogen is also used in the oxyhydrogen torch and in the atomic hydrogen torch to produce high temperatures of about 4000°C or more.

Such temperatures are required for the preparation of the synthetic rubies and sapphires used in wrist watches. Certain *fuel cells* now use hydrogen to produce electrical energy in manned space capsules (see Chapter 16).

Alkali and alkaline earth metal families   The alkali family of metals includes the elements lithium (Li), sodium (Na), potassium (K), rubidium (Rb), cesium (Cs), and francium (Fr), all of which are located in group IA of the periodic table (Fig. 5.9). The alkaline earth family, the group IIA elements, consists of beryllium (Be), magnesium (Mg), calcium (Ca), strontium (Sr), barium (Ba), and radium (Ra) (Fig. 5.9). Of these francium and radium are rare and radioactive. The families are so named because their compounds with oxygen and hydrogen, called *hydroxides,* are among the strongest bases (alkalis) known.

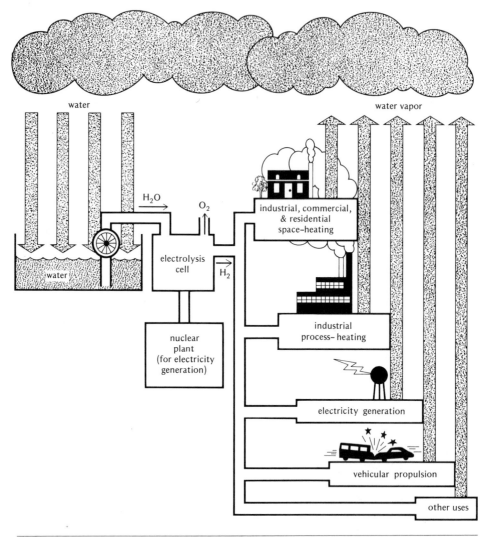

FIGURE 5.8

A hydrogen-based energy system.

Because of their great chemical reactivity, none of these metals are ever found as free elements in nature. They are found in the earth's crust mostly as components of practically insoluble minerals; the estimated percentage compositions

IA

1 H
1.0080
$1s^1$

IIA

3 Li	4 Be
6.941	9.01218
$2s^1$	$2s^2$
11 Na	12 Mg
22.9898	24.305
$3s^1$	$3s^2$
19 K	20 Ca
39.102	40.08
$4s^1$	$4s^2$
37 Rb	38 Sr
85.468	87.62
$5s^1$	$5s^2$
55 Cs	56 Ba
132.91	137.34
$6s^1$	$6s^2$
87 Fr	88 Ra
(223)	226.03
$7s^1$	$7s^2$

FIGURE 5.9

Alkali and alkaline earth families.

(Table 5.3) show that only Ca, Na, K, and Mg are fairly abundant in the earth's crust; the other alkali and alkaline earth metals are relatively scarce. Seawater contains about 3 percent by weight of alkali metal compounds, such as NaCl and KCl, and

TABLE 5.3
Abundance of alkali and alkaline earth metals in earth's crust

Element	Percent abundance, by weight
ALKALI METALS	
lithium	2×10^{-3}
sodium	2.6
potassium	2.4
rubidium	4×10^{-3}
cesium	1×10^{-4}
francium	trace
ALKALINE EARTH METALS	
beryllium	4×10^{-4}
magnesium	1.9
calcium	3.4
strontium	1×10^{-2}
barium	1×10^{-2}
radium	trace

some magnesium and calcium salts. Large quantities of alkali metal salts are also found in salt lakes and sedimentary deposits all over the world.

Unlike the common metals, such as iron, copper, and gold, with which we are so familiar, the alkali metals and the alkaline earth metals have low densities and melting points and most of them are relatively soft. Outside a stable core structure similar to that of noble gas elements, their atoms have, respectively, one (for alkali metals) or two (for alkaline earths) s electrons that are lost relatively easily. For this reason the pure elements are extremely reactive and often are kept in sealed containers or under oil to protect them from reacting with air and moisture. The loss of the outer s electrons from the atoms produces cations of $+1$ charge and $+2$ charge in the case of alkali metals and alkaline earth metals, respectively:

$$2Na + 2H_2O \longrightarrow 2Na^+ + 2OH^- + H_2$$
$$Ca + 2H_2O \longrightarrow Ca^{2+} + 2OH^- + H_2$$

The reaction of alkali metals with water proceeds with explosive violence and can sometimes be dangerous. The alkali metals dissolve in liquid ammonia to give a *solvated electron* that turns the solution blue:

$$Na + NH_3 \longrightarrow Na^+ + e^- \; (NH_3)$$

Another phenomenon associated with most of these metals (but not with Mg) is that they impart characteristic colors to an ordinary flame. In chemical laboratories *flame tests* are often used to identify the various alkali and alkaline earth elements in the form of their cations.

Both the alkali metals and alkaline earth metals are commercially prepared by the electrolysis of their molten chlorides:

$$2NaCl \xrightarrow{\text{electrolysis}} 2Na + Cl_2$$

$$CaCl_2 \xrightarrow{\text{electrolysis}} Ca + Cl_2$$

Chlorine is a by-product.

Compounds of sodium have many uses. For example, sodium hydroxide (NaOH) is used in soap, dye (coloring matter), paper, and textile industries; sodium carbonate (Na_2CO_3) is used in water softening and in soap, glass, paper, and paint industries; sodium nitrate ($NaNO_3$) is used in the explosives and fertilizer industries; and sodium cyanide (NaCN) is used in electrochemical industries. The manufacture of some of these compounds is discussed in Chapter 8. Among the alkaline earth metal compounds, large quantities of calcium hydroxide, $Ca(OH)_2$, are necessary for the manufacture of mortar and bleaching powder. Familiar household compounds derived from these two families of elements are table salt (NaCl), washing soda (Na_2CO_3), baking soda ($NaHCO_3$), and milk of magnesia [$Mg(OH)_2$].

On the biological side, sodium and potassium ions are among the indispensable constituents of animal and plant tissues. Sodium ion (Na^+) is the principal cation of the fluids outside living cells; potassium ion (K^+) is the principal cation inside the cells. Both ions play a major role in nerve impulses. The Apollo 15 crew experienced irregularities in their heartbeats presumably due to the weightlessness that caused them to eliminate K^+ ions at an unusually high rate. Astronauts on subsequent missions were therefore advised to drink potassium-spiked liquids. Potassium ions control the water balance in plants, and relatively high levels of K^+ are, therefore, needed by growing plants.

Magnesium ions are important in maintaining stability of intercellular substances and cell membranes and are required for muscle contractility. Also, magnesium ion forms an integral part of the chlorophyll molecule ($C_{55}H_{72}MgN_4O_5$ or $C_{55}H_{70}MgN_4O_6$), the green pigment in plants essential for photosynthesis. The chlorophyll serves as an antenna for receiving solar energy and passes it on for ultimate conversion to chemical energy through photosynthesis (see Chapter 7).

Calcium compounds form the main structure of bones and teeth. Strontium-90 in radioactive fallout, being chemically similar to calcium, is taken up along with calcium in the formation of green plants, milk, and bones and teeth. The presence of radioactive strontium-90 in the tissue surrounding bones curtails the production of bone marrow which is the main site of blood-cell formation. Anemia or more serious disorders may result. Cesium contamination of foods results in the same effects as strontium contamination.

In spite of the poisonous nature of beryllium (and its compounds), it has been increasingly used as structural material for missiles, space vehicles, and nuclear reactors. Finely powdered metallic beryllium is also added to increase the performance of rocket fuels because of the large amount of energy beryllium releases on combustion. Beryllium metal or insoluble beryllium compounds cause the chronic pulmonary disease, berylliosis, and soluble beryllium compounds commonly produce acute inflammation of the lungs. Consequently, the U.S. Environmental Protection Agency has included beryllium in the list of "hazardous air pollutants."

QUESTIONS

5.9 What are the representative elements?

5.10 Name as many families of representative elements as you can and indicate their location in the periodic table.

5.11 Cite specific reasons to explain why it is difficult to classify hydrogen within the periodic table.

5.12 Compare the reactivity of hydrogen toward metals and nonmetals.

5.13 What are the isotopes of hydrogen?

5.14 Why are group IA and IIA elements not found in nature in the free state?

5.15 Predict some of the properties of the elements francium and radium from their family characteristics.

5.16 Account for the presence of the large amounts of alkali metal salts in the ocean.

5.17 Name some compounds of sodium that are commonly used in your home.

5.18 Liquid ammonia forms a deep blue color when alkali metals are dissolved in it. Why?

p BLOCK ELEMENTS   The boron-aluminum family   The elements in this family are boron (B), aluminum (Al), gallium (Ga), indium (In), and thallium (Tl) (Fig. 5.10). Elemental boron is a brittle, very hard crystalline solid with nonmetallic properties. In nature, boron is found only in the combined form as borax ($Na_2B_4O_7 \cdot 10H_2O$) of which there are large deposits in the dry lakes of California. Familiar household occurrences of boron are in Pyrex glassware and in "perborate" bleaches. The element may be produced by electrolyzing molten borax or by the reaction

$$2BCl_3 + 3H_2 \xrightarrow{\text{heat}} 2B + 6HCl$$

The latter method yields high-purity elemental boron.

Except for boron, all of the members of this family are unquestionably metals. Of these, aluminum is the most common metal in the earth's crust (7.5% by weight) and one of the most important commercially produced metals. It is not found as the free element in nature but is mined as bauxite ($Al_2O_3 \cdot 3\ H_2O$) from which pure aluminum is obtained by electrolysis (see Chapter 8). Pure aluminum is rather soft and weak. Only when it contains small quantities of Cu, Mg, Mn, and Cr does aluminum develop the necessary strength for use as a structural material. Because of its light weight, aluminum has found extensive use in aircraft construction. Aluminum is also used in cooking utensils where its high thermal conductivity is an asset. Many of us come into daily contact with aluminum compounded as aluminum chlorohydrate, the active ingredient in many antiperspirants.

Elements of this group have two s electrons and one p electron in their outer energy levels. Whereas the Al^{3+} ion is quite common, the B^{3+} ion is not. Unlike the alkali and alkaline earth metal compounds, many of the compounds of the boron-aluminum family are covalently bonded in the solid state and possess an outer electron *sextet* rather than the noble gas octet configuration. For example, boron trifluoride (BF_3) has a sextet of outer electrons as seen by its Lewis structure:

$$\begin{matrix} \ddot{F} \\ \ddot{B}\!:\!F \\ \ddot{F} \end{matrix}$$

Further, boron forms a number of hydrides, called *boranes*, the electronic structures of which seem to violate all of the conventional rules. A possible structure for diborane (B_2H_6) is shown in Fig. 5.11. It has *single-electron bonds* between boron and hydrogen, unlike most covalent bonds which have at least two electrons. It is said that two of the hydrogens form "bridges" between the boron atoms. The higher boranes such as B_4H_{10}, B_6H_{10}, and $B_{10}H_{14}$ are equally surprising because there seem to be too few electrons to hold them together. These hydrides ignite spontaneously in air to give dark-colored products of unknown composition. The only important oxide of boron is boric oxide, B_2O_3, which dissolves in water to form boric acid, H_3BO_3. In this form boron is available to plants as an essential element. Although the toxicity

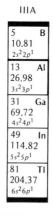

IIIA

5	B
10.81	
$2s^2 2p^1$	

13	Al
26.98	
$3s^2 3p^1$	

31	Ga
69.72	
$4s^2 4p^1$	

49	In
114.82	
$5s^2 5p^1$	

81	Tl
204.37	
$6s^2 6p^1$	

FIGURE 5.10

The boron–aluminum family.

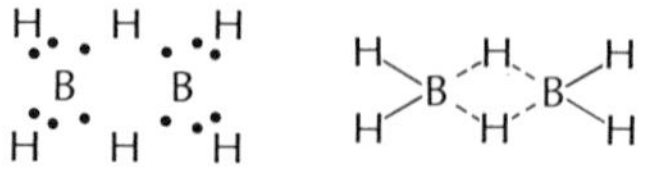

FIGURE 5.11

A possible structure for diborane (B_2H_6).

of boron is not well established, the tolerance limits of most plants to boron are very small. Levels exceeding 1 ppm of available boron in the soil are toxic to most plants. For this reason, a maximum level of 1 ppm of boron is now recommended for public water supplies. The salts of boric acid, the borates, are extremely complicated compounds containing complex anions in which boron atoms are joined together by oxygen bridges (Fig. 5.12). Borates are used in the manufacture of glass and various ceramic items. Borax, the most common of the borates is now widely used as a water softener, alone or in detergent mixtures. The presence of boron in sewage effluents may be a critical factor in the use of effluents for irrigation because of the toxicity of boron to plants.

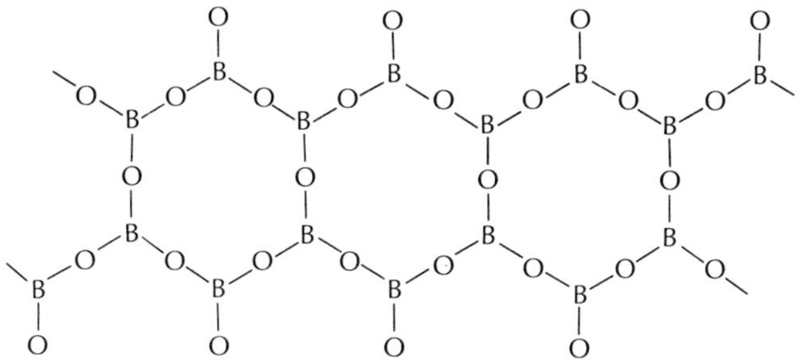

FIGURE 5.12

Borate ion.

Aluminum reacts with strong acids and strong bases evolving hydrogen:

$$2Al + 6HCl \longrightarrow 2AlCl_3 + 3H_2$$
$$2Al + 2NaOH + 6H_2O \longrightarrow 2NaAl(OH)_4 + 3H_2$$

Because of its ability to dissolve both in acids and bases, aluminum is considered to be *amphoteric*. Aqueous solutions of most aluminum salts are acid because of the hydrolysis of Al^{3+}; the ion is certainly hydrated, and its formula is probably $Al(H_2O)_6^{3+}$. Because of its small size and high charge, Al^{3+} also forms a series of quite stable *complex ions* with fluoride ion. One such complex ion AlF_6^{3-} is found in the mineral cryolite (Na_3AlF_6) which occurs in nature almost exclusively as an enormous geologic dike in Greenland. Like other +3 ions, aluminum ion may be crystallized by slow evaporation of water from aqueous solutions containing sulfate ions and singly charged cations to give double salts called *alums*. They have the general formula $MM'(SO_4)_2 \cdot 12H_2O$, where M is a singly charged cation (such as K^+, Na^+, or NH_4^+) and M' is a triply charged cation (such as Al^{3+}, Fe^{3+}, or Cr^{3+}). Ordinary alum is $KAl(SO_4)_2 \cdot 12H_2O$. Some alums find important uses in the tanning, textile, and paper industries.

The higher members of the family exhibit properties similar to those of aluminum. Monatomic ion formation becomes progressively easier with increasing atomic number within the group.

The carbon–silicon family   Of the five elements in this family, carbon (C) and silicon (Si) are nonmetals, germanium (Ge) is metalloid (semimetallic), and tin (Sn) and lead (Pb) are true metals (Fig. 5.13). Carbon occurs in the earth's crust in both the free and combined states. It is the all-important element in the molecules constituting living organisms. Silicon is found only in the combined state as oxycompounds such as silica and the silicates which are the most abundant of all the compounds in the earth's crust. Most rocks and minerals are silicates with a

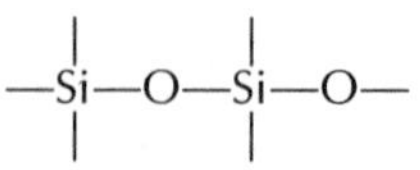

$$-\overset{\displaystyle |}{\underset{\displaystyle |}{Si}}-O-\overset{\displaystyle |}{\underset{\displaystyle |}{Si}}-O-$$

lattice derived from SiO_2. Some silicon-containing minerals are feldspar, quartz, and mica. In addition to silicon and oxygen, these minerals contain atoms of other elements such as Ca, Na, K, and Al. Elemental silicon in its crystalline form is gray or black and is prepared by the reaction

$$SiO_2 + 2Mg \xrightarrow{\text{heat}} 2MgO + Si$$

It is used as a component in transistors and in certain of the new solar batteries.

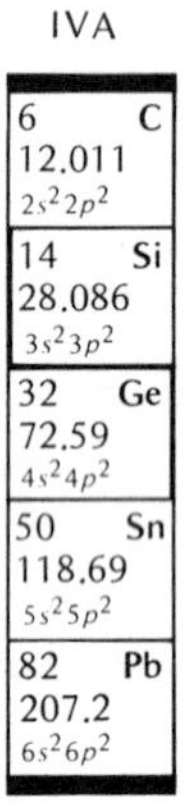

IVA

6	C
12.011	
$2s^2 2p^2$	
14	Si
28.086	
$3s^2 3p^2$	
32	Ge
72.59	
$4s^2 4p^2$	
50	Sn
118.69	
$5s^2 5p^2$	
82	Pb
207.2	
$6s^2 6p^2$	

FIGURE 5.13

The carbon–silicon family.

Carbon is characterized by the ability to exist in different *allotropic** modifications. Naturally occurring carbon is found in two allotropic forms, *diamond* and *graphite*. Diamond is a colorless solid capable of being cleaved into brilliant crystals. It is the hardest, most abrasive mineral known. By contrast, graphite is a soft black substance that actually feels greasy. In graphite the carbon atoms crystallize in a pattern of layers with hexagonal symmetry, whereas in diamond the carbon atoms crystallize with tetragonal symmetry (Fig. 5.14). The covalent bonding between carbon atoms is weaker in graphite than in diamond, thus accounting for the flakiness and lubricity of graphite. Many amorphous forms, which are random arrangements (distinct from the orderly arrangement in crystals) of carbon atoms, are also known, the common forms being charcoal, coke, carbon black, and bone black.

Carbon and silicon are quite unreactive under ordinary conditions. Unlike

---

* The existence of an element in more than one elemental form is a property called *allotropy*, and the different forms are referred to as allotropic forms.

the metals of groups I–III, their atoms have no tendency to lose outer electrons completely and form simple cations such as C^{4+} and Si^{4+}. Even though the higher members of the family are distinctly metallic in their properties, the +4 ionic state is rarely encountered. The more common ionic states among tin and lead are the +2 state (Sn^{2+}, Pb^{2+}). The greater tendency for the heavier elements to lose outer electrons is the result of the greater distance between the nucleus and outer electrons of these heavier elements.

When carbon and silicon react with other elements they do so by sharing electrons to form covalent bonds; for example, they react with oxygen to form oxides such as carbon dioxide (CO_2) and silicon dioxide (SiO_2). Of these two oxides, carbon dioxide reacts readily with water to give very weak acid solutions,

$$CO_2 + H_2O \longrightarrow H_2CO_3$$
$$\text{carbonic}$$
$$\text{acid}$$

(and this is why carbonated drinks are slightly sour tasting). The spectacular limestone caves (such as Mammoth Cave and Carlsbad and Luray Caverns) have been dissolved out of solid rock by the gentle action of the carbonic acid formed by the solution of carbon dioxide in rain water.

A most important chemical property of carbon and silicon is their tendency to form large molecules. However, carbon atoms tend to form single, double, and triple covalent bonds, whereas silicon tends to form only single bonds. Unlike silicon, carbon atoms join with each other to form a limitless variety of chains or rings of atoms. In fact over 2 million compounds of carbon are known. Many of these compounds contain only carbon and hydrogen and are called *hydrocarbons,* but the majority also contain one or more other elements such as O, N, S, Cl, Br, and P. These compounds are discussed in more detail in Chapter 9. The ability to form chains is found to a lesser degree in silicon, whereas germanium, tin, and lead exhibit little, if any, tendency to form chains.

A number of carbon-containing materials are familiar to us. In fact most of the

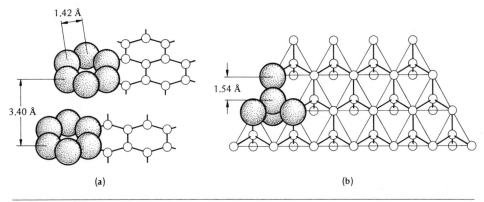

(a)   (b)

FIGURE 5.14

Structures of (a) graphite and (b) diamond. (Reproduced, by permission, from Charles W. Keenan and Jesse H. Wood, *General College Chemistry,* 4th ed., p. 560. Copyright © 1957 by Harper & Row, Inc.; copyright © 1961, 1966, 1971 by Charles William Keenan and Jesse Hermon Wood.)

food we eat and drink contains a variety of carbon compounds, and numerous others form every day from the physiological processes of living organisms (see Chapter 11). Familiar silicon-containing materials are asbestos, bricks and pottery, mortar, plaster, cement, glass, and ceramics. Asbestos is a fibrous silicate mineral used for insulation and fireproofing by the building trade. Recent findings indicate that inhalation of asbestos can produce serious lung disorders. Glass has what we call a "glassy form," which is a state in between the true crystalline solid and the liquid state. It is attributed to a random arrangement of the atoms in a distorted fashion without perfect symmetry (Fig. 5.15). The industrial production of some of these materials is discussed in Chapter 8. Among the most interesting synthetic compounds are the *silicones* which are unlike anything found in nature. Their molecules are of Si, O, C, and H atoms formed in different chain lengths and with different properties (Fig. 5.16). For example, a silicone made of short-chain molecules is an oily liquid, whereas silicones with medium-length chains behave as viscous oils, jellies, and greases. Very long-chain silicones have a rubberlike consistency. All of these compounds are resistant to chemical attack and are water-repellant and, therefore, find wide use in wood, textile, and metal industries.

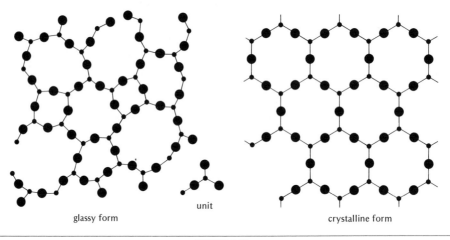

unit

glassy form   crystalline form

FIGURE 5.15

Glassy form and crystalline form of the same structural units. (Redrawn, by permission, from A. F. Wells, *Structural Inorganic Chemistry*, The Clarendon Press, Oxford, 1945, and reproduced, by permission, from Charles W. Keenan and Jesse H. Wood, *General College Chemistry*, 4th ed., p. 575. Copyright © 1957 by Harper & Row, Inc.; copyright © 1961, 1966, 1971 by Charles William Keenan and Jesse Hermon Wood.)

Because of similarities between Ca^{2+} ions and Pb^{2+} ions, bones serve as accumulation sites for lead. Most of the lead our system picks up comes from combustion of lead-containing gasoline in automobiles (see Chapter 15). Lead poisoning in humans is manifested by anemia, kidney disease, and disturbances of the central nervous system. Although the chronic effects of the ingestion of low levels of lead have not been well established, lead is considered a very undesirable component of air and water. The U.S. Public Health Service has established a maximum limit of 0.05 ppm for lead in drinking water.

The nitrogen family The elements of this group are nitrogen (N_2), phosphorus (P_4), arsenic (As), antimony (Sb), and bismuth (Bi) (Fig. 5.17). They are similar to one another in some respects but are noted more for their differences. In the nitrogen family, the stepwise change from nonmetallic to metallic character within a group is more clearly evident than in any other.

Unlike the other elements in this group, nitrogen is a colorless, odorless, and tasteless gas and is the most abundant constituent of the earth's atmosphere (78% by volume). Nitrogen compounds, especially proteins, are constituents of all living things. In nature, nitrogen is also found in the combined form as sodium nitrate ($NaNO_3$) in the mineral Chile nitrate.

Phosphorus is so active that it is found in nature only in minerals such as phosphorite $[Ca_3(PO_4)_2]$ and apatite $[Ca_5(Cl, F)(PO_4)_3]$. However, it ranks tenth among the elements in the earth's crust (0.12% by weight) and is thus the most abundant of the elements in the nitrogen family. Like nitrogen it is essential to the growth of living things. The presence of phosphorus in all living matter implies that it might be useful as a fertilizer. In fact, almost all commercial fertilizers do contain some phosphorus compounds. The phosphate ion (PO_4^{3-}) is part of the biological molecules used in energy storage and release (see Chapter 11); it is also a component of bones and teeth. Phosphates present an environmental problem known as *eutrophication* of some lakes and rivers (see Chapter 14).

Phosphorus exhibits allotropy in that it is found in at least two forms: white and red. *White phosphorus* is a soft, cream-colored solid and exists as P_4 molecules. It ignites spontaneously in air so that it is usually stored under water or kerosene. *Red phosphorus* is obtained by heating the white allotrope to 250°C in the absence of air. The molecules of red phosphorus consist of many phosphorus atoms strung together in a chain. Unlike the white variety, the red allotrope is not poisonous and will not ignite spontaneously in air, but is flammable. The most obvious chemical property of phosphorus is its reactivity with oxygen. Because of this property, phosphorus has been used extensively in the manufacture of matches (Fig. 5.18).

The elements arsenic and antimony have at least two allotropic modifications—a nonmetallic form of low density and a dense, more closely packed metallic form. Bismuth is unquestionably a metal. All these three elements are found in both the elemental and combined form in nature, the most common minerals of these

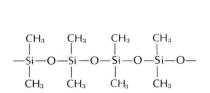

VA

7	N
14.0067	
$2s^2 2p^3$	

15	P
30.97	
$3s^2 3p^3$	

33	As
74.92	
$4s^2 4p^3$	

51	Sb
121.75	
$5s^2 5p^3$	

83	Bi
208.98	
$6s^2 6p^3$	

FIGURE 5.16

Structure of methyl silicone.

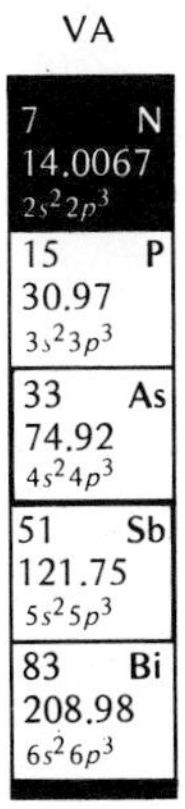

FIGURE 5.17

The nitrogen family.

being the sulfides and oxides. Arsenic also occurs with phosphate minerals and is introduced into the environment along with some phosphorus compounds. The combustion of fossil fuels introduces large quantities of arsenic into the environment, some of which eventually reaches natural waters. There is some evidence that arsenic, like mercury, may be converted to highly toxic organic derivatives by bacteria. Arsenic and its compounds have been suspected to be carcinogenic. Acute arsenic poisoning can result from the ingestion of milligram quantities of the element.

The atoms of this group have five electrons in the outer energy level, with two of them occupying an s orbital and each of the remaining three in one of the three outer p orbitals. Perhaps the most striking chemical property of the nitrogen family is the lack of reactivity of elemental nitrogen, which is so apparent in combustion, fermentation, decay, and respiratory processes. The resistance to combination with other atoms is attributed to the great affinity of one nitrogen atom for the other in N_2. It was pointed out in Chapter 4 that two nitrogen atoms share three pairs of electrons to form a triple bond in the nitrogen molecule. However, under extreme temperatures and pressures or in the presence of catalysts, nitrogen does react with other elements to form a number of compounds in which the oxidation states of nitrogen vary from -3 to $+5$. The other elements in the group have common oxidation states of $-3, +3,$ and $+5$. To acquire electrons and become negative ions is a characteristic of many nonmetals. In the nitrogen family, nitrogen has the greatest tendency to have an oxidation state of -3; bismuth, by contrast, does not form stable compounds in which its oxidation state is -3.

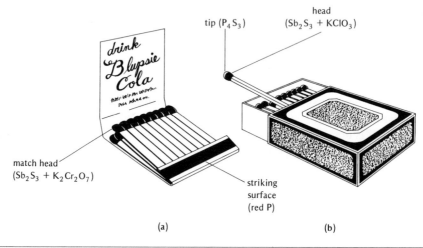

FIGURE 5.18

Use of phosphorus in matches: (a) safety matches; (b) strike anywhere matches.

Some of the important compounds of nitrogen are ammonia (NH_3), nitric oxide (NO), and nitric acid (HNO_3). Nitric oxide is known to mix spontaneously with atmospheric oxygen and form nitrogen dioxide,

$$2NO + O_2 \longrightarrow 2NO_2$$

which contributes to air pollution (see Chapter 15). Ammonia dissolves easily in water forming ammonium hydroxide or ammonia water:

$$NH_3 + H_2O \longrightarrow NH_4OH$$

It also reacts with acids forming ammonium salts some of which are compounds used extensively as fertilizers:

$$NH_3 + HCl \longrightarrow NH_4Cl$$
ammonium
chloride

$$NH_3 + HNO_3 \longrightarrow NH_4NO_3$$
ammonium
nitrate

$$2NH_3 + H_2SO_4 \longrightarrow (NH_4)_2SO_4$$
ammonium
sulfate

Ammonium salts contain the ammonium ion, NH_4^+, which acts very much like the K^+ and Rb^+ ions of the alkali metals.

The oxygen family Besides oxygen (O_2), the other elements of this family are sulfur (S_2, S_4, S_6, or S_8), selenium (Se_2, Se_8), tellurium (Te_2), and polonium (Po) (Fig. 5.19). Apart from the fact that the oxygen atom, like the other members of the family, has six electrons in its outer energy level, actually it shares very little either in its physical or chemical properties with the other members of the group. Thus, a distinctive feature among these elements is the -2 oxidation state which is, in fact, the oxidation state almost always found in oxygen compounds. Polonium is so rare that its chemistry is relatively unknown, but it is a radioactive element.

Oxygen is the most abundant element in the earth's crust, comprising, in combination with other elements, approximately half of its total mass. Elemental oxygen is found in the earth's atmosphere (20% by volume). The oxygen combined with hydrogen in water accounts for most of the total mass of our hydrosphere (the oceans, lakes, and rivers). The only allotrope of oxygen is *ozone* (O_3) which is formed when oxygen is exposed to solar (ultraviolet) radiation or an electric discharge. Unlike oxygen, which is a colorless, odorless gas essential for the survival of most forms of life, ozone has a characteristic pungent odor and is very toxic. Ozone is one of the harmful components of smog (see Chapter 15).

The most characteristic and well-known chemical reaction involving oxygen is *combustion,* the outstanding features of which are the evolution of a great quantity of heat and the generation of a flame. For example, under certain conditions oxygen and hydrogen combine to produce a oxyhydrogen flame which is quite hot. Some-

VIA

8	**O**
15,9994	
$2s^22p^4$	
16	**S**
32.06	
$3s^23p^4$	
34	**Se**
78.96	
$4s^24p^4$	
52	**Te**
127.60	
$5s^25p^4$	
84	**Po**
(210)	
$6s^26p^4$	

FIGURE 5.19

The oxygen family.

times the same reaction takes place with explosive violence. The product of the reaction is invariably water. Oxygen reacts with other elements to form *oxides.** The combustion of hydrocarbons is of great importance to our economy. For example, natural gas (CH_4) and gasoline (which contains octane, C_8H_{18}) give carbon dioxide and water on their complete combustion:

$$CH_4 + 2O_2 \longrightarrow CO_2 + 2H_2O$$
$$2C_8H_{18} + 25O_2 \longrightarrow 16CO_2 + 18H_2O$$

Unfortunately, the combustion of hydrocarbons contributes heavily to air pollution (see Chapter 15).

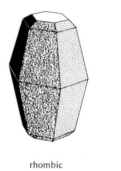 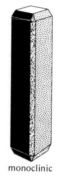

rhombic   monoclinic

FIGURE 5.20

Allotropic forms of sulfur: (a) rhombic; (b) monoclinic.

Sulfur is found as the free element in naturally occurring deposits as a yellow solid. The United States has some of the world's richest deposits of elemental sulfur. In the combined form, it occurs in nature as the sulfides of Fe, Hg, Pb, Zn, and Cu and as the sulfates of Ca (gypsum) and Ba (barite). In living organisms, sulfur occurs as an integral part of some proteins. The amount of sulfur present in the earth's crust amounts to about 0.1 percent by weight. Under normal conditions, the *rhombic* form of sulfur is the stable allotrope; above 95°C the *monoclinic* crystalline form is found (Fig. 5.20). Molten sulfur combines with hydrogen to form hydrogen sulfide (H_2S) which is a gas that smells like rotten eggs. However, most of the hydrogen sulfide in our environment is produced by the decay of organic matter and by the emissions of volcanoes. Something of the order of 200 to 300 million tons of sulfur enters the global atmosphere as H_2S from natural sources each year. There have been several acute cases of hydrogen sulfide poisoning which results in damage to the central nervous system and even fatalities. In addition to the −2 oxidation state, sulfur exhibits the oxidation states of +4 and +6 in its two compounds, sulfur dioxide (SO_2) and sulfur trioxide (SO_3), both of which contribute to air pollution (see Chapter 15). The latter readily dissolves in water to form sulfuric acid (H_2SO_4) which is perhaps the most important single industrial chemical. The manufacture of sulfuric acid is discussed in Chapter 8.

Compared to oxygen and sulfur, selenium and tellurium are relatively scarce. However, they occur naturally, both in the free form and in compounds. Whereas oxygen and sulfur are typical nonmetals with low electrical conductivities, tellurium

---

* The oxides of many metals dissolve in water to form alkaline solutions and are, therefore, referred to as *basic oxides.* The oxides of nonmetals react with water to form acidic solutions and are known as *acidic oxides.*

approaches some metals in electrical conductivity. In fact, tellurium and one form of selenium look like metals. Selenium has a rare property in that its electrical conductivity is greatly increased when exposed to light; that is, it exhibits *photoconductivity.* One of the most popular copying devices, that of the Xerox Corporation, depends on this property of selenium. It is now known that selenium in the diet in concentrations of about 3 ppm or less improves growth and helps to negate the effects of mercury in the human body. Concentrations of selenium above 5 ppm have been known to cause gastrointestinal disturbances, liver injuries, and dermatitis. The "alkali disease" among livestock occurs as a result of chronic selenium poisoning.

The halogen family The elements in group VIIA are fluorine (F_2), chlorine (Cl_2), bromine (Br_2), iodine (I_2), and astatine (At) (Fig. 5.21). They are collectively called the halogens, a name derived from the Greek *halos,* salt, and *genes,* born. Of these five elements, astatine is a radioactive element artificially produced by bombarding polonium or bismuth with alpha-particles (see Chapter 6). All the halogens are extremely reactive elements and therefore exist in nature primarily in compounds. Their most common state is as the halide ions F^-, Cl^-, Br^-, and I^-, and these ions always occur in association with positive ions such as Na^+, K^+, Mg^{2+}, and Ca^{2+}. Because the halides are usually soluble in water, the halide ions are found concentrated in the sea and in salt lakes. At moderate levels, these ions have relatively little effect upon the chemical or biological characteristics of waters. However, excessive levels of these ions can result in the deterioration of water quality due to excessive salinity (see Chapter 14).

At ordinary temperatures fluorine is a yellowish gas; chlorine, a greenish gas; bromine, a dark reddish-brown liquid; and, iodine a purplish-black crystalline solid. None possess any metallic properties. Because their atoms lack but a single p electron in the outer energy level in order to achieve a noble gas electronic structure, halogens are powerful *oxidizing agents,* that is, substances that pull electrons away from other elements or compounds. Their electronegativities are high, ranging from 2.5 for iodine to 4.0 for fluorine, which has the largest electronegativity of all the elements. In fact, fluorine is the most chemically reactive of all of the elements and

VIIA

1	H
1.0080	
$1s^1$	

9	F
18.9984	
$2s^2 2p^5$	

17	Cl
35.45	
$3s^2 3p^5$	

35	Br
79.9	
$4s^2 4p^5$	

53	I
126.90	
$5s^2 5p^5$	

85	At
(210)	
$6s^2 6p^5$	

FIGURE 5.21

The halogen family.

it is nicknamed the "tiger" of the elements. Because of its extreme reactivity, it is difficult to prepare fluorine by ordinary chemical methods. It was the French chemist Henri Moissan who first prepared fluorine by passing an electric current through fused, potassium acid fluoride (KHF_2) in a copper vessel or cell. A diagram of the apparatus is shown in Fig. 5.22. The following reaction takes place in the cell during electrolysis, with fluorine being given off at the anode and hydrogen at the cathode:

$$2HF \xrightarrow{\text{electrolysis}} H_2 + F_2$$

Fluorine reacts vigorously with water, asbestos (the well-known fireproofing material), hydrogen, and most metals at ordinary temperatures:

$$2F_2 + 2H_2O \longrightarrow 4HF + O_2$$
$$H_2 + F_2 \longrightarrow 2HF$$

Hydrogen fluoride (HF) is unique among the hydrogen halides in that it is a weak acid. However, it reacts with glass and finds practical application in the etching and frosting of glass. If you look carefully you may find that some of the reagent bottles in the chemical laboratory have the name of the reagent etched on them.

The most interesting of the compounds of fluorine are the *fluorocarbons,* which are structurally similar to hydrocarbons (Fig. 5.23). Some of these fluoro-carbons *polymerize.* A polymeric fluorocarbon familiar to us is *Teflon* [polytetra-fluoroethene, $(C_2F_4)_n$], which is almost totally inert and, therefore, has found a number of practical applications including the "greaseless" frying pans.

Unlike fluorine, the other halogens can be prepared by a number of non-electrolytic procedures. In the laboratory, they are prepared by the reaction of manganese dioxide (MnO_2) with hydrochloric acid (HCl), hydrobromic acid (HBr), or hydriodic acid (HI):

$$MnO_2 + 4HCl \longrightarrow MnCl_2 + Cl_2 + 2H_2O$$
$$MnO_2 + 4HBr \longrightarrow MnBr_2 + Br_2 + 2H_2O$$
$$MnO_2 + 4HI \longrightarrow MnI_2 + I_2 + 2H_2O$$

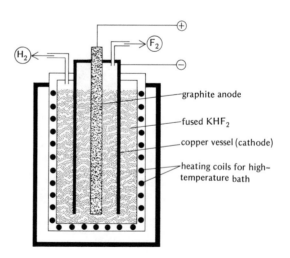

FIGURE 5.22

Apparatus for the preparation of fluorine.

Chlorine, however, is prepared commercially by the electrolysis of a concentrated solution of sodium chloride using graphite electrodes:

$$2NaCl + 2H_2O \xrightarrow{\text{electrolysis}} 2NaOH + H_2 + Cl_2$$

Bromine and iodine are manufactured by passing chlorine through a solution of a bromide and iodide salt, respectively:

$$Cl_2 + 2Br^- \longrightarrow Br_2 + 2Cl^-$$
$$Cl_2 + 2I^- \longrightarrow I_2 + 2Cl^-$$

Of all the halogens, iodine alone sublimes on heating; that is, it passes directly from the solid state to a vapor without first changing into a liquid. Advantage is taken of this fact in separating the nonvolatile impurities from crude iodine.

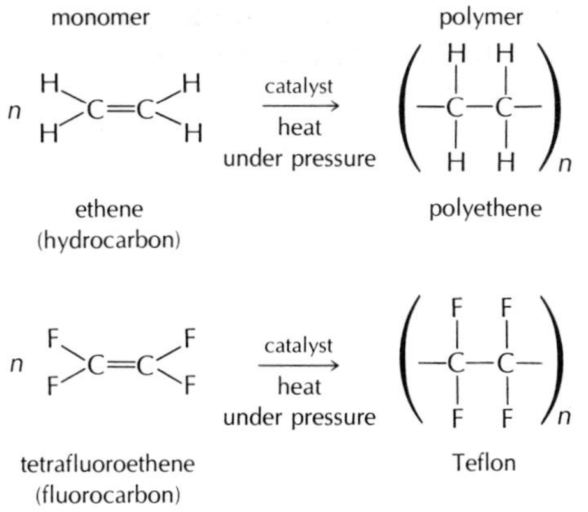

FIGURE 5.23

Comparison of hydrocarbons and fluorocarbons.

In contrast to fluorine, chlorine and the higher halogens can exist in more than one oxidation state. Thus they have an oxidation state of -1 in the halide ion (X^-), $+1$ in the hypohalite ion (OX^-), $+3$ in the halite ion (XO_2^-), $+5$ in the halate ion (XO_3^-), and $+7$ in the perhalate ion (XO_4^-). Here X stands for Cl, Br, or I.

A remarkable property of all the halogens is that they can form *interhalogen compounds* by the direct reaction of the elements, for example,

$$I_2 + Br_2 \longrightarrow 2IBr$$
$$I_2 + Cl_2 \longrightarrow 2ICl$$
$$Br_2 + Cl_2 \longrightarrow 2BrCl$$

Of considerable interest are compounds such as ClF_3, ICl_3, BrF_5, and IF_5, all of which constitute exceptions to the octet rule.

The halogens have many important uses. Their ions play important roles in the chemical reactions of our bodies. For example, Cl^- is the chief anion in extracellular fluids, and I^- is part of the thyroxin molecule produced by thyroid glands. A lack of iodine in the diet causes *goiter,* which was one of the first diseases to be recognized as stemming from a trace element deficiency. Iodized salt is now used to eliminate iodine deficiency and prevent goiter. Traces of F^- ion are effective in preventing tooth decay. However, excess fluoride causes *fluorosis,* manifested by mottled teeth and pathological effects upon bone. A level of 1 ppm is considered optimum for fluoride in water treated for the prevention of tooth decay. In areas where the water source contains high levels of fluoride, it is necessary to employ fluoride removal processes.

QUESTIONS

5.19 Which of the families of elements are located in the p block of the periodic table?

5.20 Name the p block elements that exhibit allotropy.

5.21 List some of the compounds of the p block elements that violate the octet rule.

5.22 What are boranes? Do you think they can be used as rocket fuels?

5.23 List some of the unique properties of carbon and silicon that distinguish them from their family members.

5.24 What are the common oxidation states of nitrogen? Give examples.

5.25 Suggest a method for the laboratory preparation of ozone.

5.26 Distinguish between acidic and basic oxides.

5.27 Based upon periodic relationships, predict some of the properties of the element astatine.

5.28 Can fluorine be prepared from HF by the same method used for preparing chlorine from HCl? Why?

5.29 Why are the halogens powerful oxidizing agents?

5.30 What are interhalogen compounds? Give examples.

THE NOBLE GAS FAMILY

The elements in this family are helium (He), neon (Ne), argon (Ar), krypton (Kr), xenon (Xe), and radon (Rn) (Fig. 5.24). They have been collectively called the *inert*

0

| 2 He |
| 4.0026 |
| $1s^2$ |

| 10 Ne |
| 20.179 |
| $2s^2 2p^6$ |

| 18 Ar |
| 39.95 |
| $3s^2 3p^6$ |

| 36 Kr |
| 83.80 |
| $4s^2 4p^6$ |

| 54 Xe |
| 131.30 |
| $5s^2 5p^6$ |

| 86 Rn |
| (222) |
| $6s^2 6p^6$ |

FIGURE 5.24

The noble gas family.

gases, the *rare* gases, or the *noble* gases. Since 1962, when the Canadian chemist Neil Bartlett prepared a stable compound of xenon and fluorine, the term "inert" has been questioned by many chemists. Because argon is about 1 percent by volume of atmospheric air, neither is the word "rare" fully appropriate. At present the word "noble" is preferred for these gases as well as for metals such as gold and platinum, which can react under special circumstances but are usually long-lasting.

Helium, the first member of the family, was discovered in the sun before it was found on earth. There is an abundance of helium in the sun and the name helium originates from the Greek word *helios* meaning "sun." All of the noble gases are found in varying amounts in our atmosphere, but argon is the most abundant. Helium occurs in a few natural gas wells in our country. It is formed during the radioactive decomposition of certain elements (see Chapter 6). Radon is a radioactive element, produced chiefly during the radioactive decay of radium.

Helium is the least dense of all elements except hydrogen. Because of its lack of flammability, helium is preferred to hydrogen for filling balloons. Neon is used in outdoor lighting fixtures (neon signs), giving as it does a red glow when an electric current is passed through it. Argon is used in filling incandescent light bulbs, because it does not react with the white-hot tungsten wire as oxygen or nitrogen does. Krypton and xenon, being relatively scarce and expensive, have few applications and these mostly in research work.

The limited chemical reactivity of the noble gases is attributed to their octet outer electron configuration (duo in the case of helium), which is considered to represent a condition of great stability. The formation of a limited number of compounds such as XeO_3, XeF_2, XeF_4, and XeF_6 by direct combination of the elements is explained in terms of promoting the noble gas outer electrons to higher-energy orbitals.

QUESTIONS

5.31 Are the commercial uses of the noble gases based on their properties? If so, can they be used interchangeably?

5.32 What possible oxidation states of the noble gases are known? Give examples.

5.33 Why would the noble gases be more likely to react with fluorine than with hydrogen?

5.34 What is the explanation for the formation of noble gas compounds?

THE TRANSITION (d BLOCK) ELEMENTS

Hitherto we have discussed the families of the representative elements which are in the *s* and *p* blocks of the periodic table. We shall now turn our attention to the *d* block elements or transition elements which fall between groups IIA and IIIA of the periodic table (Fig. 5.25). In the fourth period the regular addition of electrons to *d* sublevels begins with the third element in the period and continues till group IB is reached. The fifth, sixth, and seventh periods in the periodic table (see Fig. 5.5) follow a similar sequence: The added electron goes into the penultimate energy level, that is, the main energy level next to the outer one. For example, in the fourth period it is the 3*d* sublevel to which the electron is added, in the fifth to the 4*d*, and so on. As the elements in groups IB and IIB are reached the *d* sublevel is filled up with its maximum of 10 electrons so that the IB and IIB families are sometimes referred to as *neighbors* of transition elements. To justify this we need only examine the electron arrangements of the *d* block elements in the periodic table in Fig. 5.25.

IIIB	IVB	VB	VIB	VIIB	←————VIIIB————→		IB	IIB
21 Sc 44.96 $3d^14s^2$	22 Ti 47.90 $3d^24s^2$	23 V 50.94 $3d^34s^2$	24 Cr 51.996 $3d^54s^1$	25 Mn 54.94 $3d^54s^2$	26 Fe 55.85 $3d^64s^2$	27 Co 58.93 $3d^74s^2$	28 Ni 58.71 $3d^84s^2$	29 Cu 63.55 $3d^{10}4s^1$	30 Zn 65.37 $3d^{10}4s^2$
39 Y 88.91 $4d^15s^1$	40 Zr 91.22 $4d^25s^2$	41 Nb 92.91 $4d^45s^1$	42 Mo 95.94 $4d^55s^1$	43 Tc 98.91 $4d^65s^1$	44 Ru 101.07 $4d^75s^1$	45 Rh 102.91 $4d^85s^1$	46 Pd 106.4 $4d^{10}$	47 Ag 107.87 $4d^{10}5s^1$	48 Cd 112.4 $4d^{10}5s^2$
57 La 138.91 $5d^16s^2$	72 Hf 178.49 $5d^26s^2$	73 Ta 180.95 $5d^36s^2$	74 W 183.85 $5d^46s^2$	75 Re 186.2 $5d^56s^2$	76 Os 190.2 $5d^66s^2$	77 Ir 192.2 $5d^76s^2$	78 Pt 195.1 $5d^96s^1$	79 Au 196.97 $5d^{10}6s^1$	80 Hg 200.59 $5d^{10}6s^2$
89 Ac (227) $6d^17s^2$	104 Ku*	105 Ha*							

FIGURE 5.25

The transition elements.

All of the transition elements are metals, most of which are hard and high melting. They include such well-known and biologically essential elements as iron, chromium, copper, cobalt, manganese, molybdenum, and zinc. They also include the commercially important metals, silver, platinum, and gold. The transition elements are found mainly in the combined form with other elements and in varying amounts in the earth's crust. Iron is the most abundant (4.7% by weight). Many of these elements form solid solutions with other metals. Such solid solutions are called *alloys* (see Chapter 8).

In contrast to the alkali and alkaline earth metals, the *d* block elements show a certain tendency toward covalent bonding, both as the free elements and in their compounds. It is also noteworthy that most of these elements tend to show several oxidation states. Under appropriate conditions they can be made to combine with other elements, especially nonmetals, to form a variety of compounds. Many of these compounds have characteristic colors due to the excitation of *d* electrons in partially filled sublevels in the metal ion by visible light. The colors of some of the ions of the transition elements are listed in Table 5.4. Note that the color is often different for different oxidation states of the same element. The color is also modified by the formation of *hydrates,* as is the case with the so-called invisible ink. An invisible ink is a water solution of cobalt chloride ($CoCl_2$) that is pale pink in color.

TABLE 5.4

Colors of transition metal ions in dilute water solutions

Ions	Color
Mn^{3+}, Cr^{3+}, V^{2+}	violet
Ti^{3+}, MnO_4^-	purple
Cu^{2+}, Cr^{2+}	blue
Fe^{2+}, V^{3+}, Ni^{2+}	green
Fe^{3+}, Au^{3+}, Au^+, CrO_4^{2-}	yellow
Co^{2+}, Mn^{2+}	pink
Sc^{3+}, Ti^{4+}, Zn^{2+}, Cu^+	colorless

Upon warming, anything written with this solution becomes deep blue due to the formation of anhydrous $CoCl_2$. Hydrated metal ions have a tendency to behave as acids. This has a profound effect on the water environment. A good example is "acid mine water" which derives part of its acidic character from the tendency of hydrated iron to lose protons:

$$Fe(H_2O)_6{}^{3+} \longrightarrow Fe(H_2O)_5OH^{2+} + H^+$$

Another important property of the transition metals is that their ions tend to form *complex ions* by combining via covalent coordinate bonds with species that donate both the electrons. A few examples of complex ion formation are given below:

$$Ag^+ + 2NH_3 \longrightarrow Ag(NH_3)_2{}^+$$
$$Zn^{2+} + 4OH^- \longrightarrow Zn(OH)_4{}^{2-}$$
$$Pt^{2+} + 4Cl^- \longrightarrow PtCl_4{}^{2-}$$
$$Fe^{3+} + 6CN^- \longrightarrow Fe(CN)_6{}^{3-}$$

The molecules or ions that react with the transition metal ions are called *ligands*. A ligand that has two or more points of attachment to the central atom is called a *chelate* (from the Greek *chele*, meaning "claw"). The number of ligands that combines with a given metal ion is called the *coordination number* of that ion. Each of the complex ions possesses a definite geometry (Fig. 5.26). The study of complex

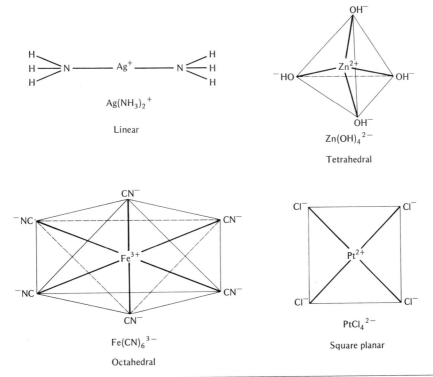

FIGURE 5.26

Geometric shapes of some complex ions.

ions is called *coordination chemistry*. Many anhydrous compounds such as $FeCl_3$ and $Cu(NO_3)_2$ are best viewed as *coordination compounds* in which the metal ions are covalently bonded to the negative species.

Closely related to the chemical activity of the transition metals is the phenomenon of *corrosion,* which is the chemical attack on a metal by its environment, usually atmospheric air in conjunction with water and various substances dissolved in water. Essentially, corrosion is a reaction in which a metal is oxidized. The rusting of iron is by far the most widespread and familiar type of corrosion in our environment.

Many transition metals form an integral part of the protein molecules known as *enzymes* or biological catalysts that are necessary to speed up the hundreds of chemical reactions in living cells (Table 5.5). Complexed iron in the form of hemoglobin, $(C_{738}H_{1166}FeN_{203}O_{208}S_2)_4$, provides a means for oxygen transport in the blood of vertebrates. The copper-containing hemocyanin, which is similar to hemoglobin, transports oxygen in the blood of many invertebrates. It is believed that the absorption of glucose by tissues depends on the interaction of chromium and the protein called *insulin.*

Although some compounds of mercury are relatively nontoxic and have been used as medicaments, most of the organic compounds of mercury are highly toxic. In fact, metallic mercury from industrial discharges into lakes and rivers settles down in sediments and is acted upon by bacteria in the absence of free oxygen to produce the highly toxic CH_3Hg^+ ion and $(CH_3)_2Hg$. Other transition metals that may pose health hazards are cadmium, nickel, vanadium, and yttrium (Table 5.6). Chemically, cadmium is very similar to zinc, easily replaces zinc in some enzymes and impairs their catalytic activity. Much of the physiological action of the toxic elements arises from their chemical similarities to the essential elements.

TABLE 5.5
Some transition elements essential to life

Element	Principal function	Deficiency effect
chromium	helps glucose absorption by tissue by interacting with the protein insulin	(unknown)
cobalt	integral part of vitamin B_{12}	anemia
copper	part of hemocyanin molecule that transports oxygen in the blood of some invertebrates	anemia and disorders of bone and connective tissue
iron	part of heme, a component of hemoglobin that is vital for oxygen transport in blood of vertebrates	anemia
manganese	part of some enzymes (biological catalysts)	abnormalities in skeletal structure
molybdenum	essential component of many enzymes, particularly those required for both nitrogen fixation and nitrate utilization	(unknown)
zinc	part of several enzymes	skin lesions

QUESTIONS

5.35 What is the relation of the transition elements to the other elements in the periodic table? Is there any difference in sublevel electronic structures of the atoms of the transition elements?

5.36 Is there a most common oxidation state of the transition metals?

5.37 Which elements resemble each other more closely: IA and IB elements or IIA and IIB elements?

5.38 What causes the coloration of the transition metal ions?

5.39 Explain what is meant by the term corrosion.

5.40 What are complex ions? Illustrate with some examples.

5.41 List the names of transition metals that are
   a. essential elements
   b. health hazards

THE INNER TRANSITION (f BLOCK) ELEMENTS

The inner transition elements or f block elements have electronic configurations corresponding to the filling of the f sublevels (Fig. 5.27).

TABLE 5.6
Some transition elements posing health hazards

Element	Source	Health effect
Cadmium	mining wastes, industrial discharges	bone fracture, high blood pressure, kidney damage, destruction of testicular tissue and red blood cells
mercury	electrical batteries, industrial discharges	nerve damage and death
nickel	steel and nonferrous metal industries, diesel and residual oils, coal, tobacco smoke, chemicals and catalysts	lung cancer [from nickel carbonyl, $Ni(CO)_4$]
vanadium	petroleum, steel and nonferrous alloys, chemicals and catalysts	no reported hazards at current levels of contamination
yttrium	coal, petroleum	carcinogenic in mice over long-term exposure

The inner transition elements whose electronic structures involve the progressive filling of the 4f sublevel in the sixth period are called the *lanthanides* or *rare earths*. These include the elements of atomic numbers 58 through 71. Except for promethium (atomic number 61) which is artificially prepared and radioactive, all other rare earths are found as compounds with other elements in the mineral *monazite*. All the lanthanide elements are soft gray metals that react vigorously with water giving off hydrogen and forming a hydroxide. In this respect the lanthanides behave like the alkali and alkaline earth metals. The +3 oxidation state is by far the most common and this leads to closely similar chemical properties making the separation of lanthanides by ordinary means rather difficult.

The inner transition elements whose electronic structures involve the progressive filling of the $5f$ sublevel in the seventh period are called the *actinides*. These include the elements of atomic numbers 90 through 103. The actinides are radioactive, and all of the elements with atomic numbers greater than that of uranium (atomic number 92) must be prepared artificially in high-energy machines (see Chapter 6). Uranium and thorium are found in nature, but only combined with other elements. All the actinides are metallic and, like the lanthanides, react readily with water with the evolution of hydrogen. However, unlike the lanthanides, they are capable of existing in a variety of oxidation states. The stability and decay of these elements are discussed in Chapter 6.

According to presently accepted practice, the actinide series ends with element 103 (lawrencium), at which point the $5f$ orbitals are filled. Element 104 (kurchatovium or rutherfordium) should, therefore, be a member of group IVB, being placed in the periodic table just under hafnium, number 72 (see Fig. 5.25).

lanthanides $\qquad$ $4f^{0-14}5d^{0-1}6s^2$

58 Ce 140.12	59 Pr 140.1	60 Nd 144.2	61 Pm (147)	62 Sm 150.4	63 Eu 151.96	64 Gd 157.25	65 Tb 158.93	66 Dy 162.50	67 Ho 164.93	68 Er 167.26	69 Tm 168.93	70 Yb 173.04	71 Lu 174.97
90 Th 232.04	91 Pa 231.04	92 U 238.03	93 Np 237.05	94 Pu (242)	95 Am (243)	96 Cm (247)	97 Bk (247)	98 Cf (247)	99 Es (254)	100 Fm (253)	101 Md (256)	102 No (254)	103 Lr (257)

actinides $\qquad$ $5f^{0-14}6p^{0-2}7s^2$

FIGURE 5.27

The inner transition elements.

QUESTIONS

5.42 Account for the fact that the inner transition elements have very similar physical and chemical properties.

5.43 Why is the separation of lanthanides by ordinary chemical methods difficult?

5.44 What are the naturally occurring actinides?

5.45 Name the synthetic actinides.

5.46 How many more elements must be discovered to complete the seventh period in a modern periodic table?

SUGGESTED PROJECT 5.2
Elements and your household

Compile a list of elements (in the uncombined form) constituting household articles and classify them into various families. Then find those household chemicals that contain the elements on your list. Cite possible routes by which these chemicals can be produced beginning with the pure elements. Based on the information available in this chapter, identify those elements and chemicals that are essential to health and those that are hazardous.

6

Nuclear transformations and our environment

It appears that we can look forward to further extension of the periodic table as a continuing result of modern-day alchemy, and thus the transuranium elements will continue to add much to our knowledge of atomic and nuclear structure and properties.

GLENN T. SEABORG (1912–   )
American chemist

About half a century ago, Rutherford concluded that the atomic nucleus is a world of its own in which a number of particles like protons and neutrons are confined in a minute volume and held together by very powerful unknown forces. What are these nuclear forces? Where does the energy that causes them originate? An answer to these questions was obtained in 1932 when the release of *nuclear energy* was achieved in the laboratory.

If we take a close look at nuclear masses, we can see that nuclei always have slightly smaller masses than the sum of the masses of the particles constituting them. As an example, consider the helium nucleus which has a mass of 4.0026 amu. It contains four nucleons out of which two are neutrons and two are protons. If we add up the masses of all these particles (see Table 4.1), we will get a mass of 4.0320 amu, which is 0.0294 amu more than the nuclear mass of helium (Fig. 6.1). It is now believed that this missing mass, called the *mass defect*, is associated with the energy required for the binding of protons and neutrons into the nucleus. The magnitude of this *binding energy*, calculated from Einstein's equation (see Chapter 3), is about 630 billion cal/mole of helium atoms. Calculations of the binding energies for the nuclei of various elements have shown that the mass losses and binding energies are greatest for nuclei of elements with mass numbers from about 20 to 150.

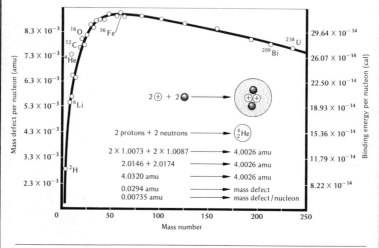

FIGURE 6.1

Mass defect and binding energy per nucleon versus mass number.

What if the nucleus of a very heavy atom (mass greater than 150) were to split into two or more nuclei of intermediate mass (about 20 to 150)? The two nuclei would weigh less than the original nucleus and the loss of mass would result in the liberation of a tremendous amount of energy. This exothermic process, called *nuclear fission*, was achieved on a laboratory scale in 1938 when the German chemist Otto Hahn succeeded in splitting uranium-235 into krypton and barium. Within 4 years a team of American scientists succeeded in harnessing the same process on a practical scale and the result was the *atomic bomb* and the *nuclear reactor*. More about these devices later.

What if two or more nuclei of light atoms (mass less than 20) were to fuse to form the nucleus of a heavier atom? Even though the resulting atom will contain all the nucleons of the lighter atoms, there would be a loss in mass that, again, would result in the liberation of a tremendous amount of energy. This exothermic process, called *nuclear fusion*, can occur and, in fact, is thought to be responsible for the tremendous amount of energy released by the sun. Nuclear fusion was first revealed in 1952 when the United States exploded its first *hydrogen bomb*. Today, much research is being carried out toward harnessing *fusion power*.

In this chapter we shall examine these nuclear transformations and their impact on our environment. First of all, let us see whether or not such transformations occur in nature.

NUCLEAR TRANSFORMATIONS IN NATURE

Following the discovery of natural radioactivity (see Chapter 4), it was believed that atoms are radioactive because they are too large and too complex to be stable. Soon this instability was traced to an unfavorable ratio of neutrons to protons in the nuclei of atoms with atomic numbers above 82. In general, that kind of radiation is emitted which tends to increase the nuclear stability.

Let us examine what happens when an α-particle (He^{2+}) is emitted from uranium-238:

$$^{238}_{92}U \longrightarrow {}^{4}_{2}\alpha + {}^{234}_{90}Th$$

Here the superscript represents the mass number (the number of protons plus the number of neutrons) and the subscript the atomic number (the number of protons). The thorium that is produced is also radioactive and ejects a β-particle according to the scheme

$$^{234}_{90}Th \longrightarrow {}^{0}_{-1}e + {}^{234}_{91}Pa$$

Because it is radioactive, the protoactinium ejects a β-particle to form $^{234}_{92}U$, which is an isotope of the original uranium-238. The $^{234}_{92}U$ isotope is also radioactive and gives off an α-particle, thus producing $^{230}_{90}Th$, which, in turn, is radioactive. The process of either α- or β-particle emission together with some γ-rays continues until a stable nonradioactive element is formed. The whole reaction sequence (Fig. 6.2) is called a *radioactive decay series*, that is, a collection of elements that are formed from a *parent* radioactive element by successive emissions of α- or β-particles. In nature there are three such series the parent elements of which are uranium-238, uranium-235, and thorium-232. Interestingly, an isotope of lead (atomic number 82) is the end product of all the three naturally occurring radioactive series.

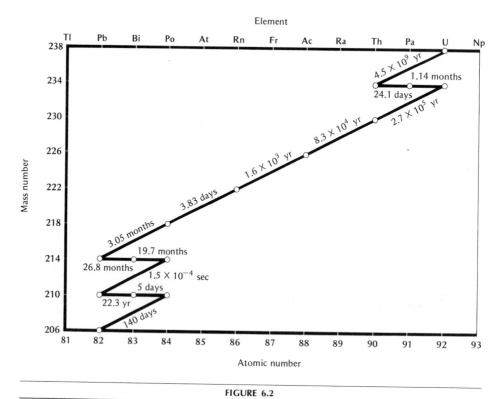

FIGURE 6.2

Naturally occurring uranium-238 decay series: α decay is represented by diagonal lines; β decay by horizontal lines.

The rate at which radioactive decay occurs gives a measure of the stability of a nucleus and is usually expressed in terms of the *half-life* of the nucleus: the time needed for half of a given sample of the isotope to decay. The many steps and the half-lives of the members of the uranium-238 series are shown in Fig. 6.2. Notice that the half-life of uranium-238 is of the order of billions of years; in contrast, that of the polonium-214 isotope is a small fraction of a millisecond. Thus, some of the transformations take place rapidly, whereas others are very slow.

How are the half-lives determined experimentally? One of the ways is to count the number of emissions in a given period of time by a known mass of the radioactive sample using a simple but versatile device known as the *Geiger-Müller counter* (Fig. 6.3). When the emitted particle passes through the window it causes ionization of some of the gas molecules in the tube. The ions thus formed conduct a pulse of current between the central wire and the tube walls. This current is used to activate a detector light or a clicker. The Geiger-Müller counter is commonly used for detecting radioactivity in our environment. Of course, there are many other devices also available for the same purpose; some of them are very sophisticated indeed.

The knowledge of the very long half-life of uranium-238 (4.5 billion years) and of the occurrence of uranium-238 together with lead-206 in certain minerals has made possible some fascinating calculations of the *minimum age of the earth*. Because lead-206 is the end product of the uranium-238 decay series, it is assumed that the lead has been formed only as the result of radioactive decay. Thus, by determining the amounts of uranium and lead present in a mineral sample and by using the known half-lives of all the intermediate elements involved in the decay, we can calculate the time required to establish the uranium-to-lead ratio found in the minerals. Such calculations for different samples indicate that the earth is about 4 to 6 billion years old.

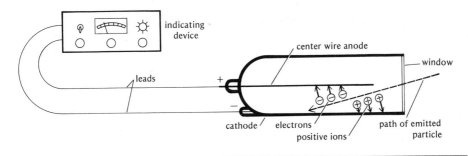

FIGURE 6.3

The Geiger-Müller counter.

Similarly, the ages of many ancient objects have been estimated using other radioactive elements. Of significance is the method developed by the American chemist Willard Libby for dating archaeological discoveries of an organic nature using the half-life of the radioactive isotope carbon-14 (5570 years). This isotope is present in the atmosphere (in the combined form as CO_2) in a constant small amount as a result of continued cosmic ray activity in the upper atmosphere. Plants consume the carbon dioxide during their growth, but when they cease to grow the carbon-14 begins to diminish through radioactive decay. By measuring this radioactive decay in ancient objects and comparing it with the radioactivity of living plants, the ages of many objects of archaeological importance have been determined; for example those of the Dead Sea Scrolls, glacial deposits, prehistoric Indian cliff dwellings, and objects found in Egyptian mummy tombs.

QUESTIONS

6.1 **What is the origin of nuclear energy?**
6.2 **Define the following terms:**
  a. mass defect
  b. binding energy
  c. half-life

6.3   Distinguish between nuclear fission and nuclear fusion.
6.4   What are the particles given off by radioactive materials?
6.5   Give two examples of nuclear transformations in nature.
6.6   What is meant by a radioactive series?
6.7   Name the parent elements in the naturally occurring radioactive series.
6.8   Describe a device for determining whether or not a substance is radioactive.
6.9   How was the minimum age of the earth calculated?

NUCLEAR TRANSFORMATIONS IN THE LABORATORY

It was originally the alchemist's dream to *transmute* base metals such as lead into gold. Following the discovery of the existence of nuclear transformations in nature, attempts were made to transmute atoms of one element into those of another artificially. Over the past five decades the *artificial transmutation* of most of the elements was achieved in many laboratories. It all began in 1919 with Rutherford's experiments on the bombardment of various gases with α-particles from radioactive sources. When nitrogen was bombarded with α radiation, Rutherford found that hydrogen and an isotope of oxygen were produced:

$$^{14}_{7}N + ^{4}_{2}\alpha \longrightarrow ^{17}_{8}O + ^{1}_{1}p$$

During the following decade, many other transmutations were accomplished using α-particles as the only projectiles. It was found that the α-particles are actually captured by the nuclei of the atoms that were bombarded prior to the emission of protons and that the energy of the emitted protons is greater than the energy of the bombarding α-particles. Perhaps, the reason for this energy increase is that the protons acquire additional energy during nuclear rearrangement.

In 1930 the German physicists W. Bothe and H. Becker observed an important reaction when they bombarded beryllium atoms with α-particles. They reported the emission of a very penetrating radiation that had the characteristics of a stream of particles but was not deflected on passing through electric and magnetic fields. Two years later, the English physicist James Chadwick showed that the particles had masses nearly equal to that of protons but were uncharged. Thus the *neutron* was discovered:

$$^{9}_{4}Be + ^{4}_{2}\alpha \longrightarrow ^{12}_{6}C + ^{1}_{0}n$$

During the years following Rutherford's discovery of artificial transmutation, physicists and chemists all over the world began developing devices, called *accelerators*, for increasing the speed of α-particles. The earliest such device, called the *cyclotron*, was invented by the American physicist Ernest Lawrence in 1930 at the University of California. The cyclotron operates on the principle that a charged particle is repelled by a like charge and attracted by an unlike charge and follows a curved path in a magnetic field. The core of the device (Fig. 6.4) is a pair of hollow and semicircular metal chambers, which are shaped like the letter D (and hence called *dees*), placed in a vacuum chamber mounted between the pole pieces of an extremely powerful magnet. These dees are supplied with powerful, rapidly alternating electric charges with the result that an α-particle entering at the center of the device is driven across the intervening space into the other dee by electrical attraction and repulsion. The alternating electrical charge increases the particle velocity and the powerful magnetic field forces the accelerated particle to follow the spiral pattern shown in Fig. 6.4. The net result is that the particle emerges from the device with a velocity approaching that of light (3×10^{10} cm/sec). Such accelerated particles have higher energy and, therefore, penetrate the nucleus more effectively than

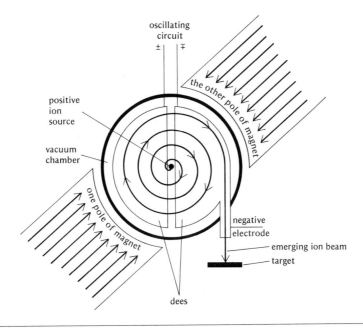

oscillating circuit

positive ion source

vacuum chamber

the other pole of magnet

one pole of magnet

negative electrode

emerging ion beam

target

dees

FIGURE 6.4

Schematic diagram of a cyclotron.

the α-particles emitted from a natural radioactive source such as the element radium. Typical transmutations achieved by the cyclotron in the early 1930s are

$$^{7}_{3}\text{Li} + ^{4}_{2}\alpha \text{ (alpha)} \longrightarrow ^{10}_{5}\text{B} + ^{1}_{0}\text{n}$$
$$^{7}_{3}\text{Li} + ^{1}_{1}\text{p (proton)} \longrightarrow ^{4}_{2}\alpha + ^{4}_{2}\alpha$$
$$^{24}_{12}\text{Mg} + ^{2}_{1}\text{d (deuteron)} \longrightarrow ^{24}_{11}\text{Na} + 2^{1}_{1}\text{p}$$
$$^{209}_{83}\text{Bi} + ^{2}_{1}\text{d (deuteron)} \longrightarrow ^{210}_{84}\text{Po} + ^{1}_{0}\text{n}$$

The cyclotron is essentially a device for accelerating positive particles. Its invention was followed by the design and construction of several other accelerators such as the betatrons for accelerating electrons, the Van de Graaff generators, and linear particle accelerators. The linear accelerator is one of the most powerful machines for particle acceleration presently available. The operation of the linear accelerator is less complicated in that it does not require a magnetic field. The principle of operation is outlined in Fig. 6.5. The largest linear accelerator in the world went into operation in 1967 at Stanford University under the sponsorship of the Atomic Energy Commission. However, the nuclear reactors, discussed later in this chapter (p. 147), are the most prolific large-scale transmuters. Neutrons liberated in these reactors are particularly useful projectiles for transmutations because neutrons have no charge and, therefore, are not repelled by the target nucleus.

TABLE 6.1
Transuranium elements[a]

Atomic number	Name	Symbol	Source of first preparation	Longest-lived isotope		More available isotope	
				Atomic mass	Half-life for α decay	Atomic mass	Half-life for α decay
93	neptunium	Np	irradiation of uranium with neutrons	237	2.2×10^6 yr	—	—
94	plutonium	Pu	bombardment of uranium with deuterons	244	76×10^6 yr	239 242	24,360 yr 3.79×10^5 yr
95	americium	Am	irradiation of plutonium with neutrons	243	7950 yr	241	458 yr
96	curium	Cm	bombardment of plutonium with helium ions	247	16.4×10^6 yr	242	162.5 days
97	berkelium	Bk	bombardment of americium with helium ions	247	1380 yr	249	314 days (β decay)
98	californium	Cf	bombardment of curium with helium ions	251	800 yr	249 250 252	360 yr 13.2 yr 2.65 yr
99	einsteinium	Es	irradiation of uranium with neutrons	254	270 days	253	20 days
100	fermium	Fm	irradiation of uranium with neutrons	257	80 days	—	—
101	mendelevium	Md	bombardment of einsteinium with helium ions	257	—	—	—
102	nobelium	No	bombardment of curium with carbon ions	255	185 sec	—	—
103	lawrencium	Lr	bombardment of californium with boron ions	256	45 sec	—	—
104	kurchatovium[b]	Ku	bombardment of plutonium with neon ions	260	—	—	—
104	rutherfordium[b]	Rf	bombardment of californium with carbon ions	257	3 sec	—	—
105	hahnium[b]	Ha	bombardment of californium with nitrogen ions	260	1.6 sec	—	—

[a] Data from Earl K. Hyde, Synthetic Transuranium Elements, Division of Technical Information, U.S. Atomic Energy Commission, Washington, D. C.
[b] Name and symbol not yet officially recognized by the International Union of Pure and Applied Chemistry.

SYNTHETIC ELEMENTS   In 1937 the first synthetic element was prepared by cyclotron bombardment of molybdenum metal with deuterons:

$$^{98}_{42}Mo + ^{2}_{1}d \longrightarrow ^{99}_{43}Tc + ^{1}_{0}n$$

The new element was named *technetium* (from the Greek *technetos,* meaning "artificial"). In the succeeding years the blank spaces in Mendeleev's periodic table (see

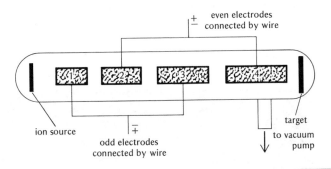

FIGURE 6.5

Schematic diagram of a linear accelerator. Notice that the electrode tubes become successively longer because of increased speeds of the particle.

Chapter 5) were filled with the new elements francium, promethium, and astatine that were synthesized in cyclotrons. In addition 13 elements heavier than uranium (atomic number 92) have been synthesized, most of them by the American chemist Glenn Seaborg and his co-workers at the University of California. These elements are called *transuranium* (beyond uranium) elements. Their methods of preparation

FIGURE 6.6

Glenn T. Seaborg (1912–), American chemist who synthesized many transuranium elements. (Photograph courtesy of Lawrence Berkeley Laboratory, Berkeley, California.)

and half-lives are shown in Table 6.1. All of them are radioactive and many are extremely unstable. For these reasons it has been suggested that these elements could have once existed on earth but were too unstable to survive the passage of time due to radioactive decay. Indeed, traces of one of the transuranium elements, namely, plutonium, have been detected in the earth.

The interior of a new type of particle accelerator that is being used in the search for new elements is shown in Fig. 6.7. Because many exciting experiments are still in progress, it is likely that elements with still higher atomic numbers will be synthesized and the periodic table will be continually extended. However, their anticipated extreme instability is suggestive of little or no practical applications to mankind in the immediate future.

FIGURE 6.7

The interior of a heavy ion linear accelerator (HILAC) used in the search for new elements at Berkeley, California. This linear accelerator was used in the first synthesis of elements 102 and 103. It is designed to accelerate heavy particles such as the ions of C, N, O, or Ne which are fired through the center of the doughnut-shaped "drift tubes." These drift tubes extend the length of the vacuum tank (90 ft long) the size of which can be judged by the apparent size of the man standing at the far end. (Photograph courtesy of the Lawrence Berkeley Laboratory, University of California, Berkeley, California.)

INDUCED RADIOACTIVITY AND RADIOISOTOPES What happens if some of the lighter stable elements are bombarded with energetic particles? The problem was initially studied in 1934 by Frédéric Joliot and Irène Joliot-Curie who observed that lighter elements such as Al, B, and Mg when subjected to α-particles emitted, besides neutrons, positively charged particles identical to electrons in mass but opposite in sign. (These new particles were later termed *positrons*.) The reaction for aluminum is

$$\ce{^{27}_{13}Al + ^{4}_{2}\alpha \longrightarrow ^{30}_{15}P + ^{1}_{0}n}$$
$$\ce{^{30}_{15}P \longrightarrow ^{30}_{14}Si + ^{0}_{1}e}$$

The phosphorus isotope produced is radioactive and decays with a distinctive half-life similar to the behavior of naturally occurring radioactive elements. This was the first example of *induced radioactivity,* and phosphorus-30 was the first artificially prepared radioactive element. Since then, over a thousand radioactive isotopes (radioisotopes) have been artificially produced because of their important uses in research, medicine, agriculture, industry, and other areas. Two very useful radioisotopes are produced by neutron bombardment reactions:

$$\ce{^{59}_{27}Co + ^{1}_{0}n \longrightarrow ^{60}_{27}Co}$$
$$\ce{^{24}_{12}Mg + ^{1}_{0}n \longrightarrow ^{24}_{11}Na + ^{1}_{1}H}$$

Cobalt-60 decays, emitting γ radiation, and has a half-life of 5.3 years; sodium-24 is a β emitter with a half-life of 14.8 hours.

One of the widely publicized uses of radioisotopes is in the treatment of human diseases. Since the discovery that γ radiation destroys living tissue, its first medical use was to control cancers that were not accessible to surgery. The naturally occurring element radium was the first emitter used in medicine; nowadays cobalt-60 has replaced radium in cancer therapy. Some of the other radioisotopes used in medicine are radioactive iodine to control overactive thyroid glands, radioactive phosphorus to treat leukemia and cancer of the bone, and radioactive sodium and gold in controlling the growth of malignant tumors.

Besides treating diseases and controlling malignancies, radioisotopes are also used for medical diagnosis. For example, a small amount of radioactive sodium is injected into the bloodstream to locate circulatory deficiencies, a dye tagged with radioactive iodine is used to locate brain tumors, and a sugar containing radioactive carbon is sometimes administered to observe the progress of diabetes. In all these instances the radioactive material is traced with a sensitive Geiger-Müller counter. Of course, there are many other possible applications of radioisotopes in medicine. Medical research is now progressing so rapidly that radioisotopes seem destined to become even more useful in the future.

In industry, radioisotopes are used for tracing leaks in gas or oil pipelines, for controlling the uniformity of sheet metal in roller mills, for detecting metal flaws in locomotives and other machinery, for evaluating dirt-removing efficiency of detergents, for exploration of mineral and petroleum resources, and so on. At present, food preservation by radiation treatment is the subject of extensive research. The process, known as *radiation sterilization,* makes use of the radiation from radioisotopes to inactivate or destroy the microorganisms responsible for the decay of foods. The doses used do not induce radioactivity in the food but do sometimes produce changes in flavor and nutritional value. The main advantage is that food thus treated can be stored without refrigeration indefinitely. Because of the difficulties involved in refrigerating food for armies entrenched in jungles, food sterilization by irradiation has been approved and used for some military rations. In addition, radiation treatment has been approved for the destruction of insects in wheat, for the prevention of sprouting in white potatoes, and for the preservation of canned bacon used for general consumption.

We have just seen that the role of radioisotopes in our society is wide and varied. Their use in industry, agriculture, and other areas is of great economic importance. However, these benefits are overshadowed by the more spectacular medical applications. It has been reported that radioisotopes are administered to over 65,000 patients annually in the United States alone. Despite all these remarkable benefits to mankind a few words of caution regarding the use of radioisotopes are desirable. We live in an environment in which we are continuously exposed to some *natural* radiation arising from radioactive minerals (such as those of radium and uranium) and from atmospheric cosmic ray activities. Our own bodies contain traces of substances in which radioactive atoms (such as carbon-14 and potassium-40) are present. All forms of radiation capable of producing ionization can tear apart almost any molecules. Although extremely small, the injury or risk is finite even at the low doses characteristic of environmental contamination. The ultimate sources of radiation injuries are unwanted changes in the molecules that constitute the body cells. Some observed consequences to body cells include swelling of the nucleus and cell degeneration, breakup of chromosomes, viscous cell fluids, delayed or abnormal cell division, abnormal metabolism, and the formation of toxic compounds in the cells. For these reasons, atomic energy commissions in every country have cautioned all personnel working with radioisotopes to take every precaution to shield themselves from harmful doses of radiation.

QUESTIONS

6.10 What is meant by artificial transmutation? Give an example.

6.11 How was the neutron discovered?

6.12 What is the principle of the cyclotron?

6.13 Can neutrons be accelerated in the cyclotron? Why or why not?

6.14 What are synthetic elements? How are they prepared?

6.15 Name one of the synthetic elements that has been detected in the earth.

6.16 Write the equations for the synthesis of
   a. the longest-lived plutonium isotope
   b. the more available isotopes of plutonium
      (Hint: see Table 6.1.)

6.17 What are positrons?

6.18 What is meant by induced radioactivity?

6.19 Name some radioisotopes used in medicine.

6.20 Describe how radioisotopes are used as tracers in medical diagnosis.

6.21 Discuss the effects of food preservation by radiation treatment.

6.22 What steps should you take to prevent radiation hazards?

ATOM-SMASHING EXPERIMENTS

In 1938 the German chemists Otto Hahn, Lise Meitner, and Fritz Strassman made the startling discovery that the impact of a neutron on a uranium nucleus caused it to undergo *fission* into medium-sized nuclei such as those of barium and krypton:

$$^{235}_{92}U + ^{1}_{0}n \longrightarrow ^{94}_{36}Kr + ^{139}_{56}Ba + 3^{1}_{0}n + energy$$

The atom-smashing process, which previously had been considered impossible, was accompanied by the release of a tremendous amount of energy, calculated to be of the order of a few trillion (10^{12}) calories per mole of uranium. Furthermore, released neutrons produced a fission that liberated even more neutrons that could, in turn, produce more fissions and thus *self-sustain* a *chain reaction*. Here was the beginning of the nuclear era with tremendous possibilities for the practical release and harnessing of *atomic* or *nuclear energy*.

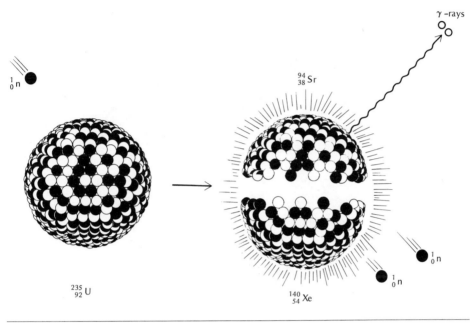

FIGURE 6.8

Fission of a uranium nucleus.

THE EXPLOSIVE REACTION   Soon after the discovery of the uranium fission reaction, it became apparent to many scientists that uranium atoms may fission in several ways and that each fission produces more neutrons. A representative fission process

$$^{235}_{92}U + ^{1}_{0}n \longrightarrow ^{94}_{38}Sr + ^{140}_{54}Xe + 2^{1}_{0}n + energy$$

is shown schematically in Fig. 6.8. In 1942 a group of scientists working under the Italian physicist Enrico Fermi in Chicago observed that a minimum amount, called the *critical mass*, of uranium-235 is necessary for the fission process to be a self-sustaining chain reaction (Fig. 6.9) and that once the critical mass is equaled or exceeded, an explosion occurs in a matter of only microseconds. With World War II in progress, the United States launched a massive effort to gather uranium-235 for an explosive device and on July 16, 1945, with less than 1 kg of uranium-235 the first atomic bomb explosion occurred at Alamogardo, New Mexico. During the next month, two similar bombs were dropped over the Japanese cities of Hiroshima and Nagasaki. Both cities were annihilated, millions of people were dead and injured, and a few days later World War II ended with the surrender of Japan. From that day the stockpiling of atomic weapons by the major world powers began and has since continued unabated.

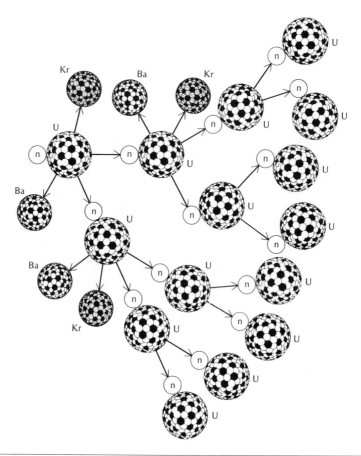

FIGURE 6.9

A nuclear chain reaction.

What is worse than the explosion, which produces a fantastic amount of heat, deadly γ radiation, and devastating shock waves, is the *fallout,* the mushroom-type cloud (Fig. 6.10) that follows the explosion. It contains the radioactive fission products which are distributed by air currents to all parts of the world. Days and weeks later, rain and snow carry the fission products back to the earth's surface and its immediate air environment, creating a hazard to all life on earth.

The basic principle of the structure of atom bombs is now well-known. In general, sufficient subcritical (less than critical) masses of uranium-235 are suddenly

FIGURE 6.10

The world's first nuclear explosion at Trinity Site in southern New Mexico on July 16, 1945. Photograph courtesy of the Los Alamos Scientific Laboratory, Los Alamos, New Mexico.

brought together to make a supercritical mass. The device shown in Fig. 6.11 makes use of the implosion wave from detonation of a common explosive such as TNT to compress rapidly loosely packed uranium-235 into a compact, supercritical mass that explodes in a fraction of a second. A heavy steel case must be used so that the neutrons and fissioning atoms are held together long enough for the chain reaction to become explosive.

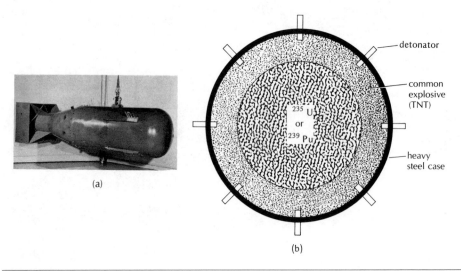

(a)

(b)

detonator

common explosive (TNT)

^{235}U or ^{239}Pu

heavy steel case

FIGURE 6.11

(a) Nuclear weapon of the "Little Boy" type—the kind detonated over Hiroshima, Japan, in World War II. This was the first type of nuclear weapon ever detonated. (Photograph courtesy of the Los Alamos Scientific Laboratory, Los Alamos, New Mexico.) (b) Basic structure of an atom bomb.

Because of the scarcity of uranium-235 in nature, it was necessary to look for other fissionable materials. It has been possible to prepare fissionable material artificially. Nonfissionable uranium-238 and thorium-232 are converted into fissionable plutonium-239 and uranium-233, respectively, by neutron capture reactions and subsequent radioactive decay:

$$^{238}_{92}U + {}^{1}_{0}n \longrightarrow {}^{239}_{92}U \xrightarrow{\text{23 min}} {}^{239}_{93}Np + {}^{0}_{-1}e$$

$$\downarrow \text{2.3 days}$$

$$^{239}_{94}Pu + {}^{0}_{-1}e$$

$$^{232}_{90}Th + {}^{1}_{0}n \longrightarrow {}^{233}_{90}Th \xrightarrow{\text{23 min}} {}^{233}_{91}Pa + {}^{0}_{-1}e$$

$$\downarrow \text{27 days}$$

$$^{233}_{92}U + {}^{0}_{-1}e$$

It is said that the second atomic bomb dropped in World War II contained plutonium-239 as the fissionable material.

THE CONTROLLED REACTION Concurrently with the development of the atomic bomb, attempts were made to carry out the self-sustaining nuclear fission reaction under controlled conditions. That is to say, attempts were made to liberate the instantaneous heat of an atomic explosion over a period of several days or weeks. In 1942, Fermi and his associates succeeded in setting up a device in Chicago, which has since become known as the *atomic pile* or *nuclear reactor,* and in achieving controlled fission. Today, the nuclear reactor is the means by which power and nonfissionable isotopes are produced for peacetime purposes.

A number of nuclear reactors have been built all over the world and more are being constructed. A typical reactor (Fig. 6.12) has a core that consists of cylindrical slugs of pure, naturally occurring, uranium oxide encased in aluminum rammed into holes in a pile of graphite blocks serving as *moderators* by slowing down the neutrons produced by fission of uranium. In addition to the graphite moderators, efficient

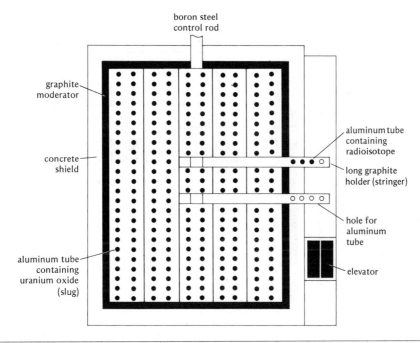

FIGURE 6.12

Schematic representation of a nuclear reactor.

neutron absorbers, such as cadmium and boron, encased in steel are sometimes pushed into the holes in graphite blocks to slow down neutron production in case the rate of fission becomes too high. The reactor core is surrounded by 4–6 ft wide concrete shield for biological protection of the surroundings. Provision is also made for circulating cooling water or some other heat-transfer medium such as cold air through the reactor. Some reactors use either ordinary water or heavy water (D_2O) both as coolant and moderator. The steam generator turns by means of a heat exchanger, a turbine that, in turn, operates a generator to produce electricity (Fig. 6.13). Nuclear power is thus obtained.

Natural uranium contains only 1 uranium-235 atom for each 140 atoms of uranium-238. Although uranium-238 does not undergo fission by slow neutrons, it readily absorbs them eventually to produce fissionable plutonium-239 (see p. 146).

Thus, if the fissionable uranium-235 core inside the reactor is surrounded by uranium-238, the otherwise wasted neutrons produce more fissionable material than is used by the reactor. Such a reactor has been nicknamed the *breeder* reactor. The existence and operation of such devices has allowed for the use of nuclear energy unrestricted by the amount of uranium-235 that exists on earth. The present-day reactors that are used for electrical power production make use of core material consisting of uranium and thorium.

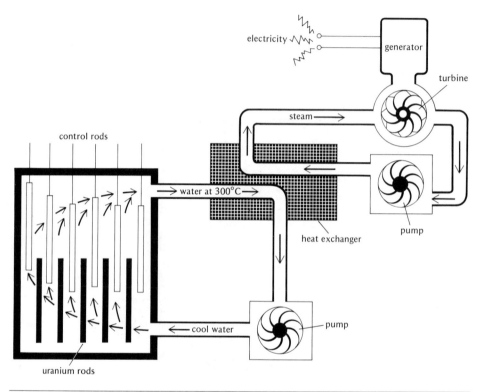

FIGURE 6.13

Basic components of a nuclear power plant.

We live in an era in which there is an ever-increasing demand for energy by an expanding population. Increased burning of coal, petroleum oils, and natural gases for obtaining the necessary energy introduces large amounts of contaminants into the atmosphere. Although this type of contamination can be avoided by the use of nuclear energy, nuclear power plants pose two grave problems. One is the problem of containment of radioactive wastes (see below); the other is *thermal pollution*, the waste heat from the generating plant heating up our environment—a discussion of this threat to our environment is deferred until Chapter 14. In spite of these somewhat serious threats the major world powers have built and continue to build numerous nuclear electric power stations. Many of them are already in operation although nuclear power is by no means cheaper than conventional hydroelectric power at this time. Even some of the smaller nations have such installations now under construction. Moreover, some of the submarines and ships that churn the waters of our oceans have been nuclear-powered for two decades. With the anticipated depletion of fossil fuels (coal, petroleum products, etc.), the question arises: Are there energy sources other than the fissionable fuel we have been discussing? An answer to this question may be seen in the discussion of *nuclear fusion* (see p. 150 of this chapter).

PROBLEM OF RADIOACTIVE WASTE DISPOSAL   The chief source of radioactive waste is the solid fission products, which are almost always the unstable neutron-rich isotopes of naturally occurring elements. In normal practice, these products are allowed to accumulate in the reactor over a period of 1 to 3 years. During this time they absorb sufficient neutrons to slow or stop the chain reaction. The extremely radioactive fuel elements are then removed and shipped in special containers to a fuel-processing plant where the fission products are separated from the remaining usable fuel. The fuel is returned to the reactor for further use. The waste fission products, which are now in liquid or slurry form, are stored in steel tanks for a year or more to dissipate thermal energy before permanent disposal in huge underground stainless steel tanks.

To a lesser extent, some radioactivity is present in spent coolant, either gas or liquid, and in any of the structural parts of a reactor that may have needed replacement. Leaks of radioactive gases at nuclear power plants have, indeed, been reported. Minute cracks in the fuel containers allow fission products to escape into the coolant. In addition, low-level, radioactive liquid wastes are formed when impurities in the coolant water and corrosion products from coolant pipes are bombarded with neutrons from the fuel elements. Much of these low-level wastes are stored in sealed containers and subsequently dropped into the ocean. This disposal technique has been questioned because of unknowns such as the lifetime of the containers used, especially when they are subject to underwater currents.

Present practice calls for solid radioactive wastes such as reactor and machinery parts to be placed in stainless steel vaults and buried underground. This practice has also been questioned because the sites may not be impervious to geological action for the time periods that are required to reduce the radioactivity to harmless levels. It has been suggested that abandoned salt mines, where the action of water in the ground is minimal, are presently the most reassuring choice site for burial of solid radioactive wastes.

Many environmentalists claim that disposal of radioactive wastes in deep wells underground or burial at sea is like sweeping them under the rug. Indeed, some of the fission products such as strontium-90 and cesium-137 are long-lived. Even the Atomic Energy Commission estimates that a storage time of about 600 years is required before the residuum of these isotopes is safe for biological exposure. Only time will demonstrate whether the present disposal techniques can in the long run isolate radioactive wastes from the biological environment. In the meantime, it is imperative that research be pursued to seek better methods of preventing this form of environmental pollution.

QUESTIONS

6.23   Write the equation for the fission of a uranium-235 atom into molybdenum-102 and tin-131 nuclei.

6.24   What is meant by a chain reaction?

6.25   How is it possible to make a fission process self-sustaining?

6.26 Distinguish between explosive and controlled nuclear fissions.

6.27 Why is the fallout from an atomic bomb worse than its explosion?

6.28 Name two artificially prepared fissionable materials.

6.29 Describe the general features of a nuclear reactor.

6.30 What are graphite blocks and boron-steel rods used for in a nuclear reactor?

6.31 How is electricity produced using nuclear reactors?

6.32 What are breeder reactors and why are they used?

6.33 Discuss the merits and demerits of nuclear power plants.

6.34 What are the problems involved in radioactive waste disposal?

FUSION AFTER FISSION?

Much more matter can be converted into energy by *fusion* of lighter nuclei into heavier ones than by fission of heavy atoms. This startling truth was revealed, in November 1952, when the United States exploded its first hydrogen device (more correctly called *thermonuclear bomb*). It made use of solid lithium deuteride, $^6_3Li^2_1D$, which, when placed around an ordinary ^{235}U or ^{239}Pu fission bomb and set off (Fig. 6.14) produced the radioactive hydrogen isotope ^3_1T (tritium):

$$^6_3Li^2_1D + ^1_0n \longrightarrow ^3_1T + ^4_2He + ^2_1D$$

At the temperature (of the order of 100 million degrees) attained by the fissioning uranium or plutonium, the tritium and deuterium nuclei fused to produce the helium nuclei:

$$^2_1D + ^3_1T \longrightarrow ^4_2He + ^1_0n + energy$$

Since 1952 several of these *fission-fusion* bombs have been tested and stockpiled by the major world powers.

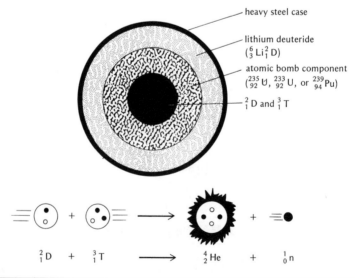

heavy steel case

lithium deuteride
$(^6_3Li^2_1D)$

atomic bomb component
$(^{235}_{92}U, ^{233}_{92}U, or ^{239}_{94}Pu)$

^2_1D and ^3_1T

^2_1D + ^3_1T $\longrightarrow$ ^4_2He + ^1_0n

FIGURE 6.14

Components of a hydrogen bomb and the nuclear fusion reaction.

Calculations show that 0.7 percent of the matter is converted into energy in the fusion of four hydrogen atoms to form one helium atom, whereas only 0.1 percent of the uranium-235 atom is converted in the fission process. If the fusion process can be controlled, as in the case of fission, the human demand for unlimited energy can be fulfilled free of radioactive waste by the unrestricted fuel supply in the form of deuterium from the oceans of the earth. However, despite active research for two decades, there has been little success in controlling these thermonuclear fusion reactions for the purpose of harnessing their energy.

What are the problems to be solved in controlling fusion reactions? First of all, temperatures of the order of 100 million degrees or more have to be achieved in a controlled fashion. At these temperatures structured atoms do not exist but only nuclei and electrons. A mixture of nuclei and electrons is termed a *plasma*. It is only in these plasmas that nuclei interact and fuse together. Second, the existing difficulties in containing the high-temperature plasma have to be overcome. Although devices such as the *magnetic bottle* (Fig. 6.15) have been constructed and experimented upon, as yet fusion reactions have not been controlled. Current research using a special kind of light device known as the *laser* appears to be promising for the future.

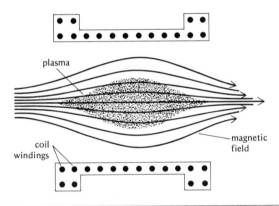

FIGURE 6.15

Diagram of a plasma held suspended in a magnetic field. (From Robert W. Medeiros, *Chemistry: An Interdisciplinary Approach*, p. 513. Copyright © 1971 by Litton Educational Publishing, Inc. Redrawn by permission of Van Nostrand Reinhold Company.)

Actually there is every reason to believe that fusion reactions will eventually be controlled. The source of nearly all our energy is the fusion reactions taking place on our sun. It is estimated that 5–6 million tons of matter (chiefly hydrogen) are converted each second into radiant energy on the sun, the principal reaction being

$$4^1_1H \longrightarrow {}^4_2He + 2^0_1e + energy$$

The energy of the other stars is of similar derivation. Clearly, nature has long preceded us in our attempts for releasing thermonuclear fusion energy. Can we equal nature's feat?

QUESTIONS

6.35 What are the advantages of nuclear fusion over nuclear fission?
6.36 Describe the components of a thermonuclear bomb.
6.37 What are the difficulties in controlling nuclear fusion for power production?
6.38 Name the principal fusion reaction that takes place on the sun and the stars.
6.39 Summarize your opinion of the effects of nuclear transformations on our environment.

SUGGESTED PROJECT 6.1

Checking for radiation (or a counting game that may save lives)

All forms of radiation capable of producing ionization can cause damage to living tissues. With the increasing use of radioisotopes in our environment both radiation hazard detection and biological shielding are vital for our protection. We have already described a radiation detection instrument called the Geiger-Müller counter. It is particularly useful in detecting β- and γ-rays.

If a portable counting device is available, check various areas of your science building and school campus for dangerous radiation. Many science laboratories and hospitals, which provide radiation therapy, have powerful sources of ionizing radiations. If there is a nuclear facility at your school or near your community it may be interesting to check the surrounding areas. In each case count and record the counts per minute. Make a report on your findings, specifying in particular the locations visited and counted. In so doing you are using your newly acquired bit of chemistry for a service to an entire community. Moreover, you can find out for yourself, and convince others too, whether or not our government has been doing one of its mandated tasks—the protection of our lives from deadly radiations.

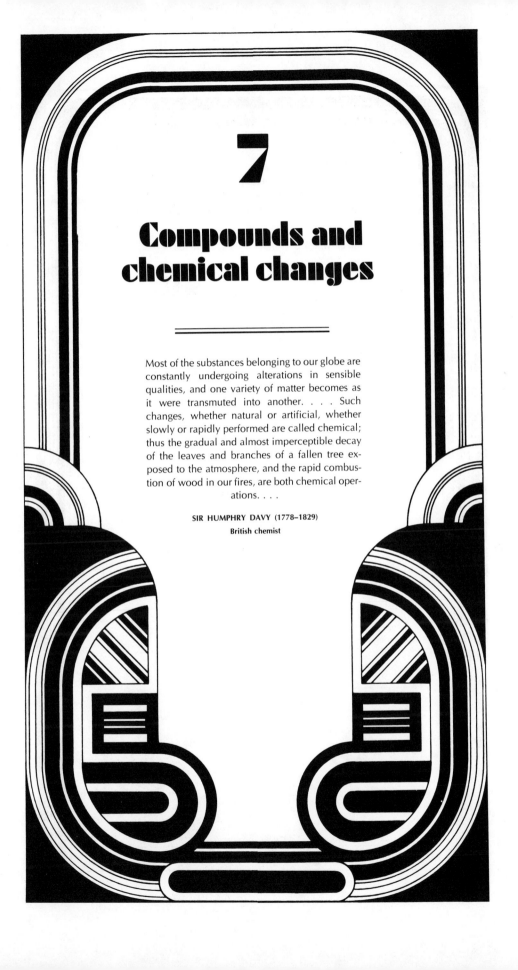

7

Compounds and chemical changes

Most of the substances belonging to our globe are constantly undergoing alterations in sensible qualities, and one variety of matter becomes as it were transmuted into another. . . . Such changes, whether natural or artificial, whether slowly or rapidly performed are called chemical; thus the gradual and almost imperceptible decay of the leaves and branches of a fallen tree exposed to the atmosphere, and the rapid combustion of wood in our fires, are both chemical operations. . . .

SIR HUMPHRY DAVY (1778–1829)
British chemist

The rich diversity of our material world arises from the millions of compounds that either occur naturally on earth or are created by chemical changes in the laboratory from a mere handful of elements. If we examine the compounds that are present in the earth's crust we

Lavoisier's experiments on conservation of mass in chemical changes

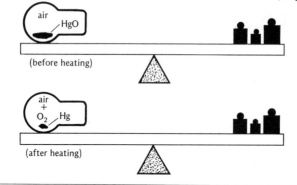

Proust's experiments on definite proportions of elements in compounds

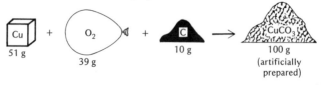

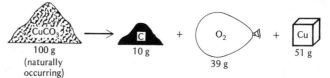

Dalton's experiments on multiple proportions

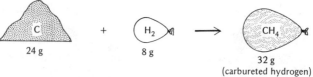

FIGURE 7.1

Early experiments on chemical changes and compounds.

can easily realize that they are primarily constituted of the elements Si, O, Cl, C, Al, Mg, Ca, Fe, Na, and K. Plants and animals, including human beings, consist almost totally of compounds made up from C, H, O, N, and to a lesser degree, Fe, Cl, I, S, and P. Clearly, only about 20 different kinds of atoms are needed to form the myriad substances of our world.

Historically speaking, it was only by the eighteenth century that chemists began observing chemical changes more carefully and measuring the resulting compounds more accurately. The French chemists Antoine Lavoisier and Joseph Proust are credited with discovering and proving the important features of chemical changes and compounds. Whereas Lavoisier provided the first concrete evidence for the conservation of mass in chemical changes that produce compounds, Proust demonstrated that a compound always contains elements in certain definite proportions (Fig. 7.1). Later, the English mathematician John Dalton showed that two elements might combine in more than one set of proportions and that each set of proportions produced a different compound (Fig. 7.1). Today, we know that there are 5–10 million known chemical compounds and all of them can be involved in numerous chemical changes.

In this chapter we shall be concerned with compounds, their composition and classification into families, and the chemical changes they undergo.

COMPOUNDS AND THEIR FAMILIES

Rough estimates show that more than ten thousand new compounds are prepared each year in the university and industrial research laboratories of the world. As new compounds are obtained, their composition must be established. In fact, the naming and symbolic representation of compounds are based on their composition which reveals their identity and serves to distinguish one compound from another.

COMPOSITION OF COMPOUNDS There are two general methods for arriving at the composition of a compound. These methods are (1) *synthesis* and (2) *analysis*. In synthesis, the compound is prepared from the elements or simple combinations of the elements; in analysis, the compound is decomposed to find the individual elements or combinations of elements of known composition. If the weights of the elements required to form a given weight of the compound can be determined, the percentage composition of the elements in the compound can be calculated.

EXAMPLE 7.1

When 2.1 g of elemental carbon was burned in oxygen, 7.7 g of carbon dioxide (CO_2) was produced. Calculate the percentage composition of the elements in carbon dioxide.

Weight of carbon = 2.1 g; weight of carbon dioxide formed = 7.7 g; weight of oxygen (by difference) = 5.6 g. Percent C in carbon dioxide:

$$\frac{\text{wt of C}}{\text{wt of } CO_2} \times 100\% = \frac{2.1 \text{ g}}{7.7 \text{ g}} \times 100\% = 27\%$$

155

Percent O in carbon dioxide:

$$\frac{\text{wt of O}}{\text{wt of CO}_2} \times 100\% = \frac{5.6 \text{ g}}{7.7 \text{ g}} \times 100\% = 73\%$$

Numerous compounds containing carbon and hydrogen exist in our environment. Their compositions are determined by burning weighed samples in a closed tube in a stream of oxygen to form carbon dioxide and water. The products are swept from the tube by the stream of oxygen into two absorbing chemicals, one of which absorbs water vapor and the other carbon dioxide (Fig. 7.2). The gain in

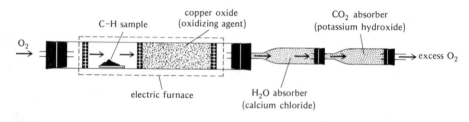

FIGURE 7.2

Apparatus for determining percent composition of a compound containing carbon and hydrogen.

weight of each of the absorbers gives the weight of water and carbon dioxide, respectively. Because water is known to be 2/18 hydrogen by weight, therefore 2/18 of the weight of water is equal to the amount of hydrogen originally present in the compound. Similarly, carbon dioxide is 12/44 carbon by weight and, therefore, 12/44 of the weight gained by the carbon dioxide absorber is the weight of the carbon originally present in the sample. From these weights, the percentage composition of the compound is calculated.

EXAMPLE 7.2

2.0 g sample of a compound containing carbon and hydrogen was burned in an apparatus similar to that shown in Fig. 7.2. The increases in weight in the water and carbon dioxide absorbers were 1.4 and 6.8 g, respectively. What is the percentage composition of the compound?

Weight of hydrogen: $\quad$ wt of water $\times \dfrac{2}{18} = 1.4 \text{ g} \times \dfrac{2}{18}$

$$= 0.155 \text{ g}$$

Percent H: $\quad \dfrac{\text{wt of H}}{\text{wt of sample}} \times 100\% = \dfrac{0.155 \text{ g}}{2.0 \text{ g}} \times 100\%$

$$= 7.7\%$$

Weight of carbon: $\quad$ wt of carbon dioxide $\times \dfrac{12}{44} = 6.8 \text{ g} \times \dfrac{12}{44}$

$$= 1.85 \text{ g}$$

Percent C: $\quad \dfrac{\text{wt of C}}{\text{wt of sample}} \times 100\% = \dfrac{1.85 \text{ g}}{2.0 \text{ g}} \times 100\%$

$$= 92.5\%$$

If the compound contains only oxygen in addition to carbon and hydrogen (for example, the sugars), the percentage of oxygen is obtained by difference. Thus, a compound of carbon, hydrogen, and oxygen that contains 40 percent carbon and 7 percent hydrogen by weight must contain 53 percent oxygen by weight.

FORMULAS OF COMPOUNDS   A *formula* is a representation of the composition of a compound. For example, the formula for water, H_2O, indicates that the compound water contains the combined elements hydrogen (H) and oxygen (O) and that there are two hydrogen atoms and one oxygen atom in one formula unit (molecule) of water. However, the formula does not show whether the atoms are combined by the sharing of electrons (covalent bonds) or by the electrostatic attraction of oppositely charged ions (ionic bonds).

If we know the percentage composition and the atomic weights of the elements involved, we can determine the *simple* or *empirical formula* of a compound. For example, let us consider the compound that has the composition of 7.7 percent H and 92.5 percent C by weight (see Example 7.2). This means that 100 g of the compound contains 7.7 g H and 92.5 g C. Because we know that 1 mole of hydrogen atoms weighs 1.0 g, therefore 7.7 g contains 7.7 g $\times$ (1 mole/1.0 g) = 7.7 moles of H atoms. Similarly, because 1 mole of carbon atoms weighs 12.0 g, therefore 92.5 g contains 92.5 g $\times$ (1 mole/12.0 g) = 7.7 moles of C atoms. We know that a mole of any element contains 6.02×10^{23} atoms (Avogadro's number; see Chapter 3, p. 44), and, hence, we infer that the compound contains an equal number of carbon and hydrogen atoms. Therefore, a molecule of the compound must also contain an equal number of atoms of these elements, and the formula must, therefore, be the one that shows this 1:1 relationship. Unfortunately, several formulas, such as CH, C_2H_2, C_3H_3, and C_4H_4, show the same 1:1 relationship. The simplest one, however, is CH and this formula is called the empirical formula—a formula that indicates the ratio of the atoms in a compound.

Unlike the formula of an ionic compound that merely indicates the ratio of the ions present, the formula for a covalent compound must show not only the composition but the actual number and kind of atoms in a molecule. Such a formula is called the *molecular formula*. It may be the same as the empirical formula or it may be a whole-number multiple of it. In order to find this whole number, the molecular weight of the compound is divided by the empirical formula weight. Because methods for experimentally determining molecular weights are known, the true molecular formula can easily be calculated.

EXAMPLE 7.3

The molecular weight of a compound containing C and H is found to be 78. If the empirical formula is CH, what is its molecular formula?

Because the empirical formula weight is 13 (C = 12, H = 1), the whole number with which the empirical formula must be multiplied is given by (78/13) = 6. The molecular formula is therefore $(CH)_6$ or C_6H_6.

How can we write formulas for compounds and remember them? In Chapter 5, it was pointed out that each element is assigned a number (or sometimes two or more) that relates to the state (or states) of combination an element may assume in a compound. These numbers, called *oxidation numbers*, have been developed from consideration of the composition of compounds and the relative electronegativities of the elements comprising the compounds. However, there are some arbitrary rules that must be followed:

1. In the uncombined state, the oxidation number of an element is zero.
2. In the combined state, the more electronegative elements usually have negative oxidation numbers and the less electronegative (electropositive) elements usually have positive oxidation numbers.
3. In a given compound, the algebraic sum of the oxidation numbers of the elements present must equal zero.

Thus, if the oxidation numbers of the combining elements are known, we can generally deduce the formula of a compound. In Chapter 5, it was shown that the oxidation numbers can be predicted from the periodic table (see Table 5.2). Although these predictions do not always include all the observed oxidation states for certain elements and all the observed differences in oxidation numbers within a given group of elements, they are often quite good in predicting the formulas of the common compounds.

EXAMPLE 7.4

By using the oxidation numbers given in Table 5.2, predict the formulas for the compounds involving (a) sodium and chlorine, (b) magnesium and oxygen, (c) calcium and nitrogen, and (d) carbon and oxygen.

a. We would expect a +1 oxidation number for sodium and a -1 oxidation number for chlorine. If the formula is Na_xCl_y, the x and y must be of the proper values to yield a zero algebraic sum of the numbers; that is, $x(+1) + y(-1) = 0$ or simply $x = 1$ and $y = 1$. Therefore, the predicted formula is NaCl. Note that the subscript 1 is never shown but is implied if there is no subscript in a formula.

b. Because the predicted oxidation states of magnesium and oxygen are +2 and -2, respectively, the proper values for x and y in the formula Mg_xO_y, obtained from the algebraic equation $x(+2) + y(-2) = 0$, are $x = 1$ and $y = 1$. Therefore, the predicted formula is MgO.

c. Because calcium has an oxidation number of +2, we would expect a -3 oxidation state for nitrogen. If the formula is Ca_xN_y, then the simplest solution for the algebraic equation $x(+2) + y(-3) = 0$ is $x = 3$ and $y = 2$. Therefore, the predicted formula is Ca_3N_2.

d. Because carbon occurs in the +2 and +4 oxidation states and oxygen in the -2 state, the expected formula in the +2 oxidation state is CO. The formula in the +4 oxidation state is obtained from the equation $x(+4) + y(-2) = 0$ whose solution is $x = 1$ and $y = 2$. Therefore, the predicted formula in the +4 oxidation state is CO_2.

In writing the formula of an ionic compound, the symbol for the element that has a positive oxidation number is always written first. However, this practice may or may not be followed in the case of covalent compounds. For example, the formulas for the compounds methane (CH_4) and ammonia (NH_3) have the symbol for the element that has a negative oxidation number written first, whereas carbon dioxide (CO_2) has the symbol for the element that has a positive oxidation number written first. Further, because some molecules may or may not be formed by the combination of the smallest number of atoms, it may not always be possible to predict the formula of a covalent compound on the basis of oxidation states alone.

NOMENCLATURE OF COMPOUNDS With the preparation of more and more compounds, it became all the more necessary to name compounds systematically. The systematic naming of compounds was initiated by Lavoisier and his contemporaries in the eighteenth century. Today, it is referred to as *chemical nomen-*

clature. Names derived according to some system of nomenclature are called *systematic names;* names that developed historically before any systems of nomenclature had been developed are called *common* or *trivial names.* Examples of such nonsystematic names that are still in use are water, ammonia, and acetic acid.

According to the rules set up by the International Union of Pure and Applied Chemistry, the systematic names are based partly on the composition of the compound and partly on the class or type of compound. For example, all compounds containing two elements (*binary* compounds) have names that show the two elements present in the order of increasing electronegativity and end in *-ide:*

Na_2O sodium ox*ide*
CaO   calcium ox*ide*
H_2O hydrogen ox*ide*
HCl   hydrogen chlor*ide*
H_2S hydrogen sulf*ide*

However, some compounds that contain more than two elements are also named in this manner (NH_4Cl, ammonium chloride; HCN, hydrogen cyanide).

Water solutions that are *acidic* have names formed by using the prefix *hydro-* with the root of the name of the element other than hydrogen followed by an *-ic* ending and the word *acid:*

HCl   *hydro*chlor*ic acid*
H_2S *hydro*sulfur*ic acid*
HCN   *hydro*cyan*ic acid*

Some metals form ternary (three-element) compounds that involve hydrogen and oxygen; they are named by stating the name of the metallic element followed by the word *hydroxide* (NaOH, sodium hydroxide; $Ca(OH)_2$, calcium hydroxide).

Many nonmetals form ternary compounds that involve hydrogen and oxygen. Of these, the compounds and their water solutions that have acidic properties are referred to as *oxyacids.* The nomenclature of the oxyacids is somewhat confusing because of the involvement of different oxidation states and because the same name is usually used to refer to both the pure oxyacid and the water solutions of the oxyacid. The nomenclature and formulas of the oxyacids that fit a certain pattern are given in Table 7.1. The names of the oxyanions can be deduced by changing the ending of the name of the corresponding acid and adding the word ion. The *-ous*

TABLE 7.1
Nomenclature and formulas of some oxyacids

General formula	Molecular formula	Name	Oxidation number of nonmetal
HXO_y	HClO	*hypo*chlor*ous* acid	+1
	$HClO_2$	chlor*ous* acid	+3
	HNO_2	nitr*ous* acid	+3
	$HClO_3$	chlor*ic* acid	+5
	HNO_3	nitr*ic* acid	+5
	$HClO_4$	*per*chlor*ic* acid	+7
H_2XO_y	H_2SO_3	sulfur*ous* acid	+4
	H_2SO_4	sulfur*ic* acid	+6
H_3XO_y	H_3PO_3	phosphor*ous* acid	+3
	H_3PO_4	phosphor*ic* acid	+5

ending of the acid name is changed to an -*ite* ending, and the -*ic* ending is changed to an -*ate* ending (NO_2^-, nit*rite* ion; NO_3^-, nit*rate* ion). When oxyacids that involve more than one hydrogen form polyatomic ions resulting from the loss of one or more hydrogens, the ions are named in the same manner as the acid anions except that the anion name is preceded by the word hydrogen that has a prefix indicating the number of combined hydrogen atoms associated with the ion (HSO_4^-, hydrogen sulfate ion; $H_2PO_4^-$, *di*hydrogen phosphate ion). Just as the subscript 1 is omitted in formulas, the prefix *mono-* is usually omitted in nomenclature.

Compounds that involve metals with two different oxidation states are named by using the root of the name of the metal with an -*ous* ending for the lower oxidation state and an -*ic* ending for the higher oxidation state of the metal (CuCl, cupr*ous* chloride; $CuCl_2$, cupr*ic* chloride). However, the best method for naming compounds of metals and nonmetals that have more than one possible oxidation state is called the *Stock system* of nomenclature. In the case of metal–nonmetal compounds, the name consists of the name of the metal followed by the oxidation number in Roman numerals within parentheses, which is, in turn, followed by the name of the anion ($FeCl_2$, iron(II) chloride; $FeCl_3$, iron(III) chloride; $FeSO_4$, iron(II) sulfate; and so forth). The nonmetal-nonmetal compounds are also named by the Stock nomenclature method; however, their binary compounds are usually named by another method that involves stating first the name of the less electronegative element followed by the root of the name of the more electronegative element with an -*ide* ending. Each part of the name is preceded by a Greek or Latin prefix indicating the number of combined atoms of that element in the compound. For example, N_2O is called *di*nitrogen oxide, SO_3 is called sulfur *tri*oxide, and P_4S_7 is called *tetra*phosphorus *hepta*sulfide. Naming compounds by this method provides a way in which the formula of the compound is completely defined by the name.

EXAMPLE 7.5

Name the following compounds by the prefix method and the Stock method of nomenclature: N_2O_5; SiO_2; OF_2; P_4O_6; Mn_2O_7.

	Prefix method	Stock method
N_2O_5	dinitrogen pentoxide	nitrogen(V) oxide
SiO_2	silicon dioxide	silicon(IV) oxide
OF_2	oxygen difluoride	oxygen(II) fluoride
P_4O_6	tetraphosphorus hexoxide	phosphorus(III) oxide
Mn_2O_7	dimanganese heptoxide	manganese(VII) oxide

CLASSIFICATION OF COMPOUNDS One of the earliest methods of classifying compounds was based on whether or not the compound contained carbon as one of the combined elements. Most compounds that contain carbon are called *organic compounds;* most of those that do not are called *inorganic compounds.* Actually, more than 90 percent of the number of known compounds are organic and usually covalent, although many ionic organic compounds exist. The organic compounds that occur in nature are found mostly in plants and animals or in materials of plant and animal origin, such as coal, natural gas, and petroleum. Carbohydrates, fats, proteins, and alcohols represent familiar classes of organic compounds. There are millions of organic compounds that do not occur in nature but have been prepared in chemical laboratories (see Chapters 9–12). Inorganic compounds constitute rocks, clay, sand, and other earthy materials. Most of the ionic compounds encountered so far have been inorganic. A few compounds such as calcium carbonate ($CaCO_3$), magnesium carbonate ($MgCO_3$) and sodium cyanide ($NaCN$) that contain carbon are classified as inorganic rather than organic because they are earthy or rocklike and are quite similar to other mineral substances.

Another method of classifying compounds is based on whether the molten compound or a water solution of it will conduct a current of electricity. If a molten compound or its water solution is a conductor of an electric current, the compound is an *electrolyte;* if not, the compound is a *nonelectrolyte.* To determine whether or not a given compound is an electrolyte, the simple apparatus shown in Fig. 7.3 may be used. Notice that one of the electrical wires is cut in two so that the current can flow only when the two ends of the cut wire are immersed in a solution of electrolyte. If the light burns, the electric current is conducted by the solution and the compound in solution is an electrolyte; otherwise, it is a nonelectrolyte. Clearly, a solution of an electrolyte contains ions that carry the electric current.

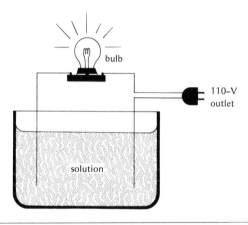

FIGURE 7.3

Apparatus to test compounds for electrolytes and nonelectrolytes.

Still another method of classifying compounds is based on the kind of bonds that hold the atoms in the compound together (see Chapter 4). If all the bonds are pairs of shared electrons, the compound is considered to be *covalent;* if one or more bonds are due to the attraction between ions of unlike charge, the compound is considered to be *ionic.* All ionic compounds that dissolve in water make an electrolyte solution, whereas some covalent compounds that dissolve in water are electrolytes and some are not. Consider, for example, the case of polar covalent compounds such as H_2O, HCl, NH_3, and $HC_2H_3O_2$ (acetic acid). In the pure liquid state these compounds do not conduct electricity, but their water solutions do conduct electricity. This behavior is attributed to the ability of these molecules to form ions in water solution:

$$HCl + H_2O \rightleftharpoons H_3O^+ + Cl^-$$
$$NH_3 + H_2O \rightleftharpoons NH_4^+ + OH^-$$
$$HC_2H_3O_2 + H_2O \rightleftharpoons H_3O^+ + C_2H_3O_2^-$$

Clearly, the ions result from a chemical reaction between the polar covalent molecules and water molecules (Fig. 7.4). In such a reaction, the covalent compound is said to ionize, and the reaction is referred to as *ionization*. In the equations, double arrows pointing in opposite directions indicate that the conditions required for a reaction to occur also allow the products to react to form the original substances; that is, the reaction is *reversible*.

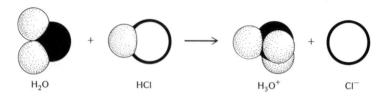

$$H_2O \qquad\qquad HCl \qquad\qquad H_3O^+ \qquad\qquad Cl^-$$

FIGURE 7.4

A hydrogen chloride molecule reacts with a water molecule to form a hydronium ion.

Consider, for example, what happens when hydrogen chloride gas is mixed with water. At the instant of mixing, only the HCl and H_2O molecules can react, because only these two are present. A very short time later some H_3O^+ and Cl^- ions will have formed, and the reverse reaction starts, slowly at first, because very few of these ions are present to react. Both reactions continue to take place simultaneously; in time, the concentrations of HCl, H_2O, H_3O^+, and Cl^- become adjusted so that both reactions are occurring at the same speed. A condition such as this is described as a *chemical equilibrium* and is indicated in equations by double arrows, the longer arrows pointing out those reactants whose concentrations are greater at equilibrium. The factors that determine the relative amounts of reactants and products at equilibrium are the nature and concentration of the reactants and the reaction temperature.

In the case of hydrogen chloride solution, the concentrations of H_3O^+ and Cl^- are relatively high and very little HCl is left in the mixture when equilibrium is reached. However, the opposite is true for water solutions of ammonia and of hydrogen acetate (acetic acid). They are rather poor conductors of electricity and are usually referred to as *weak electrolytes*. Compounds whose melts or water solutions are excellent conductors of electricity are said to be *strong electrolytes*.

In addition to being classified as to strength, electrolytes may be classified as to type. The three common types are *acids, bases,* and *salts*. Originally, acids and bases were defined in terms of properties of their water solutions (see Chapter 3). The modern definition, due to the Danish chemist J. N. Brønsted and the English chemist T. M. Lowry, suggests that acids are *proton donors* and bases are *proton acceptors*. Let us consider from the Brønsted-Lowry definition what the proton-donating species is in water solutions of strong acids such as HCl and HNO_3:

$$HCl + H_2O \rightleftharpoons H_3O^+ + Cl^-$$
$$HNO_3 + H_2O \rightleftharpoons H_3O^+ + NO_3^-$$

The proton donor is apparently the hydronium ion, H_3O^+. Even for weak acids such as acetic acid, the main proton-donating species is the hydronium ion:

$$HC_2H_3O_2 + H_2O \rightleftharpoons H_3O^+ + C_2H_3O_2^-$$

The chief proton acceptor is the hydroxide ion, OH^-, present in water solutions of bases such as $NaOH$ and NH_4OH:

$$NH_4OH \rightleftharpoons NH_4^+ + OH^-$$

The acidity or basicity of a solution depends on the concentration of hydronium or hydroxide ions. A scale called the *pH scale* has been devised to compare the acidity or basicity of solutions (Fig. 7.5). On this scale, a solution with equal concentration of hydronium and hydroxide ions has a pH of 7, which is the neutral pH point. The greater the hydronium ion concentration becomes, the lower the pH value; numbers above 7 are considered basic.

When water solutions of acids and bases are mixed together the hydronium ions of the acid and the hydroxide ions of the base combine to form water:

$$H_3O^+ + OH^- \longrightarrow 2H_2O$$

Such an equation is known as an *ionic equation* and shows as reactants or products only those particles that took part in the reaction. The reaction itself is called *neutralization*. Salts, which are ionic compounds and hence strong electrolytes, are formed during the neutralization of water solutions of acids and bases. Some common examples of salts are $NaCl$, KCl, $NaNO_3$, Na_2SO_4, Na_2CO_3, and $Ca(NO_3)_2$. All of these

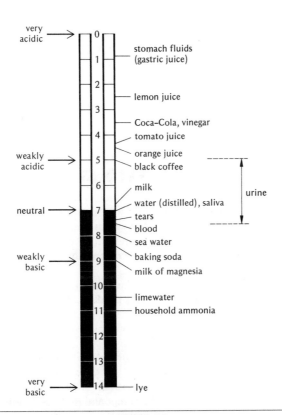

FIGURE 7.5

The pH scale and pH values of some common substances.

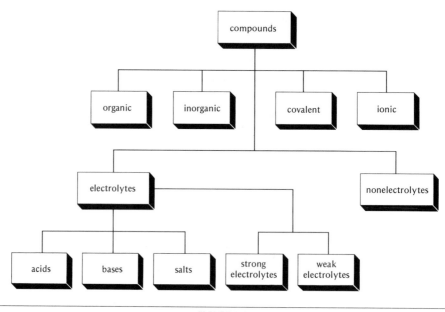

FIGURE 7.6

Families of compounds.

salts can be prepared by acid–base reactions. For example, in the case of water solutions of HCl and NaOH, the completion of the neutralization reaction leaves behind a solution of Na^+ and Cl^- ions that, when evaporated, yields the salt NaCl:

$$H^+ + Cl^- + Na^+ + OH^- \longrightarrow H_2O + Na^+ + Cl^-$$

The foregoing discussion of the families of compounds is shown in Fig. 7.6.

QUESTIONS

7.1   What are the main constituents of the compounds present in human beings?

7.2   Name Lavoisier's important observations regarding chemical changes.

7.3   List the methods for determining the composition of compounds.

7.4   Describe an experiment by which the composition of a compound of carbon and hydrogen can be determined.

7.5   A 2.0-g sample of a compound containing carbon and hydrogen gave, on complete combustion, 4.5 g H_2O and 5.5 g CO_2 as products. What is the percentage composition of the compound?

7.6   Deduce the empirical formula of the compound that has 75 percent C and 25 percent H by mass.

7.7   What is the empirical formula of a compound that contains 5.9 percent H and 94.1 percent O by weight?

7.8   Calculate the percentage by mass composition of phosphoric acid, H_3PO_4.

7.9   Calculate the molecular formula of a compound that has a molecular weight of 180 amu and the composition C = 40 percent, H = 6.7 percent, and O = 53.3 percent.

7.10   What are oxidation numbers? How are they useful?

7.11   List the expected oxidation numbers for each group of the representative elements.

7.12 By using oxidation numbers, predict the likely formulas of the compounds between
    a. cesium and bromine
    b. aluminum and sulfur
    c. phosphorus and chlorine
    d. selenium and oxygen
7.13 What is the difference between the systematic name of a compound and its trivial name? Give examples.
7.14 Name each of the following compounds by the appropriate nomenclature method:
    a. Cl_2O
    b. As_2O_3
    c. Be_3N_2
    d. H_3AsO_4
    e. SF_6
    f. K_2HPO_4
7.15 How are compounds classified?
7.16 What are acids and bases in terms of Brønsted-Lowry definitions.
7.17 What is meant by a chemical equilibrium?
7.18 Explain why a current of electricity is conducted through mineral acid solutions.

SUGGESTED PROJECT 7.1

Classifying compounds as electrolytes and nonelectrolytes
(or the ON–OFF experiment)

Construct an apparatus similar to the one shown in Fig. 7.3 and use it to test the following substances for their ability to conduct electricity:

a.   tap water
b.   distilled or deionized water
c.   a solution of one tablespoon of salt in one cup of distilled water
d.   a solution of one tablespoon of sugar in one cup of distilled water
e.   commercial vinegar (a dilute solution of acetic acid in water)
f.   household ammonia (a dilute solution of ammonia gas in water)

Make sure that you rinse (with distilled water) and dry the contact terminals between tests for each of the different solutions. Based on your observations classify the above substances as covalent or ionic compounds.

Observe the intensity of the glow in your apparatus using a sample of vinegar. Add household ammonia dropwise into the vinegar solution and observe any change in the intensity of the glow. Explain your observations taking ions into account. Can you make use of the apparatus to determine the acetic acid content of vinegar if the composition of household ammonia is known? Explain.

Caution! Avoid shocking yourself. Do not touch the bare wires while the apparatus is plugged in. Unplug the apparatus for rinsing and drying between tests.

SUGGESTED PROJECT 7.2

Your food–acidic or basic?

Compile a list of foods and classify them as acidic or basic on the basis of their tastes. If litmus paper is available, test the various food items with it and check your classification. Also, if pH paper is available, determine the extent of acidity or basicity of each of the food items. Tabulate your findings.

CHEMICAL CHANGES

Compounds are formed and made to change into new compounds as a result of chemical changes. There are millions of known chemical changes which produce perhaps as many or even more compounds. Some of these changes occur in nature, but many are brought about in the chemical laboratory. Some are so slow that they take years to occur; others are so fast that they take place instantaneously. Some liberate enormous quantities of heat, some absorb heat, and a great many occur with very little heat change. When considering chemical changes, we have to ask whether or not a reaction will occur among certain compounds if a pathway is opened up and whether or not the reaction will occur rapidly enough to be significant. Thus, in a chemical change the rate of change is as significant as the underlying energetics.

The manufacture of chemical compounds is one of the largest industrial endeavors in our world. For economic reasons, the chemical changes and the production of the chemical compounds must be accomplished in the most convenient manner with as little waste as possible. Therefore, the amount of products that can be formed from given amounts of reactants is of considerable importance. The relationship between the amounts of the reactants and the products involved in a chemical change is called the *stoichiometry* of the reaction. (*Stoichio-* is derived from the Greek *stoicheion*, meaning "element" or part of a compound; the *-metry* refers to the measuring of masses.) Because chemical equations are exact statements of the stoichiometric amounts of materials consumed and produced in a chemical reaction, we shall begin our study of chemical changes with chemical equations.

CHEMICAL EQUATIONS AND STOICHIOMETRY As we learned previously, mass is conserved in a chemical reaction, and any substance that goes into making a given compound must combine in a definite fixed proportion with other substances. Therefore, the correct formulas of the compounds must be used and the numbers of each kind of atom must be the same on both sides of the equation. In other words, the equations must be *balanced*.

How are chemical equations balanced? Often it is done by trial and error. Consider the equation for the reaction of hydrogen with oxygen to give water. Because hydrogen, oxygen, and water in their natural states have the formulas H_2, O_2, and H_2O, we can write the equation

$$H_2 + O_2 \longrightarrow H_2O$$

However, the number of atoms of each element is not the same on both sides of the equation and hence the equation is not balanced. To balance the equation we cannot change the formulas of the species, but we can change the number of each species that appears in the equation by placing numerical *coefficients* in front of the formulas. To proceed, first double the number of water molecules, which will make the total number of oxygen atoms the same on both sides of the equation:

$$H_2 + O_2 \longrightarrow 2H_2O$$

Next, double the number of hydrogen molecules,

$$2H_2 + O_2 \longrightarrow 2H_2O$$

and note that we have, finally, obtained a balanced equation.

A balanced equation can be interpreted from a molar point of view for use in stoichiometric calculations. Consider, for example, the equation representing the reaction between natural gas (CH_4) and oxygen to produce carbon dioxide and water:

$$CH_4 + 2O_2 \longrightarrow CO_2 + 2H_2O$$

How can we interpret such an equation? We could say that the equation indicates that one molecule of methane will react with two molecules of oxygen to produce one molecule of carbon dioxide and two molecules of water. We could also say that 1 mole (Avogadro's number of molecules) of methane will react with 2 moles of oxygen to produce 1 mole of carbon dioxide and 2 moles of water. Such an interpretation is called a molar interpretation, and the relationship between any two species involved in the reaction is used in the form of a *molar ratio* for stoichiometric calculations. The following example illustrates this calculation.

EXAMPLE 7.6

Iron(III) oxide reacts with carbon to produce iron metal and carbon dioxide. How many grams of metallic iron are produced from the reaction of 200 g of iron(III) oxide with sufficient carbon?

The balanced equation for the reaction is

$$2Fe_2O_3 + 3C \longrightarrow 4Fe + 3CO_2$$

The number of grams of iron can be obtained by converting the grams of Fe_2O_3 to the number of moles,

$$2(55.8 \text{ g/mole}) + 3(16.0 \text{ g/mole}) = 159.6 \text{ g/mole } Fe_2O_3$$

which can then be converted to the number of moles· of iron using the molar ratio 4 moles Fe/2 moles Fe_2O_3 and, subsequently, to the number of grams of iron metal. The arithmetic sequence should be

$$200 \text{ g} \left(\frac{1 \text{ mole } Fe_2O_3}{159.6 \text{ g}} \right) \left(\frac{4 \text{ moles Fe}}{2 \text{ moles } Fe_2O_3} \right) \left(\frac{55.8 \text{ g}}{1 \text{ mole Fe}} \right) = 139.8 \text{ g}$$

By itself, a chemical equation does not enable us to predict what the result of a reaction will be, nor does it tell us what conditions will cause the reaction in the first place. However, if we know enough about chemical behavior, we may often be able to use existing equations to extrapolate how substances will react under different conditions. For example, because we know that methane (CH_4) reacts with oxygen, it is reasonable to suspect a similar compound, such as ethane ($H_3C—CH_3$), also to react with oxygen. Further, if we know how to make methane burn smoothly without exploding, we might be able to predict how to do the same with ethane. Alternatively, such considerations may also lead us to conclude that a hypothetical reaction is impossible. Both kinds of considerations are valuable in attempts to control our environment by controlling chemical changes.

Finally, examination of a balanced chemical equation provides a basis for classifying reaction types in terms of the changes in chemical bonding that have occurred. Of course, this requires, in addition to the correct equation for the chemical change, the knowledge of the structures of the compounds involved. Two such reaction types which we have already discussed are oxidation–reduction or electron transfer reactions (see Chapter 5) and acid–base or proton transfer reactions discussed earlier in this chapter (see p. 163). Other types of reactions are characteristic of organic compounds and will be discussed in Chapter 9.

CLASSES OF CHEMICAL CHANGES   Chemical changes can be classified into two broad groups: oxidation–reduction or redox reactions in which oxidation numbers change and nonoxidative or metathesis reactions in which oxidation numbers do not change (see Chapter 5). Many redox reactions can be classified as either addition, decomposition, or displacement reactions. *Addition reactions* include all those reactions in which elements or compounds combine to form a single substance. The reaction that occurs when a flashbulb is used in a camera

$$2Mg + O_2 \longrightarrow 2MgO$$

the reaction that occurs in a fuel cell

$$O_2 + 2H_2 \longrightarrow 2H_2O$$

and the rusting of iron

$$4Fe + 3O_2 + 2xH_2O \longrightarrow 2(Fe_2O_3 \cdot xH_2O)$$

are examples of addition reactions. *Decomposition reactions* are essentially the reverse of addition or combination reactions and involve the breaking apart of a substance to form two or more simpler substances. Some examples are

$$2HgO \longrightarrow 2Hg + O_2$$
$$2KClO_3 \longrightarrow 2KCl + 3O_2$$

Displacement reactions occur when one element displaces another from its compound. The displacement of copper from copper sulfate by zinc,

$$Zn + CuSO_4 \longrightarrow Cu + ZnSO_4$$

and of hydrogen from sulfuric acid by zinc,

$$Zn + H_2SO_4 \longrightarrow ZnSO_4 + H_2 \uparrow$$

are examples of displacement reactions.

   Metathesis reactions, which do not involve any change in oxidation numbers, commonly take place between ions in solutions. These reactions are favored and usually proceed when one of the products does not exist as discrete ions in solution, as for example, when one of the products is a gas, a precipitate (insoluble solid), or a covalent compound:

$$CaCO_3 \longrightarrow CaO + CO_2 \uparrow \text{ (gas)}$$
$$NaCl + AgNO_3 \longrightarrow NaNO_3 + AgCl \downarrow \text{ (precipitate)}$$
$$HCl + NaOH \longrightarrow NaCl + H_2O \text{ (covalent compound)}$$

Many metathesis reactions that include *neutralization* and *precipitation reactions* are characterized by a simple exchange of ions and are therefore called *exchange reactions.* Other metathetical reactions may include addition and decomposition reactions, such as

$$CaO + CO_2 \longrightarrow CaCO_3 \downarrow$$
$$CaCO_3 \longrightarrow CaO + CO_2 \uparrow$$

The driving force behind the completion of these reactions is the formation of products which, being a solid or gas, escape from the immediate reaction environment and thus prevent any appreciable reverse reactions.

   The foregoing discussion of the classes of chemical changes is summarized in Fig. 7.7.

RATES OF CHEMICAL CHANGES   The rate of a chemical change or reaction is a measure of how fast reactants disappear and products appear. Experience shows us that rates of reactions vary enormously. Some reactions are so rapid that their rates are measured on time scales of microseconds (10^{-6} sec) or less; others are so slow that only the study of geologic ages will tell whether or not chemical change is occurring at all. However, reaction rates are sensitive to conditions such as temperature, the concentrations of reactants, the physical state (such as solid, liquid, gas, or solution) of the reactants, and the presence or absence of *catalysts* (materials that increase the rate of a chemical reaction without being consumed by the reaction). For example, we are familiar with the rusting of iron. It is an oxidation reaction whereby the compound iron oxide is formed. Because we know that the rate of rusting is immeasurably slow in dry climates but much faster in the presence of moisture, the reaction can be regulated and even prevented by controlling the amount of moisture. On the other hand, the oxidation of iron can be made to occur extremely rapidly by using the finely divided metal and the high temperatures available in the flame of an oxygen torch.

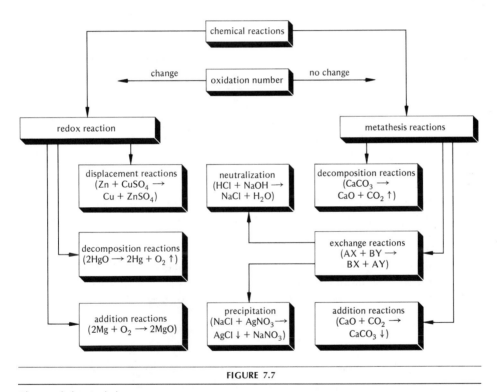

FIGURE 7.7

Classes of chemical changes.

Catalysts are powerful tools for controlling the rates of chemical changes. In fact, some chemical reactions will not occur unless catalysts are present in the reaction environment. It is believed that reactants always become bound temporarily to catalytic substances at some stage in the reaction to enable the reactants to pass through the required configuration for change into products. An important use of catalysis is in the industrial *cracking* (breaking up) of crude oil molecules into gasoline (see Chapter 9). The hydrogenation of liquid fats to form solid fats is another example of catalysis (see Chapter 11). The chemical reactions in most living organisms are speeded up by catalysts called *enzymes*. Because catalysts are *specific* in their action, numerous enzymes are involved in the reactions in living organisms. Although catalysts are employed to increase the speed of a desired chemical change, other substances called *inhibitors* are sometimes used to retard the rate of a chemical reaction. For instance, the use of certain substances in antifreezes to retard rusting, in rubber to retard aging, and in hydrogen peroxide (H_2O_2) to retard decomposition exemplifies the use of inhibitors to slow down undesirable reactions.

ENERGY CHANGES IN CHEMICAL REACTIONS Because chemical changes involve the breaking and forming of chemical bonds, energy changes may occur during chemical reactions. The energy changes that accompany reactions often take the form of heat energy. In some cases, chemical changes produce heat energy that is delivered to the immediate environment; such reactions are called *exothermic* reactions. In other cases, chemical changes occur only if the reactants are supplied with sufficient heat energy from an external source; such reactions are called *endothermic* reactions. Consider, for example, the decomposition of water into hydrogen and oxygen. The chemical change involved is endothermic, and experiments on electrolysis of water (see Chapter 4) tell us that an amount of energy equal to 68.32 kcal/mole of water (liquid, *l*) is required to change water chemically into 1 mole of hydrogen gas (*g*) and 0.5 mole of oxygen gas:

$$68.32 \text{ kcal} + H_2O(l) \longrightarrow H_2(g) + \frac{1}{2}O_2(g)$$

However, when a mixture of hydrogen and oxygen is ignited with a flame or an electric spark, the reaction proceeds rapidly and an amount of energy equal to 68.32 kcal/mole of water is given off to the surroundings:

$$H_2(g) + \frac{1}{2}O_2(g) \longrightarrow H_2O(l) + 68.32 \text{ kcal}$$

Clearly, the energy imparted to water during electrolysis can be stored in the form of hydrogen and oxygen and then released again if those two substances are allowed to react with each other to form water. The stored energy is the energy of chemical combination or *chemical energy*.

The chemical storage of energy is a natural phenomenon in our environment. For example, through the mechanism of *photosynthesis,* green plants convert solar energy to chemical energy by causing water to react with carbon dioxide, thereby producing carbohydrate and oxygen:

$$6CO_2 + 6H_2O \xrightarrow[\text{chlorophyll}]{\text{sunlight}} \underset{\text{glucose}}{C_6H_{12}O_6} + 6O_2$$

The chemical energy that carbohydrates represent is then used to maintain the life process in plants and animals, including human beings. When we burn coal or petroleum products such as gasoline or natural gas, we are using chemical energy stored millions of years ago by photosynthesis in plants.

Tremendous amounts of chemical energy derived from chemical reactions are produced for industrial and practical purposes in our society. In fact, many chemical reactions are carried out solely for the purpose of releasing energy. Today, our major source of energy is the combustion of fossil fuels such as coal, gasoline, oils, and natural gas. Naturally, combustion is an exothermic reaction in which part of the potential energy of the electrons is transformed as they move into new energy levels. The transfer may be accompanied by violent molecular motion and collisions. The energy exchanges involved in an exothermic chemical change are represented in Fig. 7.8. When the potential energy of the reactants is less than that of the products, the same figure represents an endothermic change. It is necessary, however, that a certain amount of energy is gained by the reactants from collisions before they can react to give the products. This amount of energy depends primarily on the nature of the reactants and is called the *activation energy* of the reaction. A catalyst can decrease the activation energy of the reaction; that is to say, at a given temperature the catalyzed reaction can proceed at a faster rate than the uncatalyzed reaction.

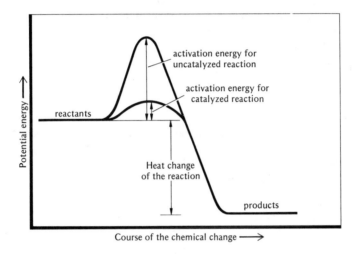

FIGURE 7.8

Energy exchanges involved in chemical changes.

CHEMICAL CHANGES IN OUR ENVIRONMENT The notion that chemical changes occur only in controlled laboratory situations is not true. In fact, there are many chemical changes that take place in our environment. Some of them, such as the photosynthesis reaction described above, are desirable and occur naturally; others are undesirable and result from the activities of an industrial society. Our population with its rapidly expanding technology is presently concerned with the undesirable chemical changes produced in the natural environment and how they affect human beings and other living things.

What are these undesirable chemical changes and how are they produced? Primarily, they are the unexpected and uncontrolled chemical reactions between substances that we produce in our environment, the products of which are considered undesirable to human health and life as a whole on the planet earth. For example, the burning of gasoline in automobile engines produces oxides of nitrogen that together with gasoline escape from automobile exhausts into our air environment. There they undergo *photochemical reactions* (that is, reactions requiring light) to initiate the complicated set of *smog* reactions and release of hazardous chemicals (see Chapter 15). The waste materials from some manufacturing processes that used to be deposited in our streams and rivers or buried under the earth were at one time considered to be unreactive and, therefore, not dangerous to health. However, during the last decade, it has been observed that some of these waste materials react with living organisms in the sea or the soil to produce highly toxic compounds (see Chapters 13–15). Clearly, compounds are prepared and chemical changes are best studied under controlled laboratory situations, rather than under the conditions of our environment where prediction and controls are difficult.

QUESTIONS

7.19   What is meant by chemical stoichiometry?

7.20   List the qualities of a balanced chemical equation.

7.21   Balance the following equations:
    a. $Al + O_2 \longrightarrow Al_2O_3$
    b. $N_2 + H_2 \longrightarrow NH_3$
    c. $C_2H_4 + O_2 \longrightarrow CO_2 + H_2O$
    d. $C + CaO \longrightarrow CaC_2 + CO$
    e. $Fe_2O_3 + H_2 \longrightarrow Fe + H_2O$

7.22   Write the balanced equation for the reaction between silicon and oxygen to produce silicon dioxide. Then calculate the number of grams of silicon required to produce 100 g of silicon dioxide.

7.23   Distinguish between redox and metathesis reactions. Give examples.

7.24   Classify each of the following reactions as redox or metathesis:
    a. $2CO + O_2 \longrightarrow 2CO_2 \uparrow$
    b. $Zn + 2HCl \longrightarrow ZnCl_2 + H_2 \uparrow$
    c. $CH_4 + 2O_2 \longrightarrow CO_2 \uparrow + 2H_2O$
    d. $2NH_3 + H_2SO_4 \longrightarrow (NH_4)_2SO_4$
    e. $H_2SO_4 + BaCl_2 \longrightarrow BaSO_4 \downarrow + 2HCl$

7.25   What is meant by the rate of a chemical change? On what factors does it depend?

7.26   Explain the role of catalysts in chemical changes.

7.27   What are inhibitors? Give examples.

7.28   Why should chemical reactions be accompanied by energy changes?

7.29   Discuss the energy exchanges involved in a chemical reaction, with particular reference to the effect of catalysts.

7.30   List some of the chemical changes occurring in our environment.

SUGGESTED PROJECT 7.3

Analyzing a chemical reaction in our environment (or observing the unobservable)

Basic to the understanding of our environment is a knowledge of what is changing and what new compounds are formed as a result of the change. A good example is the burning of a candle. Apparently the wax ($C_{46}H_{92}O_2$) upon combustion produces nothing but a little soot (carbon). Actually there are other products that come off as gases and escape unobserved.

    Place a glass inverted over a burning candle. After a few minutes, introduce a lighted match inside the glass. What happens? Cool the outside of the glass by placing ice cubes over it. What happens inside the glass? Write a balanced equation showing the products you have just recognized.

8

Large-scale chemical transformations and our environment

The chemical industry has been too preoccupied
with making *things*. It is now time for us to apply
properties to solving problems.

EDWARD J. BOCK
President, Monsanto Company

Many of the materials we use in everyday life are produced by large-scale chemical transformations. Such materials include iron and steel products, fabricated items of aluminum, copper, glass and ceramics, and numerous chemicals used as food preservatives, fertilizers, detergents, and drugs. The branch of chemistry concerned with the manufacture of these and other materials on a commercial scale is known as *industrial chemistry*. It is also involved in the manufacture of prepared foods. In this chapter, we shall discuss the industrial chemistry of basic metals and some chemicals, and examine the effects of chemical industries on our environment.

METALS AND ALLOYS

Some 80 of the 105 elements now known are metals. Whether used in relatively pure form or alloyed with one another, metals have been the materials used to constitute the cornerstone of modern technology. The reason for their usefulness lies in their physical and mechanical properties, such as hardness and strength, malleability and ductility, and ability to conduct heat and electricity, which make them superior to all other classes of material for numerous applications. The unique properties of metals stem from the fact that the atoms within them are held together by a special type of bond called the *metallic bond.* If we look at the electronic structures of a number of metals (see Chapter 5), we realize that they have outer *s* electrons and no outer *p* electrons. The outer *s* electrons in a metal have a tendency to become delocalized. They diffuse themselves throughout the body of the metal. In short, the metal is considered as a crystal consisting of positive metal ions arranged in a regular three-dimensional pattern and enveloped by a cloud of free, delocalized *s* electrons (Fig. 8.1). The cloud of free electrons that permeates the lattice structures of metals accounts for the unique properties and usefulness of many metals.

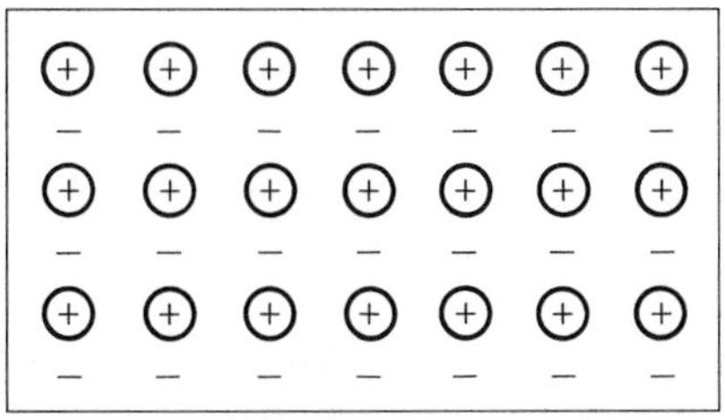

FIGURE 8.1

Cross section of the crystalline lattice in metallic sodium showing the delocalized bonding electrons.

Metals occur in the earth's crust or in seawater mostly as compounds, except perhaps for some of the less active *coinage* metals such as Cu, Ag, and Au which are found also as free elements. The abundances of metals in seawater and in the earth's

TABLE 8.1
Abundance of the metals

Metal	Abundance (% by weight)	
	In seawater	In earth's crust
aluminum	—	7.51
iron	—	4.71
calcium	0.04	3.39
sodium	1.06	2.63
potassium	0.04	2.40
magnesium	0.13	1.93
titanium	—	0.58
manganese	—	0.09
barium	—	0.04
strontium	0.01	0.02

crust are shown in Table 8.1. Except for magnesium from seawater and some manganese from the floor of the oceans, all metals are produced commercially from natural materials, called *minerals,* that are found in the earth's crust. Those materials that can be used as a source for the production of pure metals are called *ores.* The most common ores are stable compounds such as oxides, sulfides, halides, carbonates, sulfates, and silicates of the metals. Of these the silicates are the most abundant minerals, but they are of relatively little value at present because of the difficulty in decomposing them chemically. However, with the depletion of the currently available deposits of simpler ores that are easier and cheaper to process, methods will have to be devised for separating metals from their naturally occurring silicates.

The extraction of metals from their ores comprises the chemistry of *metallurgy.* In general, there are three steps in the production of most metals: concentration of the ore, its decomposition to the metal, and the refining, purifying, or alloying of the metal.

The concentration of ores is in part an art and in part a science. It essentially involves getting rid of the worthless rocks, called *gangue,* that are contained in the mined ores. After crushing and grinding the ore, physical means such as washing with a turbulent stream of water or flotation with a detergent or foaming agent are used to separate the desired mineral. Sometimes magnetic attraction is also used. If the ore cannot be sufficiently concentrated by these means, it is roasted in air to drive off volatile impurities and to burn off organic matter. Roasting in air usually converts sulfides and carbonates to oxides:

$$2FeS + 3O_2 \longrightarrow 2FeO + 2SO_2$$
$$CaCO_3 \longrightarrow CaO + CO_2$$

To remove the last of the gangue, a *flux* is added to the ore and the mixture heated in a furnace. At high temperatures the flux combines with the gangue to make a molten liquid called *slag,* which, being insoluble in the molten metal, can be removed. If the gangue is an acidic oxide such as silica (SiO_2), a cheap basic oxide, such as lime (CaO), is used for the flux:

$$SiO_2 + CaO \longrightarrow CaSiO_3$$

The calcium silicate ($CaSiO_3$) formed is low melting and constitutes the slag. If the gangue is basic, for example, calcium or magnesium carbonate, the flux will naturally be a cheap acidic oxide such as silica.

Once the ore is concentrated, the preparation of metal involves chemical *reduction.* Indeed, the concept of oxidation and reduction developed from metallurgical operations. Recall from Chapters 5 and 7 that oxidation is the loss of electrons and that reduction is the gain of electrons. The chemical methods used to reduce a given metal from its oxidation state in the ore to the elemental state are summarized in Table 8.2. The chosen method depends on the nature and type of the metal compound present in the ore. Because many concentrated ores contain oxides of the less active metals, roasting in air is all that is needed for reduction. Oxides of moderately active metals are reduced by carbon and carbon monoxide. However, carbon tends to form carbides with certain metals such as Cr and Mn, so that it cannot be employed for the reduction of all oxide ores. But it is used when possible, because it is both cheap and convenient. If compounds of metals are not satisfactorily reduced by carbon, a more active metal such as Al, Mg, Na, or Ca is used as the reducing agent. Reduction by hydrogen is more expensive than reduction by carbon and is used only when carbon is not suitable. The cost of electric power is of prime importance in using the electrolytic reduction method. Metals produced commercially by any of the above methods usually contain considerable amounts of impurities. In some applications the pure metal is required. Probably the most widely used refining and purifying process is the electrolytic process.

The number of useful metals is sufficiently large that space does not permit us to discuss them all in this text. Therefore, in the following discussion we shall limit ourselves to the metallurgy of the three most important metals — iron, copper, and aluminum. As we shall see, these metals are extracted from ores concentrated from the minerals found in the earth's crust. In addition, we shall discuss the extraction of magnesium from seawater because in the future, as the earth's richer ores are depleted, we may be looking to the sea as a source of metals.

IRON AND STEEL   In our civilization one of the most useful metals is iron. The iron and steel (a relatively purer form of iron) industries provide us with many of the materials for the industries which we need for the comforts and conveniences of our everyday living. In the United States, iron ores such as hematite (Fe_2O_3), magnetite (Fe_3O_4), limonite ($2Fe_2O_3 \cdot 3H_2O$), and low-grade oxide ores, such as taconite, are plentiful in the states of Minnesota, Alabama, Wyoming, Utah, and California. Other sources of the world's iron are in Canada, Chile, Sweden, Russia, and England among other countries.

Iron oxides are converted to metallic iron on a large scale quickly and economically using a mammoth chimney called a blast furnace (Fig. 8.2). The solid material fed into the top of the blast furnace consists of a mixture of an oxide of iron (Fe_2O_3), coke (C), and limestone ($CaCO_3$). A blast of heated air is forced into the furnace near the bottom to initiate the following reactions:

$$2C + O_2 \longrightarrow 2CO + \text{heat energy}$$
$$Fe_2O_3 + 3CO \longrightarrow 2Fe + 3CO_2 + \text{heat energy}$$
$$CaCO_3 \longrightarrow CaO + CO_2$$
$$CaO + SiO_2 \longrightarrow CaSiO_3$$

TABLE 8.2
Chemical methods for the reduction of ores to metals

Reduction method	Metals	Example
heating the elemental ore in air	Pt, Ag, Au	
heating the sulfide ore in air	Hg	$HgS + O_2 \xrightarrow{\text{heat}} Hg + SO_2$
heating the oxide ore with carbon	Fe, Co, Ni, Pb, Sn, Zn	$ZnO + C \xrightarrow{\text{heat}} Zn + CO$ $ZnO + CO \longrightarrow Zn + CO_2$
heating the halide ores with active metal	Ti, U, Cr	$TiCl_4 + 2Mg \xrightarrow{\text{heat}} 2MgCl_2 + Ti$ $UF_4 + 2Ca \xrightarrow{\text{heat}} 2CaF_2 + U$
heating the oxide ore with hydrogen	W, Ni	$WO_3 + 3H_2 \xrightarrow{\text{heat}} W + 3H_2O$
electrolysis of fused chlorides and other salts	alkali and alkaline earth metals, aluminum, and lanthanide metals	$2NaCl \xrightarrow{\text{electrolysis}} 2Na + Cl_2$ (fused)

Silica, which is an impurity, reacts with the calcium oxide (formed by the decomposition of limestone) to form the slag of calcium silicate. Consequently, as the blast furnace operates, two molten layers collect at the bottom of the furnace. The lower layer, which is mostly liquid iron, is denser than the upper layer, which is essentially molten slag. From time to time the furance is tapped at the bottom and the molten iron drawn off.

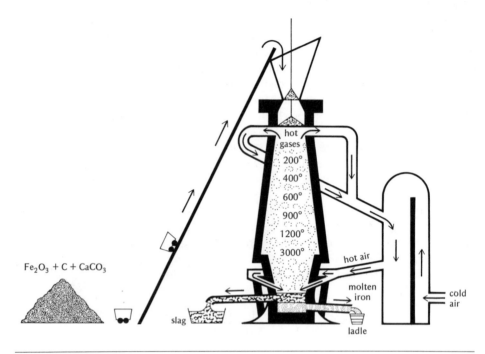

$Fe_2O_3 + C + CaCO_3$

FIGURE 8.2

The blast furnace.

The iron that is obtained from the blast furnace is called *pig iron* and contains a fair amount (4–5%) of dissolved carbon and often smaller amounts of other impurities. In this form it is brittle and of little use. Most of the pig iron is converted to steel without being allowed to solidify. The United States produces nearly 200 million tons of steel each year, which is roughly one-fourth of the world's production. Steel is an alloy of iron with a relatively small amount (less than 1.5%) of carbon. In order to convert pig iron to steel, the carbon content must be reduced. This is done by burning out the excess carbon with oxygen. About 90 percent of the steel manufactured in this country makes use of either the open-hearth furnace or the modern basic oxygen converter.

The *open-hearth furnace* (Fig. 8.3) is built of firebricks lined with oxide refractories such as MgO or CaO. The charge consists of molten pig iron, scrap steel, limestone, and mill scale (Fe_3O_4) or iron ore (Fe_2O_3). Preheated fuel gas and air enter through separate jets at one end of the furnace. Combustion takes place above the charge and the iron oxide oxidizes carbon to CO, silicon to SiO_2, phosphorus to P_4O_{10} and sulfur to SO_2. Heat decomposes limestone to CaO which then reacts with SiO_2 or P_4O_{10} to form a slag. The carbon monoxide bubbles out and burns, and exhaust gases sweep out through brick checkerworks. When these checkerworks have become heated, the direction of air and gas flow is reversed so that the checkerworks now serve to heat the incoming gases, and the combustion products sweep out through a similar structure at the other end. Thus the gas flow is switched from end to end, conserving otherwise wasted heat and making possible a higher temperature in the furnace. The entire process for a hearth of 100 to 150 tons capacity takes from 8 to 10 hours. Toward the end of an operation, a scavenger such as Al, Mg, or Mn is added to remove excess oxygen and the desired amount of carbon is added in the form of coke. The melt is allowed to stand until oxygen removal is complete, the carbon is uniformly distributed, and all slag has risen to the surface. The slag is removed and the steel is poured or tapped into a ladle for casting.

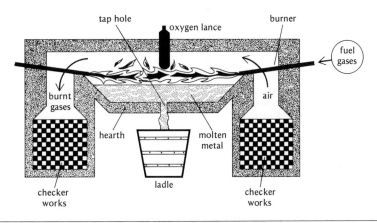

FIGURE 8.3

The open-hearth furnace.

The *basic oxygen converter* (Fig. 8.4) is an egg-shaped steel vessel lined with refractory materials. The charge consists of 30 percent scrap steel and the rest molten pig iron. By blasting in nearly pure oxygen from a water-cooled lance, lowered from the top to within a few feet of the molten iron, nearly all of the impurities such as C, Si, P, and S, as well as some of the iron, are oxidized. A flux consisting of limestone, fluorspar (CaF_2), and mill scale (Fe_3O_4) is added and converts the oxides of impurities to slag by reactions similar to those of the open-hearth furnace. The oxygen blast lasts for about 20 minutes, and in one operation about 150 tons of highly pure iron results. This iron is then poured into a ladle and C, Mn, and alloying metals are added to meet specifications. The process produces a high grade of steel much more rapidly and at a lower cost than the open-hearth process. Therefore, it is anticipated that this process will soon become the principal steel-making method in this country.

Stainless steels are alloys of iron that contain other metals, such as nickel, chromium, or cobalt. They are manufactured by melting together basic oxygen or open-hearth steel or steel scrap and the alloying elements in an electric furnace

(Fig. 8.5). The resulting steels are resistant to corrosion (a reaction with the oxygen and water of the air which transforms a metal into its oxide or hydroxide) and are, therefore, known as "stainless" steels. They find wide use in households and industrial plants alike.

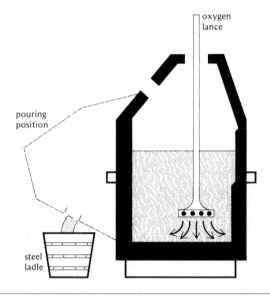

FIGURE 8.4

Basic oxygen converter.

A limited amount of pig iron is converted to *cast iron,* which contains about 2 percent miscellaneous impurities (chiefly carbon) and is used for the manufacture of stoves, steam or hot water radiators, and other objects where low cost is more important than high strength or toughness. Objects made from it are cast into the

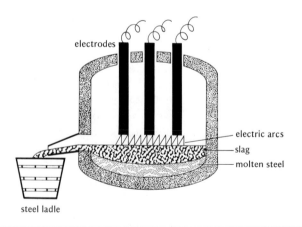

FIGURE 8.5

Electric furnace for stainless steel manufacture.

desired shape in a mold. Iron free of almost all carbon, silicon, phosphorus, and sulfur and containing only traces of slag is called *wrought iron*. It is soft and easily workable, yet tough and resistant to corrosion, and is, therefore, used for manufacturing such items as chains, bolts, wires, nails, and pipes.

COPPER Copper, together with its alloy bronze, was the first metal to be used by man for his tools and weapons. In fact, an entire age in human civilization was based upon bronze, the alloy of tin with copper. Because of its relative lack of chemical reactivity, copper is sometimes found in nature as the free element. Numerous deposits of native copper occur in northern Michigan. Most copper, however, is obtained from minerals such as chalcopyrite ($CuFeS_2$), chalcocite (Cu_2S), and covellite (CuS). Because the copper content of these ores is around 1–2 percent, the powdered ore is first concentrated by a flotation process (Fig. 8.6).

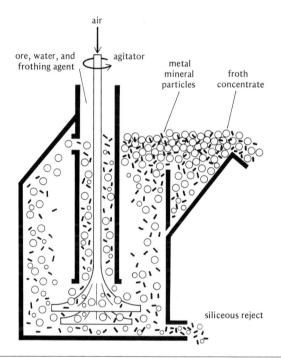

FIGURE 8.6

A flotation device for concentrating ores. (Redrawn, by permission, from Mark M. Jones, John T. Netterville, David O. Johnston, and James L. Wood, *Chemistry, Man and Society*, p. 242. Copyright © 1972 by W.B. Saunders Company.)

Extraction of copper from copper sulfide ore involves roasting in air to oxidize some of the copper sulfide and any iron sulfide present:

$$2Cu_2S + 3O_2 \longrightarrow 2Cu_2O + 2SO_2$$
$$2FeS + 3O_2 \longrightarrow 2FeO + 2SO_2$$

At the higher temperatures involved in the roasting process, some copper is also formed:

$$Cu_2S + 2Cu_2O \longrightarrow 6Cu + SO_2$$

TABLE 8.3
Common alloys

Name	Alloying metals	Special properties	Uses
FERROUS ALLOYS			
chrome vanadium steel	2–10% Cr, 0.2% V	high tensile strength	tools, auto springs
Duriron	15% Si, 0.85% C	acid resistant	plumbing, tanks, and tubing in chemical industries
high-speed steel	18% W, 7% C, 4% Cr, 0.3% Mn	temper at high temperature	cutting tools for metal lathes
manganese steel	10–18% Mn	abrasion resistant	crushers, shovels, etc. in mining industries
molybdenum steel	1–3% Mo	heat and wear resistant	auto axles
nickel steel	5% Ni	hard, tough	guns, axles, wire cable
stainless steel	18% Cr, 8% Ni,	corrosion resistant	auto parts, kitchen-ware, panels, etc.
NONFERROUS ALLOYS			
Aluminum alloys casting alloy	8% Cu	casts well	household appliances
dural	3.5–4.5% Cu, 0.3–1% Mg, 0.4–1% Mn, 0.3–1% Si	heat treatable, light but tough and strong	aircraft parts
Magnalium	5–30% Mg	light weight	household appliances and instruments
Bismuth alloys Wood's metal	25% Pb, 12.5% Sn, 12.5% Cd	very low melting point	low-melting alloy, fuses, etc.
Copper alloys brass, naval	39% Zn, 1% Sn	saltwater resistant	ship parts and trim
brass, red	15% Zn	corrosion resistant	hardware, radiator cores, etc.
brass, yellow	33% Zn	can be spun or pressed	cartridges, musical instruments
bronze, aluminum	10% Al	corrosion resistant	gilt paint
bronze, ordinary	10% Sn	corrosion resistant	valves, rods, statuary, architectural ornaments, etc.
bronze, phosphor	4.8% Sn, 0.2% P	hard, tough	spring metal
nickel silver	18% Ni, 18% Zn	corrosion resistant	silverware, plating, resistance wire, etc.
Gold alloys gold, white	3.5% Cu, 16.5% Ni, 5% Zn	strong, hard	jewelry
Lead alloys Linotype metal	16% Sb, 5% Sn		printing

The product of this operation is called *matte* and consists of a mixture of copper metal and sulfides of copper, iron, other ore constituents, and iron oxide slag. The molten matte is then heated in a converter lined with silica materials. When air is blown through the molten material in the converter, the remaining copper sulfide is converted to copper oxide and then to the metal; the iron is converted to a slag according to the reaction

$$FeO + SiO_2 \longrightarrow FeSiO_3$$

The copper thus obtained is impure and is called *blister copper*. It is purified electrolytically to obtain 99.95 percent pure copper suitable for use as an electrical conductor. In the electrolytic purification of copper, the crude copper forms the anode which is placed in a water solution of copper sulfate; the cathode is made of pure copper. As the electrolysis proceeds, copper is oxidized at the anode, moves through the solution as Cu^{2+} ions, and is deposited on the cathode:

$$Cu \longrightarrow Cu^{2+} + 2e^- \quad \text{(at the anode)}$$
$$Cu^{2+} + 2e^- \longrightarrow Cu \quad \text{(at the cathode)}$$

Besides its electrical uses, copper also finds important applications in photo-engraving and electrotyping. The alloys of copper (Table 8.3) find widespread use

TABLE 8.3 (Cont.)
Common alloys

Name	Alloying metals	Special properties	Uses
	NONFERROUS ALLOYS		
Magnesium alloys Dowmetal A	8% Al, 0.2% Mn	light, can be extruded	aircraft parts
Dowmetal H	6% Al, 3% Zn, 0.3% Mn	very light, can be cast	aircraft parts, household appliances
Nickel alloys Monel	30% Cu	bright, corrosion resistant	kitchen fixtures, work surfaces in food industries
Nichrome	15–20% Cr	high melting point, electrical resistance	heating elements
René 41	19% Cr, 11% Co, 10% Mo, 5% Ti + Al		sheathing of space capsules, jet engines
Silver alloys sterling silver	7.5% Cu		silverware
Tin alloys babbitt	10% Sb, 5% Cu		bearings
pewter	7% Cu, 6% Bi, 2% Sb		metal dishes
solder, soft	40% Pb	low melting point	soldering metal surfaces

in households and industrial plants alike. Coins in many countries are made of either pure copper or copper alloys. Our pennies are almost pure copper, nickels contain about 75 percent copper, and the so-called silvers are about 10 percent copper.

ALUMINUM   Aluminum is the most abundant metal in the earth's crust. Because of its relatively high reactivity, aluminum is found only in the combined state together with other elements. Some of the abundant aluminum minerals are feldspar, mica, and clay. The principal ore from which the metal is extracted is *bauxite*, $Al_2O_3 \cdot 2H_2O$, large deposits of which are found in Arkansas. Bauxite usually contains Fe_2O_3, TiO_2, silicates, and other impurities so that it is first treated with a hot sodium hydroxide solution that dissolves Al_2O_3 as sodium aluminate:

$$Al_2O_3 + 2NaOH \longrightarrow 2NaAlO_2 + H_2O$$

The impurities are filtered off and the solution is seeded with Al_2O_3 to precipitate the hydrated oxide, $Al_2O_3 \cdot 2H_2O$. After it is filtered off, the hydrated oxide is converted to the oxide Al_2O_3 by heating in a rotary kiln. The purified oxide, called *alumina*, thus obtained is electrolyzed for the extraction of metallic aluminum.

The electrolytic manufacture of aluminum was originally accomplished in 1886, independently by the French chemist Paul Héroult and by the American chemist Charles Hall. The same process is used commercially today. Anhydrous Al_2O_3 dissolved in molten cryolite, Na_3AlF_6, is electrolyzed in an iron tank lined with carbon which serves as the cathode (Fig. 8.7). Large blocks of carbon suspended in the tank serve as anodes. Molten aluminum collects at the carbon cell lining and is tapped from time to time. The process is continuous and yields 99.5 percent pure aluminum. Ultrahigh purity aluminum is obtained by repeated electrolysis.

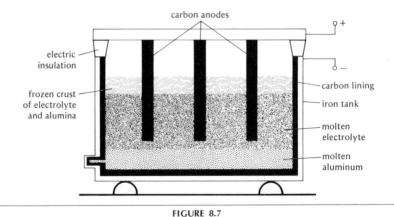

FIGURE 8.7

Electrolytic cell for the manufacture of aluminum.

Pure aluminum is weak and is usually alloyed with a few percent of metals such as Cu, Mg, or Mn to give it the properties desired for structural materials and cooking utensils. At the present time the world production of aluminum is about 10 million tons per year. Of this, over half a million tons goes into the manufacture of food and beverage cans in the United States alone. Even though aluminum has only about two-thirds of the electrical conductivity of copper, its lighter weight has resulted in its having replaced copper for cross-country high-voltage transmission lines.

MAGNESIUM Magnesium is the lightest structural metal in common use. Its "ores" include seawater, which has an Mg^{2+} concentration of 0.13 percent, and dolomite, a mineral with the composition $CaCO_3 \cdot MgCO_3$. Seawater is at present the most important source of magnesium: There are more than 6 million tons of magnesium in every cubic mile of seawater.

The extraction of magnesium from the sea is simple in principle but requires careful chemical control and follows the scheme shown in Fig. 8.8. It begins with the precipitation of magnesium hydroxide by the addition of lime (prepared by burning oyster shells) to seawater pumped into a vat. The magnesium hydroxide is filtered and reacted with hydrochloric acid to form a concentrated magnesium chloride solution. The water is then evaporated, and the dried magnesium chloride is melted at about 700°C and electrolyzed in a huge steel tank so constructed that the products, magnesium and chlorine, are separated as soon as they are formed. The steel tank itself constitutes the cathode, and carbon or graphite rods serve as anodes. The magnesium metal is formed on the steel cathodes and rises to the top where it is

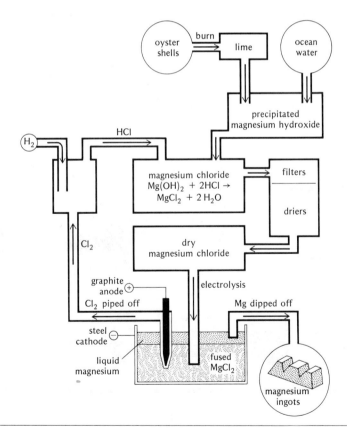

FIGURE 8.8

Flow sheet for the extraction of magnesium from seawater.

dipped off periodically. Chlorine gas, which is a by-product, is formed on the graphite anodes and is piped off.

ALLOYS Alloys are mixtures or combinations of metals that have properties different from the component metals. Although only some 40 metals are commercially important, the possibility of alloying makes available many times this number of metallic materials, often with a great variety of characteristics. An *amalgam* is an alloy of metal with mercury.

Alloys are usually prepared by melting the component metals together. If the metals are mutually insoluble in one another, the resulting alloy may be a *heterogeneous mixture* of the metals. If the metals are partially or wholly soluble in one another, a *homogeneous solid solution* may form. Sometimes the metal atoms in an alloy may combine in a definite ratio forming an *intermetallic compound.*

The physical properties of an alloy may be intermediate to, greater than, or less than the corresponding values of the component metals. For example, the melting point of an alloy is usually lower than the melting point of the component metals. The melting point will vary, of course, with the composition of the alloy. Alloys are usually harder and much more resistant to corrosion than the parent metals, but poorer conductors of heat and electricity.

Some of the more common alloys, their composition, special properties, and uses are listed in Table 8.3 (see p. 182).

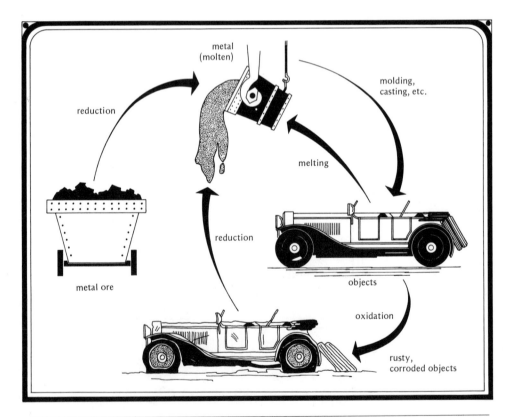

FIGURE 8.9

The metal cycle. (Redrawn, by permission, from John W. Hill, *Chemistry for Changing Times,* p. 69. Copyright © 1972 by Burgess Publishing Company.)

RECYCLING OF METALS Our demand for metals threatens to exhaust the earth's natural ore supplies. Our disposal of corroded and useless metal objects threatens to pollute the environment. Although iron and steel rust, disintegrate, and eventually disappear, aluminum resists corrosion and the widespread use of aluminum for food and beverage cans has caused some concern over its ultimate disposal.

The most promising solution to these problems appears to be the recycling of metals (Fig. 8.9). For example, our iron ore reserves are conserved and environmental pollution lessened if we can reuse old automobiles and broken-down farm and industrial machinery. In fact, many of our iron and steel industries use scrap steel in their manufacturing processes. Methods for transforming waste metals into useful products are also being worked out in various laboratories. Wasted aluminum cans, for example, are being converted into *alum* which finds important uses in the tanning, textile, and paper industries, and in water purification. However, energy dissipation and pollution will eventually limit even the amount of recycled metal in circulation.

QUESTIONS

8.1 What is meant by large-scale chemical transformation?

8.2 Name all the metals that are used as household materials.

8.3 Make a list of the unique properties of metals that make them superior to other materials.

8.4 What is a metallic bond?

8.5 Name some metals that you would expect to find free in nature and some that you would not.

8.6 What metal is most abundant in
   a. seawater
   b. the earth's crust

8.7 Describe the common methods of concentrating an ore. Which of them were probably used by gold prospectors in the West?

8.8 Why does the winning of a metal from its ore almost always require chemical reduction?

8.9 What is the difference between iron and steel?

8.10 Why is lime necessary for the production of iron?

8.11 What advantages has the basic oxygen process over the basic open-hearth method? What disadvantages has it?

8.12 How do pig iron, cast iron, wrought iron, and steel differ from each other?

8.13 Why is steel much more resistant to corrosion than iron?

8.14 What is blister copper and how is it purified?

8.15 In terms of access to needed raw materials and power requirements, what would be the most suitable location economically of aluminum and magnesium plants?

8.16 What are alloys? Name the different types of alloys, giving examples.

8.17 List the properties that make an alloy more useful than a pure metal.

8.18 For what uses might a pure metal be preferred to an alloy?

SUGGESTED PROJECT 8.1
Electroplating

Electroplating of base metal articles is usually done to make them more attractive (and to protect them against corrosion). Design cells for electroplating:

a. silver on knives, forks, and spoons
b. chromium on automobile parts
c. gold on jewelry items

Write chemical equations to show the reactions involved.

SUGGESTED PROJECT 8.2

Alloys

Prepare a list of the alloys that are found in your home. Use a Handbook of Chemistry to determine their compositions. Compare their physical properties with those of the component metals.

SOME CHEMICAL BUILDING MATERIALS

In Chapter 5 it was pointed out that silica and the silicates, which are abundant in the minerals of the earth, are the basic constituents of the building materials such as glass, cement, ceramics, and refractories (Table 8.4). In fact, our oldest factories were built for the manufacture of building materials such as bricks, mortar, plaster, and cement. It is now clear that the covalent silicon–oxygen bonds are responsible for the strength of many of these building materials.

GLASS The art of glass-making is probably six thousand years old and has played an important role in the development of civilization. Ordinary soft glass (sometimes called *lime glass*) is a mixture of sodium and calcium silicates and is manufactured by heating together highly purified silica, sodium carbonate, and calcium carbonate in a furnace at 1400–1500°C. Both carbonates react with the sand, forming the mixture of sodium and calcium silicates that constitutes the glass:

$$Na_2CO_3 + SiO_2 \longrightarrow Na_2SiO_3 + CO_2$$
$$CaCO_3 + SiO_2 \longrightarrow CaSiO_3 + CO_2$$

TABLE 8.4
Common chemical building materials

Material	Composition	Method of preparation
quicklime	CaO	heating limestone ($CaCO_3$) to 900°C
slaked lime	$Ca(OH)_2$	mixing CaO and water
whitewash	$Ca(OH)_2$	solid suspension of $Ca(OH)_2$ in water
mortar	$Ca(OH)_2 + SiO_2$	mixing CaO, SiO_2, and water
plaster of Paris	$(CaSO_4)_2 \cdot H_2O$	heating gypsum ($CaSO_4 \cdot 2H_2O$) to 130°C
cement	silicates of Al, Ca, Fe	heating $CaCO_3$ and clay to 900°C
concrete	cement + sand + gravel + water	mixing cement, sand, gravel, or rock with water
brick	$Al_2O_3 \cdot 2SiO_2 \cdot 2H_2O$	heating clay to 1200°C
firebrick	predominantly Al_2O_3	heating clay and Al_2O_3 to 1500°C
tile	$Al_2O_3 \cdot 2SiO_2 \cdot 2H_2O$	heating clay to 1200°C
glass	$Na_2O \cdot CaO \cdot 5SiO_2$	heating Na_2CO_3, $CaCO_3$, and SiO_2 to 1400°C

The glass is kept molten for several days to permit the CO_2 to escape so that gas bubbles and streams would not be present in the finished product. Portions of the molten mixture are formed into the desired shapes on automated equipment or manually by a glassblower. On solidifying, the glass objects are usually *annealed* to remove stresses that could cause the glass to break easily. Annealing is accomplished by heating the glass to a temperature just below its softening point, then allowing it to cool slowly to room temperature over a period of several days.

Bottles are made by machine from molten soft glass and sheets by drawing a layer of molten glass through a slit and over a conveyor. Plate glass is first drawn as a sheet, allowed to solidify, and then the faces are ground and polished to parallel surfaces. Shatterproof glass consists of two sheets of plate glass bonded by a layer of adherent plastic. Window glass is made in a manner similar to that used for bottles, except that sodium sulfate is substituted for part of the sodium carbonate. Does it surprise you to hear that the formula of modern window glass is very similar to that of an Egyptian glass made about 1400 B.C.?

Borosilicate glasses, of which Pyrex is an example, are made by substituting boron and aluminum oxides for the calcium oxide and increasing the proportion of silica. Such glasses have higher softening points than ordinary glass and very low thermal expansion coefficients so that they can be heated or cooled quickly without shattering. Therefore, Pyrex glass finds wide use in the manufacture of household and laboratory wares. When borosilicate glass is treated with hot acid to remove alkali and reheated, a glass of very high softening temperature and an even lower thermal expansion coefficient is obtained. Such glass is known under the trade name Vycor.

Besides ordinary and Pyrex glasses, hundreds of different colored glasses are made by adding small proportions of metal oxides. For example, cobalt oxides produce purple and blue glasses; iron, chromium, and copper oxides produce green glasses; and selenium or gold produces red glass. More often, however, the metal oxides are added to make glasses with specially desired properties. Thus, lead glasses that have one part SiO_2 and four parts PbO are capable of absorbing large amounts of X-rays and γ-rays; cadmium glasses that contain large concentration of CdO are capable of absorbing neutrons; and, high concentrations of pure arsenious oxide (As_2O_3) in glass makes it transparent to infrared radiation.

CEMENT A cement is a material used to bind other materials together because of its rocklike strength due to silicon–oxygen bonds. The Romans are credited with the first use of cements. Their cements were natural mixtures of lime (CaO) and volcanic ash which, after burning, would set under water. Modern cement, known as *Portland cement*, was first made in the early 1800s in England. It is made by roasting a powdered mixture of limestone, sand, or aluminosilicate minerals such as kaolin, clay, or shale and iron oxide at a temperature of 800–900°C in a rotating kiln. As the materials pass through the kiln, they lose water and carbon dioxide and ultimately form a "clinker" in which the materials are partially fused together. The clinker is reground, mixed with a small amount of gypsum ($CaSO_4 \cdot 2H_2O$) to retard setting, and bagged for shipment.

The approximate composition of Portland cement is 60–67 percent CaO, 17–25 percent SiO_2, 3–8 percent Al_2O_3, 1–5 percent Fe_2O_3, and small amounts of MgO, Na_2O, K_2O, and $MgSO_4$. Although the major compounds in cement have been identified as calcium silicates, calcium aluminates, and calcium alumino-ferrites, the reactions occurring during the setting of cement are quite complex and largely unknown. It is presumed that the setting involves the low-temperature hydration reactions of these compounds with water and, subsequently, at the surface, with the water and carbon dioxide in air.

Over 500 million tons of cement are manufactured each year in the United States. Most of it is mixed with sand, gravel, or rock of the proper size distribution to produce *concrete* for building construction. Although concrete is highly noncompressible, it lacks tensile strength and, therefore, it is reinforced with steel in construction works that are subject to tension. Besides Portland cement, a large variety of special purpose cements that have specially designed adhesive and binding properties are available on the market.

CERAMICS   Since well before the dawn of recorded history, ceramic materials have been made from mixtures of various finely divided minerals and rocks that form a strong rocklike mass when heated to a high temperature. The most important ingredient is clay ($Al_2O_3 \cdot 2SiO_2 \cdot 2H_2O$), a naturally occurring mineral formed by the weathering of certain feldspars ($KAlSi_3O_8$). The term ceramic is now used to include brick, terra-cotta, tileware, stoneware, china, porcelain, or any other product made by molding and baking clay.

When the right amount of water is added the clay becomes plastic as a result of hydration (the binding of ions by the charged end of water molecules; see Chapter 5) and can be formed into objects of desired shapes. After drying, the objects are fired in a kiln at about 1200°C. Fusible silicates partially melt into a glasslike material that bonds the clay particles into a homogeneous mass, called *bisque*. The bisque is finally glazed by applying a low-melting, glasslike material in powdered form to the surface. It is then fired again, but this time to higher temperatures.

In the past few decades the techniques developed with natural clay have been applied to a wide range of other inorganic materials. For example, nearly pure alumina (Al_2O_3) and zirconia (ZrO_2), used as bases for ceramic materials, make them excellent electrical or thermal insulators. Magnetic ceramics or ferrites, containing iron compounds, are used as memory elements in computers. A new class of materials, the *glass ceramics,* has also been developed. First glass is heated under controlled conditions until a very large number of tiny crystals has developed in it and then it is cooled, forming a material that is much more resistant to breaking than normal glass. Materials produced in this way are generally opaque and are used for cooking utensils and kitchenware—they include materials marketed under the trade name Pyroceram. Clearly, the special properties of these materials develop from controlled heat treatment which is indispensable in modern ceramic technology.

QUESTIONS

8.19   Name three common building materials and their chemical composition.

8.20   What gives strength to materials such as bricks, mortar, and cement?

8.21   How is ordinary glass manufactured?

8.22   What is meant by annealing and how is it accomplished?

8.23   What are the differences in the compositions of ordinary glass, window glass, and borosilicate glass?

8.24   Make a list of the special glasses that are familiar to you. What added ingredients produce the desired special properties?

8.25   How is Portland cement manufactured?

8.26   Write equations for the reactions believed to be involved in the setting of cement.

8.27   What is the main ingredient of ceramic materials?

8.28   How is glass ceramic produced?

SUGGESTED PROJECT 8.3

Industrial chemistry of calcium

Limestone ($CaCO_3$) is the main industrial raw material from which many building materials are prepared. Illustrate this statement by outlining a scheme for the preparation of some of the materials used in building

a.   your house
b.   your school

SOME BASIC CHEMICALS

Many chemical industries all over the world are concerned mainly with the large-scale production of a few basic chemicals such as sulfuric acid, ammonia, hydrochloric acid, sodium hydroxide, sodium carbonate, and chlorine from naturally occurring materials on earth. These basic chemicals, in turn, are used by many other industries for producing the variety of chemical products that we use in everyday life. For example, sulfuric acid, which is produced as a basic chemical, finds use in the production of chemical products such as rayon, paper, detergents, dyes, paints, and plastics. In this section we shall discuss a number of useful inorganic materials that are considered to be basic chemicals and are manufactured on a large scale.

SULFUR AND SULFURIC ACID   Sulfur in underground mineral deposits is mined by the *Frasch process* which utilizes superheated steam to melt the sulfur and compressed air to raise the molten sulfur to the surface of the earth (Fig. 8.10). The

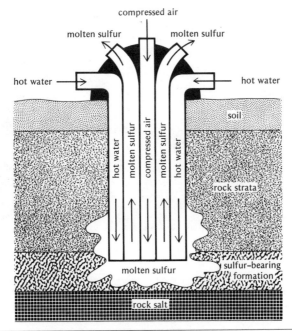

FIGURE 8.10

Diagrammatic representation of the Frasch process for the mining of sulfur.

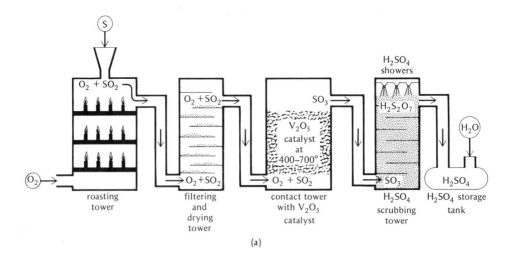

roasting tower

filtering and drying tower

contact tower with V_2O_5 catalyst

H_2SO_4 scrubbing tower

H_2SO_4 storage tank

(a)

(b)

FIGURE 8.11

(a) The contact process for manufacturing sulfuric acid. (b) A sulfuric acid production unit at the Cleveland Works of the du Pont Company's Industrial Chemicals Department. The unit can be operated around the clock from a central control room by a relatively small crew of employees. (Photograph courtesy of E. I. du Pont de Nemours & Company.)

sulfur thus mined is burned in air to give mostly sulfur dioxide gas:

$$S + O_2 \longrightarrow SO_2$$

which is used for the manufacture of sulfuric acid. In the United States most sulfuric acid is prepared by the *contact process* (Fig. 8.11). Sulfur dioxide gas is reacted with oxygen in the presence of finely divided vanadium pentoxide, which acts as a catalyst for the production of sulfur trioxide:

$$2SO_2 + O_2 \underset{400°C}{\overset{V_2O_5}{\rightleftharpoons}} 2SO_3$$

Note the reversible character of the reaction. Above a temperature of about 450°C the reverse reaction becomes appreciable. However, too low a temperature means an excessively long wait until attainment of equilibrium so that a compromise temperature of about 400°C is generally used. At this temperature, about 98 percent yield of SO_3 is obtained and the process is economically feasible. The SO_3 is absorbed in H_2SO_4 to form $H_2S_2O_7$ which then is diluted with water to form H_2SO_4:

$$SO_3 + H_2SO_4 \longrightarrow H_2S_2O_7$$
$$H_2S_2O_7 + H_2O \longrightarrow 2H_2SO_4$$

By this method about 25 million tons of sulfuric acid are prepared each year in the United States for use in the manufacture of fertilizers, in the petroleum industry, and in the production of steel.

AMMONIA AND NITRIC ACID   Elemental nitrogen from the atmosphere combines with hydrogen to form ammonia when heated under pressure and passed over a catalyst. In the *Haber process* (Fig. 8.13a) the mixture of gases is passed over iron oxide heated to 500°C:

$$N_2 + 3H_2 \underset{500°C}{\overset{Fe_2O_3}{\rightleftharpoons}} 2NH_3$$

The reaction is reversible but economically feasible. Because of its relatively high boiling point compared to other gases, the ammonia can be condensed out of the mixture. In the United States over 10 million tons of ammonia are manufactured annually by this process and used as starting material for the production of other nitrogen chemicals. An example is *nitric acid,* HNO_3, which is manufactured by the *Ostwald* process (Fig. 8.13b). Ammonia is burned at about 800°C over a platinum catalyst to obtain nitric oxide:

$$4NH_3 + 5O_2 \overset{Pt}{\longrightarrow} 4NO + 6H_2O$$

The nitric oxide reacts readily with atmospheric oxygen to form nitrogen dioxide which readily dissolves in water to form nitric acid:

$$2NO + O_2 \longrightarrow 2NO_2$$
$$H_2O + 3NO_2 \longrightarrow 2HNO_3 + NO$$

About one half of the ammonia and nitric acid manufactured in the United States is converted into ammonium nitrate (NH_4NO_3) for use as fertilizer.

PHOSPHORUS AND PHOSPHORIC ACID   Elemental phosphorus is extracted from the naturally occurring phosphate mineral, $Ca_3(PO_4)_2$. An electric furnace is used to heat a mixture of the phosphate ore, silica, and coke (Fig. 8.14). At a

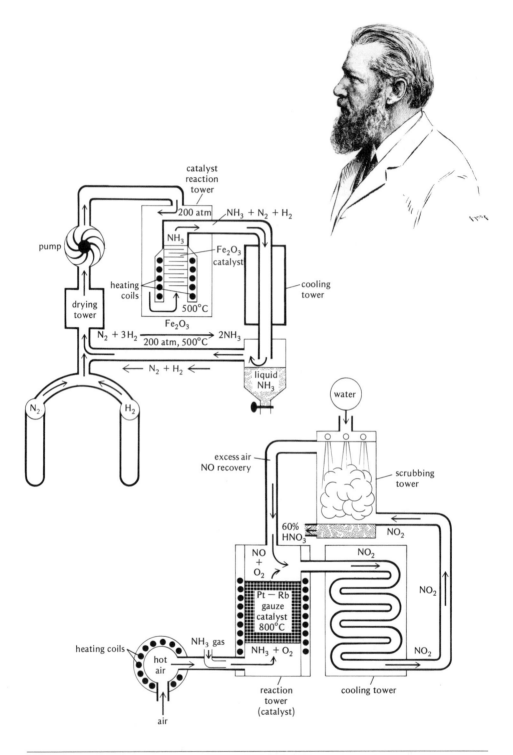

catalyst
reaction
tower

200 atm $NH_3 + N_2 + H_2$

pump

NH_3 Fe_2O_3 catalyst

drying tower

heating coils

cooling tower

500°C

Fe_2O_3

$N_2 + 3H_2 \xrightarrow{\text{200 atm, 500°C}} 2NH_3$

$\longleftarrow N_2 + H_2 \longleftarrow$

N_2 H_2

liquid NH_3

water

excess air
NO recovery

scrubbing tower

60% HNO_3 NO_2

NO + O_2 NO_2

Pt — Rb gauze catalyst 800°C NO_2

heating coils NH_3 gas

hot air $NH_3 + O_2$ NO_2

air

reaction tower (catalyst)   cooling tower

FIGURES 8.12 AND 8.13

Wilhelm Ostwald, German physical chemist of the early twentieth century who pioneered in catalysis and developed a method that, in several modifications, is now used for the manufacture of nitric acid. Manufacture of (above) ammonia (Haber process) and (below) nitric acid (Ostwald process).

temperature of 1500°C a reaction occurs in which elemental phosphorus is produced:

$$2Ca_3(PO_4)_2 + 6SiO_2 + 10C \longrightarrow P_4\uparrow + 10CO + 6CaSiO_3$$

The phosphorus is then condensed from the gaseous vapor and purified. It is transformed into phosphoric acid first by oxidation with air to give P_4O_{10} which is then hydrated by absorption into hot phosphoric acid containing about 10 percent water:

$$P_4 + 5O_2 \longrightarrow P_4O_{10}$$
$$P_4O_{10} + 6H_2O \longrightarrow 4H_3PO_4$$

Phosphoric acid is made on an industrial scale by reacting pulverized phosphate ore with concentrated sulfuric acid:

$$Ca_3(PO_4)_2 + 3H_2SO_4 \longrightarrow 2H_3PO_4 + 3CaSO_4\downarrow$$

Water is added to the mixture, the precipitated calcium sulfate filtered off, and the filtrate evaporated until a syrupy liquid containing 85 percent H_3PO_4 is obtained. Most of the phosphoric acid thus produced is transformed to phosphate fertilizers such as $Ca(H_2PO_4)_2$ and $(NH_4)_3PO_4$ that are readily assimilated by plants.

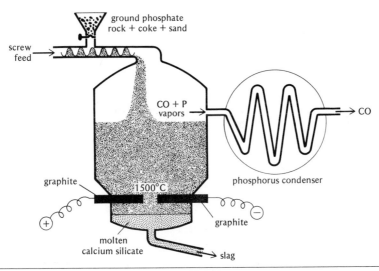

FIGURE 8.14

Electric arc furnace for the manufacture of phosphorus. (From Robert W. Medeiros, *Chemistry: An Interdisciplinary Approach*, p. 472. Copyright © 1971 by Litton Educational Publishing, Inc. Redrawn by permission of Van Nostrand Reinhold Company.)

THE CHLOR–ALKALI INDUSTRIES Sodium hydroxide (caustic soda) is manufactured by the electrolysis of a concentrated solution of sodium chloride in water:

$$2NaCl + 2H_2O \xrightarrow{\text{electrolysis}} Cl_2\uparrow + H_2\uparrow + 2NaOH$$

The by-products of the electrolysis are chlorine (at the anode) and hydrogen (at the cathode) which can be used for the production of hydrochloric acid:

$$H_2 + Cl_2 \longrightarrow 2HCl$$

The cell must be so constructed that mutually reactive products such as H_2 and Cl_2 or NaOH and Cl_2 are kept apart. This is done by separating the anode and cathode compartments by a porous asbestos diaphragm. Several versions of the diaphragm cell are in use; one of them is shown in Fig. 8.15. It uses an asbestos diaphragm through which the sodium chloride solution soaks and comes in contact with the perforated steel cathode where hydrogen gas is liberated. The bubbles of gas are unable to penetrate the asbestos and pass instead through the perforations in the cathode into the outer chamber. In the other type of cell (Fig. 8.15), a thin layer of mercury flowing slowly through the cell is used as the cathode. The sodium liberated at the cathode forms an amalgam (alloy) with the mercury and is carried from the cell. It finally reacts with water to produce H_2 and NaOH, freeing the mercury for reuse. The use of this type of cell is now believed to be responsible for much of the mercury pollution in certain streams into which the waste solutions from this process are dumped.

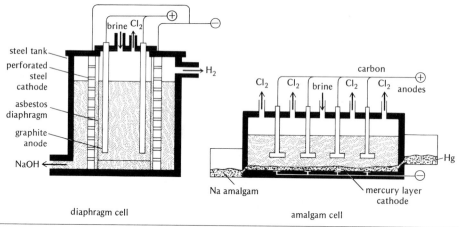

diaphragm cell   amalgam cell

FIGURE 8.15

Electrolytic cells used in the chlor–alkali industries.

Some of the sodium hydroxide produced is transformed into sodium hydrogen carbonate (baking soda):

$$NaOH + CO_2 \longrightarrow NaHCO_3$$

This large-scale conversion is more a matter of economic necessity than anything else.

THE SOLVAY PROCESS Sodium carbonate (washing soda) and sodium bicarbonate (baking soda) are manufactured by the Solvay process. In the process, ammonia and carbon dioxide are added under pressure to a solution of sodium chloride:

$$NaCl + NH_3 + CO_2 + H_2O \longrightarrow NaHCO_3\!\downarrow + NH_4Cl$$

The sodium bicarbonate that is precipitated is filtered and dried. It may then be transformed into sodium carbonate by heating, which drives off water and carbon dioxide:

$$2NaHCO_3 \xrightarrow{\text{heat}} Na_2CO_3 + H_2O + CO_2$$

The Solvay process is an example of how the chemical industry endeavors to manufacture its products from the cheapest possible starting materials and to recycle the reacting substances whenever possible. For instance, the sodium chloride is obtained inexpensively from underground salt deposits and from the ocean. The carbon dioxide is produced at minimal cost simply by heating limestone ($CaCO_3$) which is abundant on earth. The most costly item in the Solvay procedure is ammonia. It is initially obtained by the Haber process (see p. 193), but is then recycled according to the following procedure:

$$CaO + 2NH_4Cl \longrightarrow 2NH_3\!\uparrow + CaCl_2 + H_2O$$

That is, the ammonium chloride (produced as a by-product in the step where the $NaHCO_3$ is manufactured) is heated with the calcium oxide (arising from the decomposition of the limestone) to regenerate the ammonia. The carbon dioxide generated at the time of transformation of $NaHCO_3$ into Na_2CO_3 is also recycled and reused in the initial reaction for producing $NaHCO_3$. The entire process is illustrated by the flow diagram in Fig. 8.16.

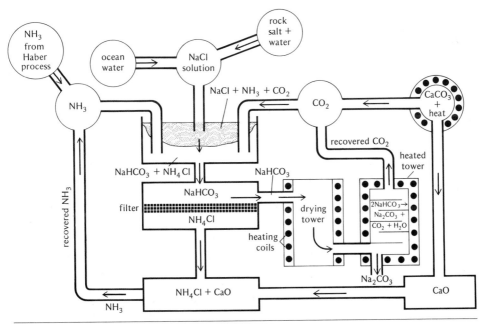

FIGURE 8.16

Flow diagram of the Solvay process.

QUESTIONS

8.29 What are basic chemicals? Name any four.

8.30 Describe the contact process for the manufacture of sulfuric acid.

8.31 What artificial methods are employed to make atmospheric nitrogen useful?

8.32 List some important uses of phosphorus chemicals.

8.33 How is caustic soda manufactured?

8.34 What are the by-products of the chlor–alkali industry?

8.35 Compare and contrast the electrolysis of molten NaCl with the electrolysis of an aqueous solution of NaCl.

8.36 Discuss the economics of chemical manufacturing using the Solvay process as an example.

EFFECT OF CHEMICAL INDUSTRIES ON OUR ENVIRONMENT

Although the chemical industries are helping to meet our needs and wants by large-scale transformations of material things into desirable products, we should not ignore that some undesirable products from the very same industries are discharged into our air, water, or soil environment as waste materials. For example, some of the industrial operations described in this chapter produce poisonous gases such as carbon monoxide, sulfur oxides, and nitrogen oxides that escape into the atmosphere and contaminate the air we breathe (Table 8.5). It is estimated that each year the metal industries in our country allow about 6 million tons of carbon monoxide and 4 million tons of sulfur oxides to escape into the air. Further, each year our sulfur chemicals industries discharge about 600,000 tons of sulfur oxides and the nitrogen

TABLE 8.5
Major industrial pollutants and their health hazards[a]

Pollutant	Industry	Level of production (10^6 tons/yr)	Tolerance level (ppm)	Health hazard
carbon monoxide	iron and steel	6.0	32	headache, fatigue, drowsiness, coma, respiratory failure, death
	petroleum refineries	2.2		
	paper pulp	0.8		
sulfur dioxide	smelters (chiefly copper)	3.9	>5 (0.5 in plants)	irritation of respiratory tract, lung diseases
	petroleum refineries	2.1		
	sulfuric acid plants	0.6		
	coking operations	0.6		
nitrogen oxides	nitric acid plants	0.2	>5 (0.25 in plants)	inflammation of lung tissue, death
particulates	iron and steel	1.91	375 $\mu g/m^3$	respiratory problems with particles less than 0.5 μ in diameter
	sand, stone, rock	0.87		
	cement	0.87		
	paper pulp	0.72		

[a] Data from the *Second Annual Report of the Council of Environmental Quality*, Washington, D. C., August 1971; and other sources.

chemicals industries about 200,000 tons of nitrogen oxides into the atmosphere. In humid areas, the oxides react with water vapor producing *acid mists* that are potent irritants to human beings. A study published recently shows significantly higher rates of lung cancer, emphysema, bronchitis, and heart disease in heavily industrialized areas.

The pollution from chemical industries is by no means confined to gaseous contaminants. Approximately 7 million tons of *particulate* (particle) matter are emitted into the air each year in the United States alone. A breakdown shows that the metal, cement, and ceramic industries alone contribute about one-half of the total particulate pollution. The pollutants include also a number of trace metals that are toxic and may pose health hazards in the environment (see Chapter 15). Beryllium and asbestos (a substance used for insulation by the building trade) are examples of airborne particulates that have become matters of concern. In 1971, the U.S. Environmental Protection Agency announced that both of these pollutants had been added to the list of "hazardous air pollutants." Besides the health hazards, particulate pollution cuts down on the amount and type of solar radiation that reaches the earth's surface and influences the formation of clouds, rain, and snow by acting as nuclei upon which water condensation can take place. Thus, worldwide climatic changes may be related to particulate pollution.

As well as polluting the air, many of our chemical industries discharge wastes into our streams and soils. For example, let us consider the chlor–alkali industry (see previous section, p. 195). It is the largest industrial user of mercury. Theoretically, no mercury is consumed during the process of producing chlorine or caustic soda. Actually, however, loss of mercury does occur—some is lost to the products, some leaves with effluent water, and some is lost through air ventilation systems. It is estimated that about 400 tons of mercury are thus lost each year and that most of it ultimately shows up in the waste-water discharges from the plants. Since 1970, however, the chlor–alkali industries have been able to reduce these discharges by 90 percent, mainly by recirculating waste water and installing lagoons and settling ponds in which the mercury collects.

Many inorganic salts, mineral acids, finely divided metals or metal compounds discharged into our streams and soil by industries increase the acidity, salinity, and toxicity of our natural waters and soil. The effects of these changes lead to destruction of aquatic life and agricultural crops, to structural damages by excessive corrosion, and sometimes to cumulative poisoning of humans.

Finally, many chemical industries contribute to the serious problem of thermal pollution by their practice of using water as a coolant. Most water used as coolant is returned, with the added heat, to the original sources. The addition of excess heat to a body of water may have as many adverse effects as chemical pollutants. For example, as the temperature increases the dissolved oxygen levels on bodies of water decreases, threatening the survival of plant and animal life that depends on the ability of the water to maintain certain minimal concentrations of this vital oxygen. However, several industries have lately incorporated cooling towers into their operations to remove heat from cooling water before returning it to the natural water supply. Although this innovation may save aquatic life for the time being, it, nevertheless, heats up our air environment. Indications are that waste heat disposal will continue to be a problem for many years.

No doubt the chemical industries are largely responsible for the high standard of living most of us enjoy. On the other hand, they are equally responsible for the accelerated changes in our environment. It is gratifying to note that many chemical industries are moving into the areas of pollution abatement for protection of our environment. Clearly, industrial chemistry should be and probably will be limited by the two relatively new demands: not only must it produce useful materials but it must also provide for the return of the useless and dangerous materials to nature in a desirable form—that is, the eventual recycling of materials. In a reasonable economic package, there must be enough demand for the desired materials to offset the cost of neutralizing the harmful effects of undesired and dangerous by-products.

QUESTIONS

8.37   Name some of the major industrial pollutants.
8.38   How are acid mists produced? Write equations.
8.39   Make a list of the more important toxic metals and their reported health effects.
8.40   How can thermal pollution by chemical industries be prevented?

SUGGESTED PROJECT 8.4

Assessing chemical industries in your state

Prepare a list of the chemical industries in your home state. Determine the respective chemicals and pollutants that each industry produces. If the levels of production of the pollutants are known, evaluate what sort of health hazards are involved.

9

Organic matter

. . . the atoms were gamboling before my eyes.
This time the smaller groups kept modestly in
the background. My mental eye, rendered more
acute by repeated visions of the kind, could now
distinguish larger structures of manifold conformation: long rows, sometimes more closely
fitted together, all twining and twisting in snake-
like motion. But look! What was that? One of the
snakes had seized hold of its own tail, and the
form whirled mockingly before my eyes. As if by
a flash of lightning I woke. . . .

FRIEDRICH KEKULÉ (1829–1896)
German chemist

Any carbon-containing material whose original source is living organisms, either plant or animal, is called *organic matter*. Wood, coal, and petroleum fuels, cotton and wool fabrics, and fruits, vegetables, milk, and meat are all examples of organic matter. Clearly, it is the food, the fuel, and the fabric of human existence. In the form of dead and decaying animals and plants, sewage, and garbage, it provides a culture for the growth of disease organisms and may rob natural water of its oxygen, rendering it unfit for consumption or recreation. Unquestionably, organic matter plays a major role in our environment. In this chapter we shall be concerned with the carbon compounds that form the organic matter of our environment.

ORGANIC CHEMISTRY

The branch of chemistry concerned with carbon-containing compounds is known as *organic chemistry*. The term organic is derived from the original thought that all carbon-containing molecules had to be directly produced by living organisms. Today, however, the term includes not only those molecules in presently living organisms and those derived from organisms once alive, but also carbon-containing molecules that are synthesized or altered in the laboratory. Thus, organic compounds may or may not be derived from living systems.

Although the abundance of elemental carbon in the crust of the earth is only 0.08 percent by weight, a few million compounds of carbon have so far been isolated. By contrast, the total number of known compounds that do not contain carbon is less than 200,000. Why the great numerical preponderance of carbon compounds? The answer lies in the electronic structure of the carbon atom ($1s^2 2s^2 2p^2$) which has an exactly half-filled outer energy level (with four electrons) capable of forming four strong covalent bonds not only with atoms of other elements but with other carbon atoms as well. The ability of like atoms to bond together into chains or rings is called *catenation*.

Carbon is unique among elements in the extent and variety with which it exhibits catenation. A consequence of this is that among the millions of carbon compounds are many that can be classified together into series, each of which contains structurally similar molecules. Such a series is called a *homologous series*. Many homologous series exist, the simplest being the one that is called the *paraffin* series or the *saturated hydrocarbons* (*alkanes*) found in crude oil. In this series all carbons are joined to other carbons by single bonds and the rest of the carbon valencies are occupied with a maximum possible number of hydrogen atoms. Table 9.1 lists several members of this series. When describing chemical reactivities, any one member can serve as a prototype for others of the same series.

Another consequence of catenation is that a great number of structural arrangements are possible with carbon, each structure representing a different compound with different physical and chemical properties. Consider, for example, the organic compound butane which has the molecular formula C_4H_{10}. It can be assigned a *straight-chain* structure or a *branched-chain* structure:

n-butane
(straight-chain)

isobutane
(branched-chain)

Because both structures have the formula C_4H_{10}, they are said to constitute two *structural isomers* of the hydrocarbon butane; the straight-chain isomer is named normal butane (abbreviated *n*-butane or n-C_4H_{10}) and the branched-chain isomer is named isomeric butane (abbreviated isobutane or iso-C_4H_{10}). The conversion of one compound to an isomer is known as an *isomerization reaction*. Although such reactions appear deceptively simple, the actual mechanisms of change are often complicated. Usually the conversion from one form to the other involves two or more steps in which the molecules are actually taken apart and reassembled. As the number of carbon atoms increases, so does the number of possible isomers; for instance, the hydrocarbon decane ($C_{10}H_{22}$) has 75 possible isomers, and the hydrocarbon tricontane ($C_{30}H_{62}$) has more than 4 billion possible isomers. However, you will recall that saturated carbons are tetrahedrally bonded with a 109° angle (see Chapter 4), so that the actual geometry of the molecules is zig-zag in spite of the carbon atoms linking in serial fashion (Fig. 9.1).

TABLE 9.1
Some members of the alkane homologous series

Name	Molecular formula	Structural formula
methane	CH_4	$\begin{array}{c} H \\ | \\ H-C-H \\ | \\ H \end{array}$
ethane	C_2H_6	$\begin{array}{cc} H & H \\ | & | \\ H-C-C-H \\ | & | \\ H & H \end{array}$
propane	C_3H_8	$\begin{array}{ccc} H & H & H \\ | & | & | \\ H-C-C-C-H \\ | & | & | \\ H & H & H \end{array}$
butane	C_4H_{10}	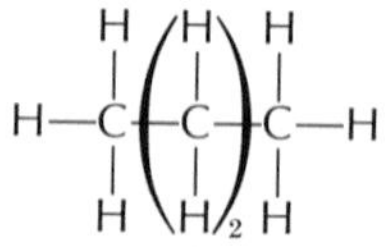 $H-C{-}(C)_2{-}C-H$
pentane	C_5H_{12}	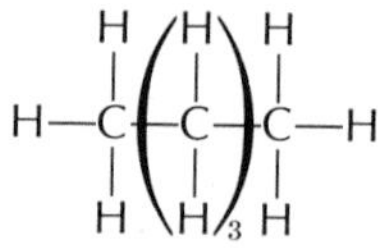 $H-C{-}(C)_3{-}C-H$
hexane	C_6H_{14}	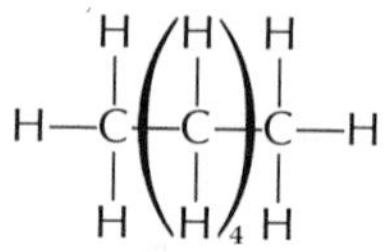 $H-C{-}(C)_4{-}C-H$
heptane	C_7H_{16}	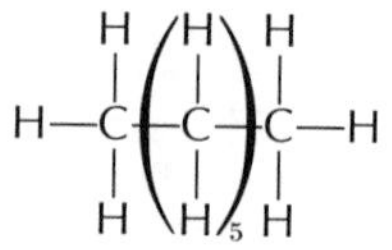 $H-C{-}(C)_5{-}C-H$
octane	C_8H_{18}	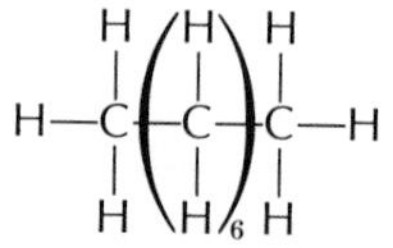 $H-C{-}(C)_6{-}C-H$
⋮	⋮	⋮
alkane	C_nH_{2n+2}	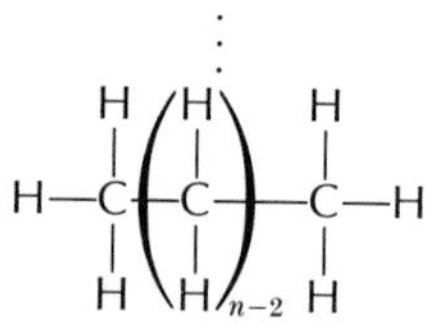 $H-C{-}(C)_{n-2}{-}C-H$

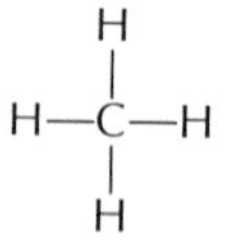

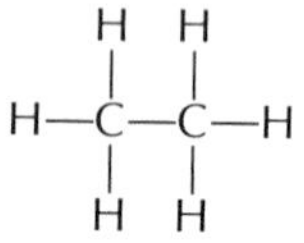

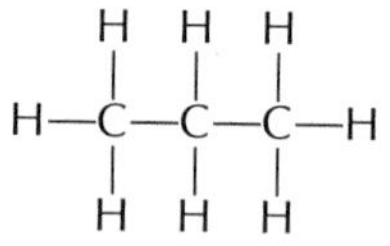

It is also possible to have atoms of elements other than hydrogen (such as O, P, S, and halogens) attached to the carbon skeleton as well as *functional groups,* such as hydroxyl (OH), amino (NH_2), carboxyl (COOH), and nitro (NO_2) (Table 9.2). A functional group is a combination of atoms in a molecule that exhibits a consistent

TABLE 9.2
Typical organic functional groups

| Name | FUNCTIONAL GROUP | | Class name of organic molecules formed |
| | Formula | | |
	Structural	Condensed	
amino	$\overset{H}{\underset{H}{-N}}$	$-NH_2$	amines
carbonyl	$\diagdown C=O$	$\diagdown CO$	aldehydes, ketones
carboxyl	$-\underset{O}{\overset{\|}{C}}-OH$	$-COOH$ *or* $-CO_2H$	acids
ester	$-\underset{O}{\overset{\|}{C}}-O-R$	$-COOR$ *or* $-CO_2R$	esters
halide	$-X$	$-X$ (X = F, Cl, Br, or I)	organic halides
hydroxyl	$-OH$	$-OH$	alcohols
nitro	$-N\overset{O}{\underset{O}{\diagdown}}$	$-NO_2$	nitro compounds
sulfhydryl	$-SH$	$-SH$	mercaptans

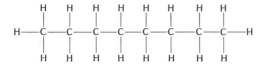

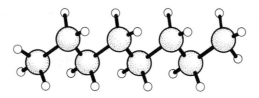

FIGURE 9.1

Zig-zag shape of "straight-chain" molecules.

kind of reactivity (function) the same as an atom, and the other parts of the molecule have little or no modifying influence on this function. Saturated organic compounds (for example, the alkanes) undergo *substitution reactions* that always involve replacement of an atom or group of atoms in a molecule by a functional group. For example, one or more hydrogen atoms in a methane molecule may be replaced by atoms of chlorine, bromine, or other functional groups:

$$H-\underset{\underset{H}{|}}{\overset{\overset{H}{|}}{C}}-H + Cl-Cl \longrightarrow H-\underset{\underset{H}{|}}{\overset{\overset{H}{|}}{C}}-Cl + H-Cl$$

methane   methyl chloride

The carbon–carbon double bond, which we have already introduced in Chapter 4, is also a functional group. One of its typical reactions is the *addition reaction*. Bromine or other reagents will react with any molecule containing C=C:

$$\underset{H}{\overset{H}{>}}C=C\underset{H}{\overset{H}{<}} + H-Cl \longrightarrow H-\underset{\underset{H}{|}}{\overset{\overset{H}{|}}{C}}-\underset{\underset{H}{|}}{\overset{\overset{H}{|}}{C}}-Cl$$

ethene   ethyl chloride
(ethylene)

The hydrogenation reactions by which liquid fats are converted into solid fats are also examples of addition reactions.

Because of the existence of a homologous series, often a number of closely related molecules, all of which have a common structural feature, serve as a focus for similar reactivity. Thus, when a saturated molecule contains the grouping —C—OH, it is classified as an *alcohol* and undergoes the reactions typical of that group. The generic formula for a homologous series containing a common functional group usually is written with the symbol R representing the rest of a hydrocarbon chain. These groups are also known as hydrocarbon *radicals* (hydrocarbon molecules that are short of one hydrogen atom) and are named according to the names given to the complete hydrocarbon molecule. For example, a compound containing the group —CH_3 is named a methyl compound (CH_3Cl is methyl chloride), and a compound containing the group—C_2H_5 is named an ethyl compound (C_2H_5Cl is ethyl chloride). In general, when the group is a residue from a straight- or branched-chain saturated hydrocarbon, it is classed as an *alkyl* group. Table 9.3 lists the hydrocarbon radicals that we shall take as examples in this chapter.

TABLE 9.3
Some hydrocarbon radicals

Name	Formula	Structure
methyl	CH_3—	CH_3—
ethyl	C_2H_5—	CH_3—CH_2—
n-propyl	C_3H_7—	CH_3—$(CH_2)_2$—
isopropyl	C_3H_7—	CH_3—CH— $\underset{CH_3}{\overset{|}{}}$
butyl	C_4H_9—	CH_3—$(CH_2)_3$—

The carbon skeletons of organic compounds may have single, double, and/or triple bonds in them or they may exist as rings (Fig. 9.2). If we consider the different positions that various atoms or functional groups can occupy on the many different possible structural arrangements of carbon, we can understand why an enormous number of organic compounds is theoretically possible. However, we are fortunate that there is a systematic method for organizing the study of the large number of carbon compounds. The method involves classifying organic compounds into groups according to the specific elements that they contain and the functional groups in the molecule. In this chapter, we shall discuss some of the more note-worthy groupings of organic compounds and their environmental impact.

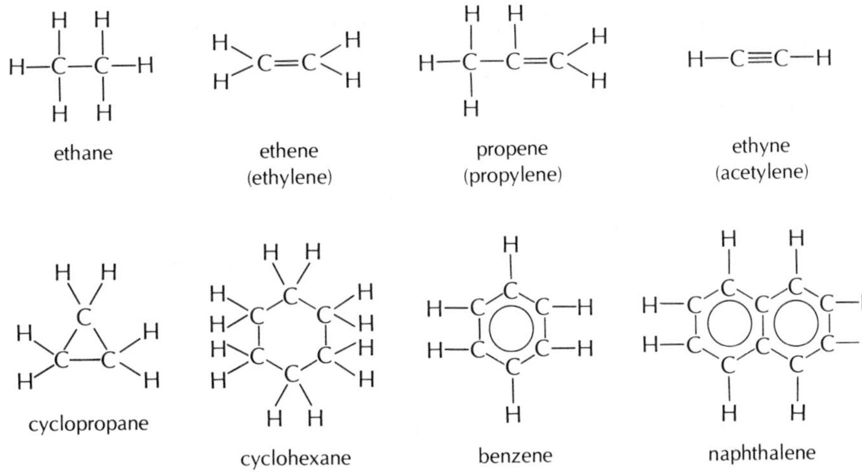

ethane   ethene (ethylene)   propene (propylene)   ethyne (acetylene)

cyclopropane   cyclohexane   benzene   naphthalene

FIGURE 9.2

Some organic compounds possessing single, double, and triple bonds and ring structures.

QUESTIONS

9.1   What is organic chemistry?

9.2   How would you account for the great numerical preponderance of carbon-containing compounds?

9.3   Define the following terms:
a. catenation
b. homology
c. isomerism
d. functional group

9.4   Distinguish between addition reactions and substitution reactions.

9.5   What is an alkane? Give two examples.

9.6   Name the following groups:
a. $-NH_2$
b. $-COOH$
c. $-SH$
d. CH_3-CH-
         $|$
        CH_3

(CH$_2$)$_n$ (CH$_2$)$_n$

FIGURE 9.3

The catenanes.

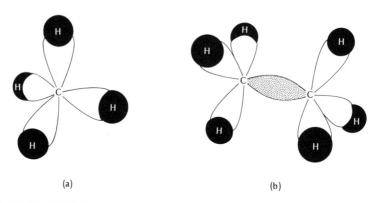

(a)   (b)

FIGURE 9.4

Schematic representation of the bonding orbitals in (a) methane and (b) ethane.

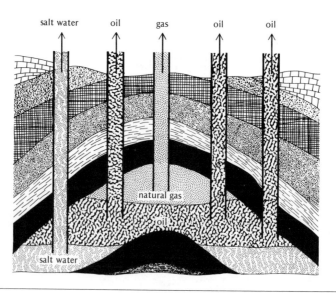

FIGURE 9.5

Diagram of a typical oil trap and natural gas well.

COMPOUNDS OF CARBON AND HYDROGEN

The compounds formed when only hydrogen atoms are bonded to the carbon skeleton are known as *hydrocarbons*. Of the hydrocarbons, the simplest member is methane which has only one carbon atom with four hydrogens bonded to it in a tetrahedral structure. As the number of carbon atoms increases in the chain, the compounds ethane, propane, butane, and so on are produced in succession (see Table 9.1). As the number of carbon atoms increases, the chains can have side branches yielding isomeric compounds, that is, different arrangements of the same number of carbon and hydrogen atoms. There are still other hydrocarbons in which the carbon atoms exist in rings that may or may not have side chains of carbon atoms. In some cases, these rings will be fused together to form a multiring system. A group of hydrocarbons that have been discovered recently have rings linked through each other. They are called the *catenanes* (Fig. 9.3). These complex molecules should be viewed in three-dimensional space rather than in planes because of the tetrahedral orientation of the bonds from carbon.

For our purposes we can classify the hydrocarbons broadly into three categories: (1) *aliphatic* hydrocarbons characterized by the absence of any rings in their structures; (2) *alicyclic* hydrocarbons, comprising all hydrocarbons with ring structures except for benzene and its derivatives; and (3) *aromatic* hydrocarbons, containing the six-member benzene rings. Aliphatic and alicyclic hydrocarbons, in turn, may be subdivided into *saturated* and *unsaturated* hydrocarbons. Saturated aliphatic hydrocarbons are called *alkanes* and they are saturated with hydrogen atoms; that is, they contain the maximum number of hydrogen atoms possible for a given carbon skeletal structure in only single bonds. On the other hand, unsaturated hydrocarbons contain double or triple carbon-to-carbon bonds: Those that contain one or more double bonds are referred to as *alkenes* or *olefins* and those that contain carbon-to-carbon triple bonds are called *alkynes* or *acetylenes*.

ALKANES The relative number of carbon and hydrogen atoms in the alkanes is expressed by the general formula C_nH_{2n+2}. Thus the simplest alkane ($n = 1$) has the formula CH_4 (methane), the second member of the series ($n = 2$) has the formula C_2H_6 (ethane), and so on. The carbon atoms in alkane molecules are joined by sp^3 bonds (see Chapter 4) which are equivalent, that is, identical in reactivity and stability (Fig. 9.4). In the formation of methane, each sp^3 orbital is thought to interpenetrate the s orbital of a hydrogen atom. That is to say, each hydrogen atom bonded to carbon involves an s orbital of hydrogen and an sp^3 hybrid orbital of carbon. Because the overlapping atomic orbitals give the greatest probability of finding the electrons along a line between the two nuclei, the newly formed C—H bonds are designated as *sigma bonds*. In an ethane molecule (C_2H_6), two carbon atoms are joined by a sigma bond formed by the overlap of an sp^3 orbital from each carbon atom; the six hydrogen atoms, three to each carbon, are also bonded to carbon by sigma bonds as in methane (Fig. 9.4). As a matter of fact, the carbon–carbon bonding and the carbon–hydrogen bonding in all alkanes are thought to involve sigma bonds of the same type shown for ethane.

The alkane hydrocarbons are found chiefly in natural gas and petroleum, both of which are thought to have originated from plant and animal sources. *Natural gas* is made up of about 50–94 percent methane, some ethane, propane, and butane. It is often encountered in drilling wells for petroleum (Fig. 9.5). *Petroleum* is a complex liquid solution of gaseous, liquid, and solid hydrocarbons with small amounts of nitrogen and sulfur compounds. About 10 million barrels of crude petroleum are produced each day in our country and refined by *fractional distillation* into batches with different boiling ranges (Table 9.4). In the process (Fig. 9.6), crude oil is pumped continuously through a furnace, where it is heated, and then is passed on to a "flash chamber" that vaporizes the lower-boiling components due to a lowering in pressure. The vapors enter a "bubble tower" where the higher-boiling components flow downward from plate to plate and are condensed at the bottom of the tower.

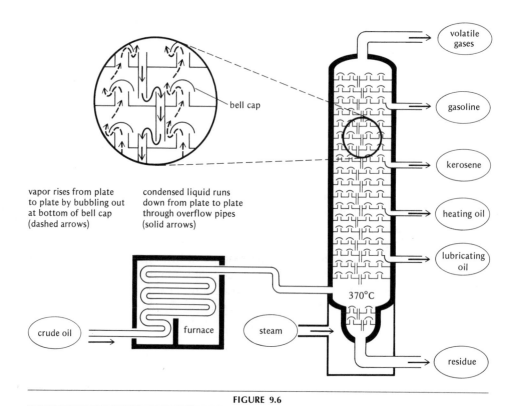

vapor rises from plate to plate by bubbling out at bottom of bell cap (dashed arrows)

condensed liquid runs down from plate to plate through overflow pipes (solid arrows)

FIGURE 9.6

Diagram of a fractionating tower for petroleum distillation. The cutaway view shows how the vapor and liquid phases are kept in contact with each other so that the condensation and distillation occur throughout the column. (Reproduced, by permission, from Charles W. Keenan and Jesse H. Wood, *General College Chemistry*, 4th ed., p. 596. Copyright © 1957 by Harper & Row, Inc.; copyright © 1961, 1966, 1971 by Charles William Keenan and Jesse Hermon Wood.)

The more volatile lower-boiling components ascend bubbling through the liquids that have condensed on the plates and collect in the upper portion of the tower. Distillation thus carried out causes the separation of petroleum into fractions that have rather definite composition and boiling-point range. However, each fraction still consists of a mixture of hydrocarbons. The low-distilling volatile fractions (petroleum ether, gasoline, and kerosene) consist of alkanes with 5 to 16 carbon atoms in their molecules. The crude oil remaining after the volatile hydrocarbons have been removed may be composed mainly of the higher alkanes or of alicyclic hydrocarbons which are isomeric with alkenes (see p. 215).

All the alkanes are generally nonreactive toward many common chemical reagents such as aqueous acids and bases, and aqueous solutions of oxidizing agents. However, they do readily burn in air,

$$CH_4 + 2O_2 \longrightarrow CO_2 + 2H_2O$$

and their combustion is of great technological importance (see Chapter 16). They also undergo photochemical reaction with halogens in the presence of sunlight:

$$\underset{\text{methane}}{CH_4} \xrightarrow{Cl_2} \underset{\substack{\text{methyl}\\\text{chloride}}}{CH_3Cl} \xrightarrow{Cl_2} \underset{\substack{\text{methylene}\\\text{chloride}}}{CH_2Cl_2} \xrightarrow{Cl_2} \underset{\text{chloroform}}{CHCl_3} \xrightarrow{Cl_2} \underset{\substack{\text{carbon}\\\text{tetrachloride}}}{CCl_4}$$

At one time chloroform was used as an anesthetic and carbon tetrachloride as a dry-cleaning agent and fire extinguisher. But both chloroform and carbon tetrachloride are highly toxic and their use has been banned for some time.

ALKENES The relative number of carbon and hydrogen atoms in the alkenes (or olefins) is expressed by the general formula C_nH_{2n}. The simplest alkene ($n = 2$) has the formula $H_2C\!=\!CH_2$ (ethene or ethylene), and the next member of the monoalkene (one double bond in the molecule) series has the formula $CH_3\!-\!CH\!=\!CH_2$ (propene or propylene). Isomers contingent upon the position of the double bond as well as upon the arrangement of the carbon skeleton are possible with higher alkenes.

In the formation of ethene, it is thought that one s and only two p orbitals hybridize to form three equivalent sp^2 bonding orbitals. (These orbitals extend toward the corners of an equilateral triangle having the carbon nucleus at its center. Two hydrogen atoms are then bonded to a carbon atom, each by a sigma bond resulting from the overlap of a hydrogen s orbital and a carbon sp^2 orbital. The two carbon atoms are bonded in part by the overlap of the remaining sp^2 orbitals of each.) The remaining p orbital of each carbon atom contains one electron and lies

TABLE 9.4
Hydrocarbon fractions from petroleum distillation

Fraction	Composition (range of C atoms)	Boiling point range (°C)	Uses
gas	C_1–C_5	<30	fuel, polymerized to gasoline
petroleum ether	C_5–C_7	30–90	solvent
gasoline	C_5–C_{12}	40–200	motor fuel
kerosene	C_{12}–C_{16}	175–275	fuel; illuminant
gas oil, fuel oil, diesel oil	C_{15}–C_{18}	250–400	fuel for furnaces and diesel engines; raw material for petrochemicals
lubricating oils, greases, Vaseline	>C_{16}	>350	lubricants
paraffin (wax)	>C_{20}	melts (50–55°C)	candles; matches; waterproofing
pitch and tar		residue	artificial asphalt
petroleum coke		residue	fuel; electrodes

above and below the plane of the sp^2 bond orbitals with the axis of its lobes perpendicular to this plane; the two axes are parallel to one another. Because of the nearness of the two carbon atoms, the p orbitals overlap laterally (along the edges rather than head-on) to form a second bond, called a pi bond, between the two carbon atoms. Thus, the carbon atoms in ethene can be thought of as being joined by a double bond consisting of a sigma bond and a pi bond (Fig. 9.7). Measurements of bond energies reveal that the energy of the double bond in ethene is less than twice that of the sigma C—C bond in ethane, indicating that the pi bond is a weaker bond than the sigma bond. The bonding system that we have discussed for ethene is thought to exist for one pair of carbon atoms in the molecules of all ethene homologs, the remaining atoms being bonded as in the alkanes.

The alkenes do not occur abundantly in nature. Both ethene and propene are gases at room temperature and are produced in large quantities by catalytic *cracking* processes in oil refineries:

$$C_2H_6 \longrightarrow C_2H_4 + H_2$$
$$C_3H_8 \longrightarrow C_3H_6 + H_2$$

The reaction is known as an *elimination reaction,* a reaction in which some molecule is always broken down into smaller units. Sizeable quantities of ethene and propene thus produced are used in the manufacture of polyethylene and polypropylene, both important commercial plastics (see Chapter 10).

Like all hydrocarbons, alkenes undergo combustion to give carbon dioxide and water:

$$C_2H_4 + 3O_2 \longrightarrow 2CO_2 + 2H_2O$$

The double bonds of alkenes are quite reactive toward a number of substances, the reaction occurring across the double bonds in so-called addition reactions:

$$H_2C{=}CH_2 + HCl \longrightarrow H_3C{-}CH_2Cl$$
ethyl chloride

$$H_2C{=}CH_2 + Br_2 \longrightarrow BrH_2C{-}CH_2Br$$
ethylene dibromide

$$H_2C{=}CH_2 + HOCl \longrightarrow H_2C{-}CH_2$$
$$\overset{|}{O}H \ \overset{|}{C}l$$
ethylene chlorohydrin

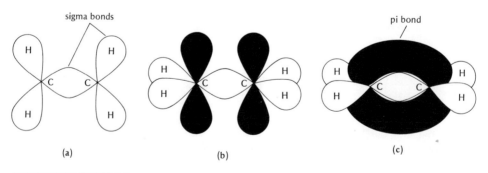

FIGURE 9.7

Schematic representation of an ethene (ethylene) molecule: (a) sigma bond between carbons and sigma bonds between hydrogen and carbon; (b) p orbitals not involved in sigma bond formation; and (c) a single pi bond, formed from p orbitals, has two regions of electron density.

Ethyl chloride is widely used as a local anesthetic, ethylene dibromide as an anti-knock fluid in gasolines, and ethylene chlorohydrin is an intermediate in the manufacture of a number of important industrial chemicals. The important substances produced from ethene are summarized in Fig. 9.8.

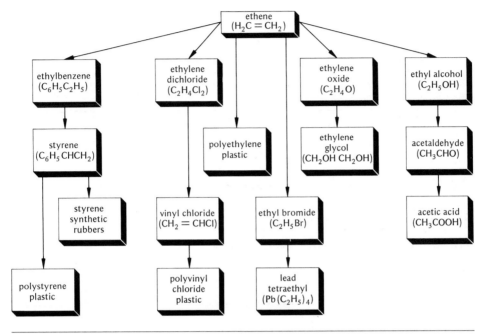

FIGURE 9.8

Industrial uses of ethene (ethylene).

ALKYNES   The relative number of carbon and hydrogen atoms in the alkynes (or acetylenes) is expressed by the general formula C_nH_{2n-2}. The simplest alkyne ($n = 2$) has the formula $HC{\equiv}CH$ (ethyne or acetylene). It is by far the most commercially used member of the series.

   In the formation of ethyne, it is thought that one s and one of the three p orbitals hybridize to form two equivalent sp bonding orbitals. (These orbitals have axes that lie at 180° to each other. The two carbon and hydrogen atoms are then bonded by sigma bonds; however, the sigma bond is only a part of the bonding system joining the two carbon atoms.) Because only one p orbital was used to form the sp hybrid, each carbon atom possesses two ordinary p orbitals, each orbital containing one electron. The carbon atoms are close enough to each other for the p orbitals to overlap laterally to form two pi bonds. Furthermore, the two pi bonds are thought to interact with each other to form an electron cloud that is cylindrically symmetrical around the line joining the two carbon nuclei (Fig. 9.9).

   Alkynes do not occur abundantly in nature. They are obtained in large amounts by elimination reactions using the alkanes as starting materials:

$$C_nH_{2n+2} \longrightarrow C_nH_{2n-2} + 2H_2$$

Ethyne is a gas and is produced by reacting calcium carbide with water:

$$CaC_2 + 2H_2O \longrightarrow C_2H_2 + Ca(OH)_2$$

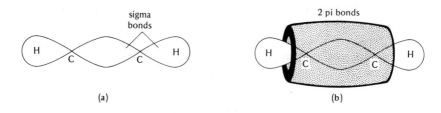

FIGURE 9.9

Schematic representation of an ethyne (acetylene) molecule: (a) sigma bonds between carbons and between hydrogen and carbon; (b) two pi bonds, formed by the lateral overlap of two $2p$ orbitals, forming a cylindrical electron cloud.

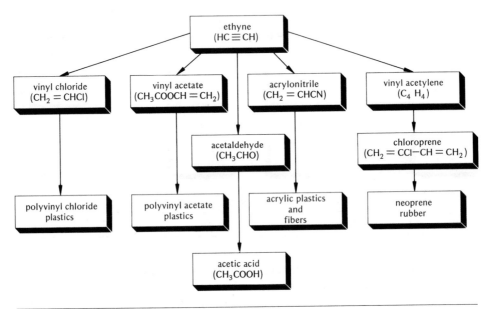

FIGURE 9.10

Industrial uses of ethyne (acetylene).

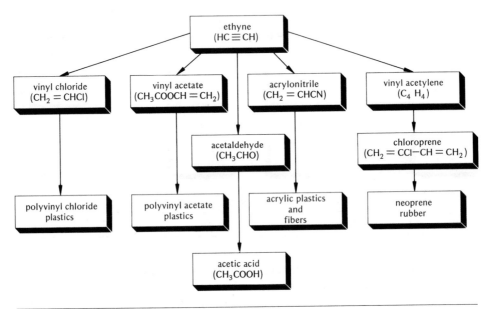

cyclopropane   cyclobutane   cyclohexane

FIGURE 9.11

Graphic formulas of representative alicyclic hydrocarbons.

The calcium carbide, in turn, is formed by heating lime and coke together in an electric furnace:

$$CaO + 3C \longrightarrow CaC_2 + CO$$

Ethyne burns in air with a luminous flame. When burned in oxygen it produces an intensely hot flame (oxyacetylene flame) which is used for welding and cutting steel. It undergoes addition reactions similar to those of the alkenes. The industrial uses of ethyne are summarized in Fig. 9.10.

ALICYCLICS Alicyclic hydrocarbons are saturated or unsaturated cyclic hydrocarbons containing carbon rings, but have most of the properties of the aliphatic chain compounds. They are isomeric with the alkenes but different in that electron pairs are utilized in building a ring rather than in forming the alkene double bonds. The graphic formulas shown in Fig. 9.11 represent some typical alicyclic hydrocarbons. Clearly, the bond angles between carbon–carbon bonds of the smaller rings, such as cyclopropane and cyclobutane, represent considerable deviations from the normal saturated carbon bond angle of 109°. Thus mechanical strains are built into the molecular structures of the small-ring compounds. These strains give the molecules some of the properties of unsaturated hydrocarbons. For example, cyclopropane adds hydrogen much as propene does:

The ring opening results in relief of the strain.

Six-membered rings and larger rings produce the most stable substances. Whereas the smaller rings have carbon atoms that lie in the same plane, *all* saturated ring compounds with six or more atoms in the ring skeleton exist as nonplanar structures. Therefore, the cyclohexane ring exists not as a flat structure, but actually in the puckered forms shown in Fig. 9.12.

Cyclopropane is a gas that is extensively used as a general anesthetic for major operations. Its popularity for this purpose is due in large measure to its relative lack of toxic side effects.

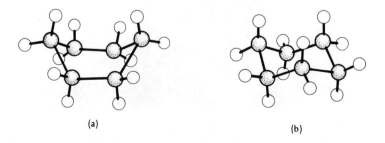

(a)

(b)

FIGURE 9.12

Puckered structures of cyclohexane: (a) boat form; and (b) chair form.

AROMATICS Aromatic hydrocarbons are cyclic hydrocarbons that usually contain very stable six-membered rings. There are many homologous series of aromatic hydrocarbons, the most important ones being derived from the simplest member, *benzene* (C_6H_6). Benzene was first isolated in 1825 from a whale oil by-product by the English chemist Michael Faraday. However, it was not until 1865 that a satisfactory structure for benzene and in fact the existence of rings of atoms was proposed. The German chemist Friedrich Kekulé, having dreamed that a snake seized its own tail forming a swirling ring (see opening quotation of this chapter), proposed the hexagonal ring structure for benzene:

This alternate double-bond arrangement accounts for the fourth valence of each carbon atom but is incorrect because all carbon-to-carbon bonds in the ring are really equivalent. The modern interpretation is to consider that each carbon atom uses sp^2 hybrid orbitals for sigma bonds with two neighboring carbon atoms and one hydrogen atom. The remaining electron of each carbon atom is in a *p* orbital and is used in pi-bond formation to give a multicenter pi-bonding (doughnut-shaped) system that contains six electrons and lies above and below the plane of the six-membered ring (Fig. 9.14). The modern representation of benzene ring bonding is

FIGURE 9.13

Friedrich Kekulé, German chemist of the late nineteenth century who postulated the existence of rings of atoms and the ring structure for benzene.

where the circle represents the pi-bonding six electrons. Each corner of the hexagon is assumed to be occupied by a carbon atom with a hydrogen atom attached. If some other atom or group is shown attached to a corner it is assumed that the carbon is still there but that the hydrogen has been replaced by the other unit.

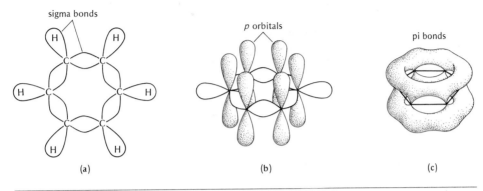

sigma bonds

p orbitals

pi bonds

(a)   (b)   (c)

FIGURE 9.14

Schematic representation of a benzene molecule. (a) Sigma bonds involving sp^2 and s orbitals; (b) p orbitals (shaded) of carbon not involved in sigma-bond formation; (c) the overlap of the p orbitals, forming the pi electron clouds above and below the benzene ring.

Benzene is a classic example of a compound forming a resonance hybrid (see Chapter 4). The degree to which the molecule is stabilized by resonance is such that it is very difficult to rupture the benzene ring with most chemical reagents. The pi bonds in benzene do not add atoms or groups as readily as do the pi bonds in chain compounds. The most important class of reactions involving benzene is substitution in which one or more hydrogen atoms on the ring are replaced by other atoms or functional groups without rupturing the ring structure:

$$\bigcirc + Cl_2 \longrightarrow \bigcirc^{Cl} + HCl$$

chlorobenzene

Aromatic substitution reactions are influenced by the fact that the benzene ring is electron rich (nucleophilic) through its pi-electron orbitals and tends to combine with electron-poor (electrophilic) structures. More than one functional group may be substituted onto a benzene ring:

salicylic acid   aspirin   trinitrotoluene (TNT)

It is also possible to transform one type of aromatic functional group into another without affecting the benzene ring:

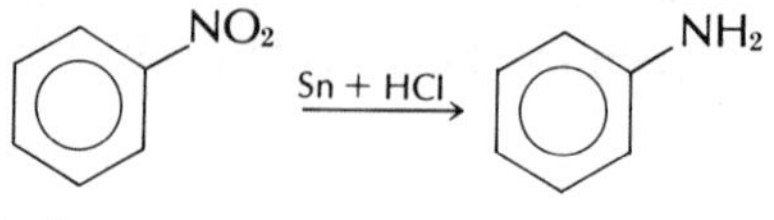

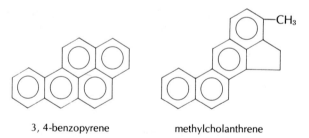

nitrobenzene   aniline

Aniline is an essential starting material for the manufacture of a great number of commercially useful products (see p. 234).

Other aromatic hydrocarbons have chains substituted on the benzene ring:

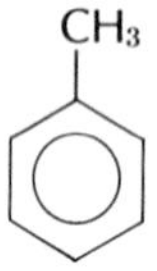

toluene   o-xylene   ethylbenzene   styrene
(C_7H_8) (C_8H_{10}) (C_8H_{10}) (C_8H_8)

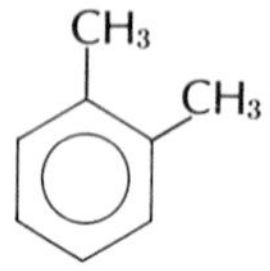

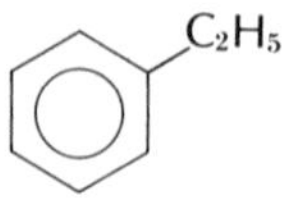

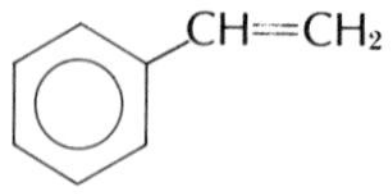

Still others consist of several fused rings having carbon atoms in common:

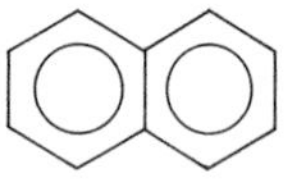

naphthalene   anthracene
$(C_{10}H_8)$ $(C_{14}H_{10})$

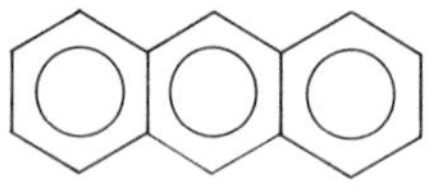

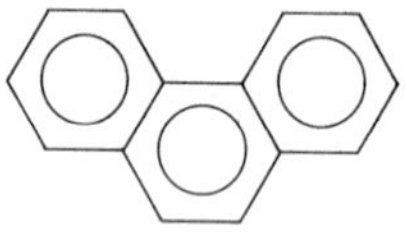

phenanthrene
$(C_{14}H_{10})$

A number of the higher fused-ring hydrocarbons, such as 3,4-benzopyrene and methylcholanthrene,

3, 4-benzopyrene   methylcholanthrene

are thought to be the *carcinogenic* (cancer-causing) agents associated with heavy cigarette smoking.

Coal has long been an indirect source of aromatic compounds. The *destructive distillation* of coal, that is, the heating of coal in a closed vessel in the absence of air (Fig. 9.15), is used in industry for production of coke, a fuel used in the iron and steel industry. It leaves a black viscous liquid called *coal tar*. This tar, on fractional distillation, yields aromatic compounds such as benzene, toluene, naphthalene, and phenol. However, our increased demand for benzene (about 9 billion lb used annually by industries that manufacture the dyes, drugs, plastics, synthetic fibers and rubbers, insecticides, and detergents we need in everyday life) has necessitated its production from petroleum products. In the latter case, the gasoline fraction from petroleum refining is used. Huge distilling columns are used to separate alicyclic hydrocarbons such as cyclohexane, methylcyclopentane, and methylcyclohexane from the gasoline. These cyclic compounds are then fed into the so-called·reform towers in which they are catalytically converted at moderately high temperatures and pressures into benzene and toluene with the liberation of hydrogen:

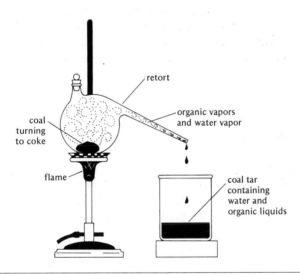

cyclohexane   benzene

This type of chemical change is known as *aromatization*. Huge quantities of benzene are thus manufactured.

FIGURE 9.15

Destructive distillation of coal. (Redrawn, by permission, from Mark M. Jones, John T. Netterville, David O. Johnston, and James L. Wood, *Chemistry, Man and Society,* p. 295. Copyright © 1972 by W. B. Saunders Company.)

QUESTIONS

9.7   What are catenanes?

9.8   How can you distinguish alkanes, alkenes, and alkynes from each other?

9.9   Discuss the bonding differences apparent in the different types of hydrocarbon molecules.

9.10   What is meant by fractional distillation? Explain by using petroleum distillation as an example.

9.11   What is meant by an elimination reaction? Give two examples.

9.12   How do pi bonds differ from sigma bonds?

9.13   How is acetylene manufactured? Name some important uses of acetylene.

9.14   List the differences between alicyclic and aromatic hydrocarbons.
9.15   Name five important derivatives of benzene and describe their uses.
9.16   What is meant by aromatization? How is it effected?

SUGGESTED PROJECT 9.1

Petroleum refining and gasoline manufacture

The importance of petroleum and petrochemicals in our society is a matter of general knowledge. However, many of us are not aware of the processes used in the separation of crude oil into fractions that have properties which render them suitable for our use.

If there is an oil refinery in your neighborhood you should visit the refinery installations and laboratories to gain first-hand knowledge of the manufacture of gasoline and other petroleum products. (Note that your visit to the refinery is best arranged through your school.) Prepare a report of your visit, emphasizing the chemistry involved in the manufacture of gasoline. Design an apparatus using crude oil that can be fractionated on a smaller scale in a chemical laboratory.

ORGANIC HALOGEN COMPOUNDS

The compound formed when one hydrogen atom from a saturated hydrocarbon is replaced by a halogen is known as an *alkyl halide*. To name a few typical alkyl halides we may mention methyl chloride (CH_3Cl), methyl bromide (CH_3Br), and ethyl chloride (C_2H_5Cl). They are made by direct substitution of halogen into an alkane in the presence of ultraviolet light. The reaction is hard to control and tends to give a mixture of products. Because the alkyl halides are very active, they are used for the synthesis of many other compounds. Methyl bromide is used as a fumigant, ethyl chloride as a local anesthetic for minor surgery, and the higher alkyl halides are useful as good solvents.

Many polyhalogen compounds of the hydrocarbons are equally interesting and useful. A group of compounds, known as *Freons*, contain fluorine as well as chlorine:

$$
\underset{\text{Freon-11}}{\overset{\displaystyle Cl}{\underset{\displaystyle Cl}{Cl-C-F}}}
\qquad
\underset{\text{Freon-12}}{\overset{\displaystyle Cl}{\underset{\displaystyle F}{Cl-C-F}}}
\qquad
\underset{\text{Freon-22}}{\overset{\displaystyle Cl}{\underset{\displaystyle F}{H-C-F}}}
\qquad
\underset{\text{Freon-114}}{\overset{\displaystyle F\ \ \ F}{\underset{\displaystyle Cl\ \ Cl}{F-C-C-F}}}
$$

They are used as refrigerants and also as the dispersing gases in aerosol cans. Ethylene dibromide ($C_2H_4Br_2$) is used as a constituent of the antiknock additive for motor gasoline. (It produces volatile lead compounds that prevent engine deposits from tetraethyllead, $Pb(C_2H_5)_4$, in the fuel; however, the vapors thus produced contribute significantly to air pollution.) The analogous chlorine compound, ethylene dichloride ($C_2H_4Cl_2$) is an excellent solvent for some plastics and an ingredient in the manufacture of synthetic rubber. Tetrafluoroethene (C_2F_4) on polymerization yields Teflon, one of the newer plastics widely used because of its high stability and excellent resistance to chemical attack (see Chapter 5).

Aryl (aromatic) halides are equally important. Chlorobenzene (C_6H_5Cl) and its relatives are widely used in the synthesis of other organic compounds. The polychlorinated hydrocarbon dichlorodiphenyltrichloroethane, known as DDT,

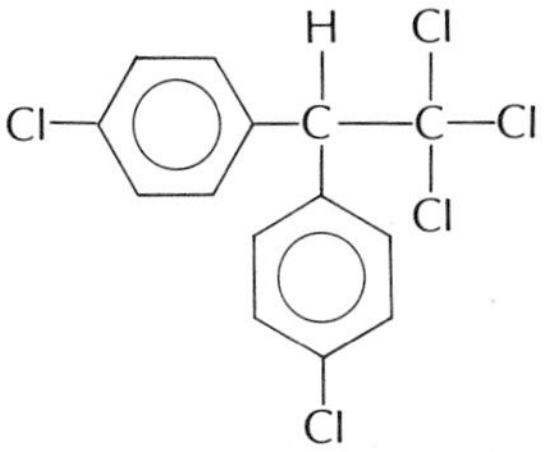

DDT

first saw extensive application as a *delouser* among the troops in World War II. It was also used to kill typhus-bearing lice and malaria-bearing mosquitoes, often being sprayed from airplanes. We shall examine the use of DDT and other chlorinated hydrocarbons as insecticides later in Chapter 12.

QUESTIONS

9.17 What are alkyl halides? Give examples.
9.18 Name some commercially important alkyl halides and their applications.
9.19 What is Teflon? How is it manufactured?
9.20 Write the formula of DDT. List the various functional groups present in its molecule.

COMPOUNDS OF CARBON, HYDROGEN, AND OXYGEN

Alcohols and phenols (R—OH), ethers (R—O—R), aldehydes (R—$\overset{\overset{\textstyle O}{\|}}{C}$—H), ketones (R—$\overset{\overset{\textstyle O}{\|}}{C}$—R), carboxylic acid (R—$\overset{\overset{\textstyle O}{\|}}{C}$—OH), and esters (R—$\overset{\overset{\textstyle O}{\|}}{C}$—OR′) are the important classes of organic compounds containing carbon, hydrogen, and oxygen only. Just as there are aliphatic, alicyclic, and aromatic series of hydrocarbons, there are also alcohols, ethers, aldehydes, ketones, carboxylic acids, and esters within each of these series.

ALCOHOLS, PHENOLS, AND ETHERS **Alcohols** The alcohols, like the alkanes and other families of compounds, compose a homologous series that varies more-or-less regularly in the properties of its compounds. Alcohols have the generic formula R—OH. Methanol (CH_3OH) is an example of an aliphatic alcohol, cyclohexanol ($C_6H_{11}OH$) an example of an alicyclic alcohol, and benzyl alcohol (C_7H_7OH) an example of an aromatic alcohol:

CH_3—OH		
methanol	cyclohexanol	benzyl alcohol
(aliphatic)	(alicyclic)	(aromatic)

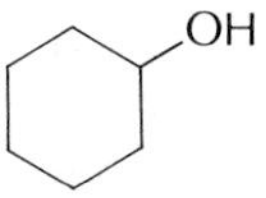

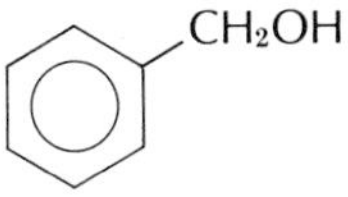

Methanol was originally known as *wood alcohol* because it was prepared by the destructive distillation of wood, that is, by heating wood chips in a closed vessel in the absence of air. Today, most of it is manufactured by a synthetic process that involves reacting carbon monoxide with hydrogen at 350°C in the presence of a mixed metal oxide catalyst:

$$CO + 2H_2 \xrightarrow[350°C]{ZnO-Cr_2O_3} CH_3OH$$

Nearly 4 billion lb of methanol are produced each year in the United States; over one-half of it is used in the production of the formaldehyde used in the manufacture of plastics (see Chapter 10). Smaller amounts are used in the production of other chemicals, jet fuels, and antifreeze mixtures as well as a denaturant (a poison added to ethanol to make it unfit for beverages). Methanol is quite poisonous and causes blindness in less than lethal doses. Many deaths and injuries resulted from methanol present in bootleg liquor.

Ethanol (C_2H_5OH) was originally called *grain alcohol* because it was prepared by the distillation of fermented sugar made from grains such as corn and barley:

$$(C_6H_{10}O_5)_n + nH_2O \longrightarrow nC_6H_{12}O_6$$
grain starch

$$2C_6H_{12}O_6 \xrightarrow{yeast} 4C_2H_5OH + 4CO_2$$

Fermentation takes place through certain enzymes (biological catalysts) that are found in yeast or mold. A solution of 95 percent ethanol and 5 percent water can be recovered by fractional distillation of the fermented liquor. Most ethanol used as industrial solvent and starting material for other chemical products is produced by reacting ethene first with sulfuric acid and then the product with water:

$$C_2H_4 + H_2SO_4 \longrightarrow C_2H_5OSO_3H$$
$$C_2H_5OSO_3H + H_2O \longrightarrow C_2H_5OH + H_2SO_4$$

Synthetic ethanol thus prepared is denatured (poisoned) by additives such as benzene and methanol to render it unsuitable for use in beverages.

Alcoholic beverages are usually prepared by the fermentation of grain or other starchy or sugary materials. The fermentation continues until there is about 12 percent ethanol by volume in the mixture. A higher alcohol content in beverages is achieved by distillation of the fermented mixture. Whiskey, gin, and brandy are liquors produced by distillation followed by aging or addition of other substances to provide the final product. The alcohol content of liquors is indicated by the term *proof* which has a numerical value twice that of the percentage by volume of alcohol present. Most whiskies are marked 80–100 proof, meaning thereby that they contain 40–50 percent alcohol by volume.

Another alcohol familiar to most people is *isopropyl alcohol* ($CH_3{-}\overset{\displaystyle CH_3}{\underset{|}{C}}HOH$). It is prepared by the hydration of propylene (C_3H_6) and is sold as *rubbing alcohol.*

When more than one hydroxyl group is present in a molecule the compounds are called *polyhydric* alcohols. Examples of such alcohols are

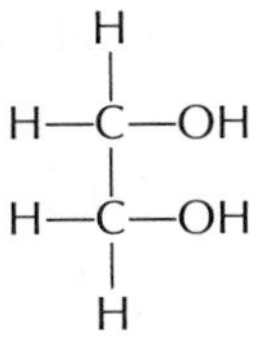

$$
\begin{array}{cc}
\text{H} & \text{H} \\
| & | \\
\text{H}-\text{C}-\text{OH} & \text{H}-\text{C}-\text{OH} \\
| & | \\
\text{H}-\text{C}-\text{OH} & \text{H}-\text{C}-\text{OH} \\
| & | \\
\text{H} & \text{H}-\text{C}-\text{OH} \\
& | \\
& \text{H}
\end{array}
$$

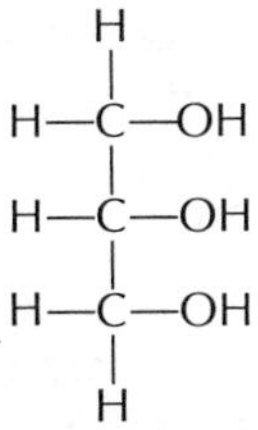

ethylene glycol   glycerol (glycerin)
(*di*hydric alcohol)   (*tri*hydric alcohol)

Glycols are the most-used radiator antifreeze and are sold under trade names such as Prestone and Zerex. Glycerol is a by-product of the soap industry (see p. 230) and has many uses in the manufacture of drugs and cosmetics, in the production of nitroglycerin, plastics, and numerous other substances.

Many compounds of biological interest have the alcohol functional group. Some vitamins, sugars, birth control agents, and cholesterol are alcohols. We shall discuss them in Chapter 11.

Phenols   When the hydroxyl group characteristic of alcohols is attached directly to a benzene ring, as, for example, in the compound C_6H_5OH, the properties of the compound are markedly different from the properties of alcohols. Such compounds are called *phenols*. The simplest phenol is a much more acidic compound than alcohols and is sometimes called *carbolic acid*. It is produced on a commercial scale by heating chlorobenzene at high pressures with an aqueous solution of sodium hydroxide:

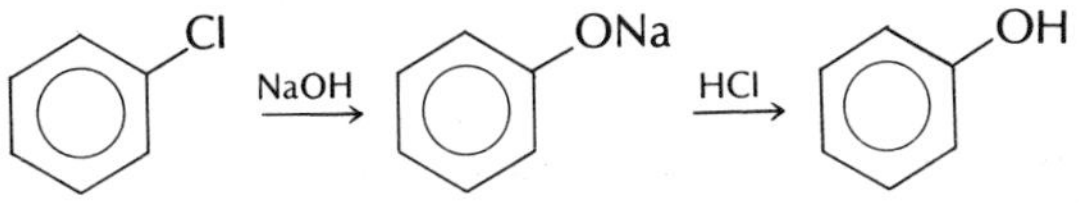

Phenol is used as an antiseptic and as a starting material in the manufacture of dyes and plastics.

The cresols (methylphenols) are excellent antiseptics and are used in consumer products such as Lysol. There are three isomeric cresols:

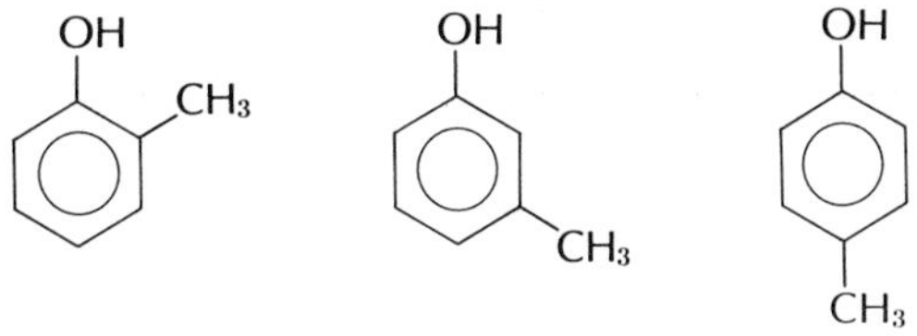

ortho-methylphenol *meta*-methylphenol *para*-methylphenol
   (*o*-cresol)   (*m*-cresol)   (*p*-cresol)

Because the phenolic hydroxy group is the principal functional group, the cresols are named on the basis of the position of the CH_3 group with respect to the OH group. In general, doubly substituted aromatic compounds are named according to the following rules: If the second substituent is adjacent to the principal one, the prefix *ortho*- is used; if it is one carbon atom removed from the first the prefix *meta*- is used; and, if it is opposite the first (two carbon atoms away), the prefix *para*- is used.

Ethers   Ethers have the general formula R—O—R and can be prepared by careful elimination of water from two molecules of alcohol, as indicated by the enclosing dashes in the equation

$$CH_3\overline{[OH + H]}OCH_3 \longrightarrow CH_3OCH_3$$

dimethyl ether

Diethyl ether $[(C_2H_5)_2O]$ or simply ether is the most familiar of all ethers and is manufactured by heating ethanol with sulfuric acid at 140°C:

$$2C_2H_5OH \xrightarrow[140°C]{H_2SO_4} C_2H_5OC_2H_5 + H_2O$$

Diethyl ether has few toxic effects and, therefore, it is widely used as an anesthetic and solvent. Although ethers are less reactive than the alcohols, they have a tendency to react with atmospheric oxygen to form the highly explosive *organic peroxides:*

$$CH_3CH_2OCH_2CH_3 + O_2 \longrightarrow CH_3\underset{\underset{OOH}{|}}{C}HOCH_2CH_3$$

Formation of such peroxides is usually inhibited by placing an iron nail in bottles containing ethers. Commercial ether contains minute amounts of organic chemical inhibitor.

Typical aromatic ethers are diphenyl ether and methyl phenyl ether (anisole):

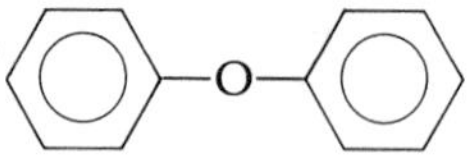

diphenyl ether

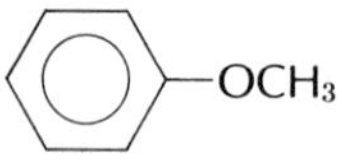

methyl phenyl ether
(anisole)

Both are used in the manufacture of synthetic perfumes.

ALDEHYDES AND KETONES   Both aldehydes and ketones are characterized by the presence of a carbonyl group ($>C=O$) in their molecules. In aldehydes, one or both of the groups attached to the carbonyl is a hydrogen atom,

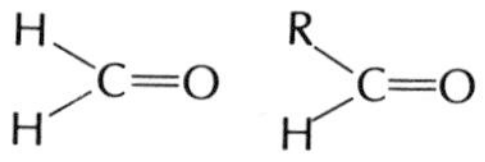

whereas in ketones, both of the groups are hydrocarbon residues

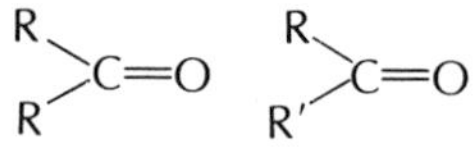

The carbonyl group of aldehydes and ketones, like the carbon–carbon double bond of alkenes, is capable of undergoing addition reactions:

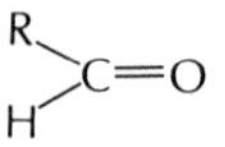

$$\underset{H}{\overset{R}{>}}C=O + H_2 \xrightarrow{Pt} RCH_2OH$$

alcohol

Further, aldehydes are very easily oxidized to acids having the same number of carbon atoms:

$$2 \underset{H}{\overset{R}{>}}C{=}O + O_2 \longrightarrow 2R{-}\overset{\overset{\displaystyle O}{\|}}{C}{-}OH$$

Ketones are not oxidized as easily as the aldehydes. If they are oxidized, they form acids with fewer carbon atoms than the ketones. Oxidation of a ketone involves breaking of carbon-to-carbon bonds, which are stronger than the carbon-to-hydrogen bonds broken in aldehyde oxidation. This difference has been used to distinguish between aldehydes and ketones.

Formaldehyde $\left(H{-}\overset{\overset{\displaystyle H}{|}}{C}{=}O \right)$, the simplest of the aldehyde homologous series, is manufactured by the air oxidation of methanol at 250°C in the presence of a mixed metal oxide catalyst:

$$2CH_3OH + O_2 \xrightarrow[250°C]{Fe_2O_3 + MoO_3} 2H{-}\overset{\overset{\displaystyle H}{|}}{C}{=}O + 2H_2O$$

Pure formaldehyde is a gas with a penetrating odor and is often handled as a 40 percent solution in water, called *formalin*. It is used as a disinfectant, a preservative of biological specimens, and in the manufacture of the plastic known by the trade name Bakelite.

Acetaldehyde $\left(CH_3{-}\overset{\overset{\displaystyle H}{|}}{C}{=}O \right)$, can be made cheaply from acetylene and water:

$$C_2H_2 + H_2O \xrightarrow[HgSO_4]{H_2SO_4} CH_3{-}\overset{\overset{\displaystyle H}{|}}{C}{=}O$$

However, it is manufactured commercially by the air oxidation of ethanol in the presence of a metal catalyst:

$$C_2H_5OH + O_2 \xrightarrow{Ag} CH_3{-}\overset{\overset{\displaystyle H}{|}}{C}{=}O + H_2O$$

A liquid, acetaldehyde is used largely as an intermediate for the synthesis of other organic compounds. One of these is *chloral hydrate*, commonly known as "knock-out drops" (because they are sometimes used to render people unconscious). It is made by first reacting acetaldehyde with chlorine to form chloral and then adding water to produce the hydrate:

$$CH_3{-}\overset{\overset{\displaystyle H}{|}}{C}{=}O \xrightarrow{Cl_2} \underset{\text{chloral}}{CCl_3{-}\overset{\overset{\displaystyle H}{|}}{C}{=}O} \xrightarrow{H_2O} \underset{\text{chloral hydrate}}{CCl_3CH(OH)_2}$$

Chloral hydrate is one of the few known stable organic compounds containing two hydroxyl groups on the same carbon.

A typical aromatic aldehyde is *benzaldehyde* $\left(C_6H_5{-}\overset{\overset{\displaystyle H}{|}}{C}{=}O \right)$. It is more popularly known as *oil of bitter almonds*. A substituted benzaldehyde,

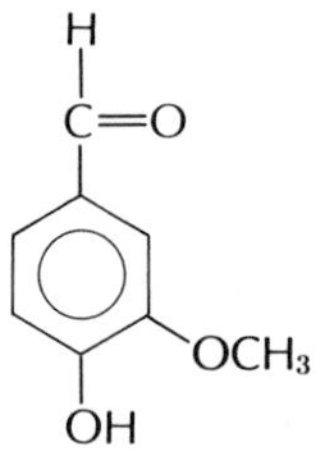

$$
\underset{\substack{\text{(benzene ring with)} \\ \text{OCH}_3 \\ \text{OH}}}{\overset{\overset{\text{H}}{\underset{}{\mid}}}{\underset{}{\overset{\text{C}=\text{O}}{}}}}
$$

called *vanillin,* is used in vanilla flavorings.

Acetone, or dimethyl ketone $\left(\text{CH}_3-\overset{\overset{\text{O}}{\|}}{\text{C}}-\text{CH}_3\right)$, is the most important member of the ketone group. It was at first produced by the *pyrolysis* (heating) of calcium acetate:

$$
\text{Ca}(\text{C}_2\text{H}_3\text{O}_2)_2 \xrightarrow{\text{heat}} \text{CH}_3-\overset{\overset{\text{O}}{\|}}{\text{C}}-\text{CH}_3 + \text{CaCO}_3
$$

Acetone is manufactured by the oxidation of isopropanol and is used extensively as a solvent and in the manufacture of the plastic Lucite. Other aliphatic ketones find application in the production of dyes, explosives, and special types of plastics. Aromatic ketones, such as benzophenone and acetophenone,

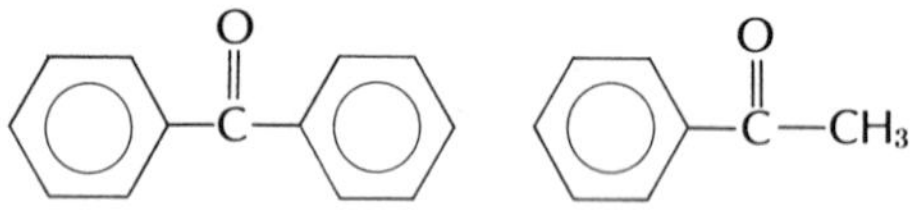

| diphenyl ketone | phenyl methyl ketone |
| (benzophenone) | (acetophenone) |

are used for making synthetic perfumes. Acetophenone is also a hypnotic (sleep inducer) and is sold under the trade name Hypnone.

CARBOXYLIC ACIDS AND THEIR DERIVATIVES Organic compounds that contain the functional group $-\overset{\overset{\text{O}}{\|}}{\text{C}}-\text{OH}$ are known as *carboxylic acids.* They are acids because of their tendency to release protons to other compounds, such as water, that function as bases:

$$
\text{R}-\overset{\overset{\text{O}}{\|}}{\text{C}}-\text{OH} + \text{H}_2\text{O} \rightleftharpoons \text{R}-\overset{\overset{\text{O}}{\|}}{\text{C}}-\text{O}^- + \text{H}_3\text{O}^+
$$

However, they are very weak acids compared to the mineral acids HCl and HNO$_3$. Nevertheless, they react with bases to form salts:

$$
\text{R}-\overset{\overset{\text{O}}{\|}}{\text{C}}-\text{OH} + \text{NaOH} \longrightarrow \text{R}-\overset{\overset{\text{O}}{\|}}{\text{C}}-\text{ONa} + \text{H}_2\text{O}
$$

With alcohol they form an important class of compounds called *esters:*

$$
\text{R}-\overset{\overset{\text{O}}{\|}}{\text{C}}-\text{OH} + \text{R}'\text{ OH} \rightleftharpoons \underset{\text{ester}}{\text{R}-\overset{\overset{\text{O}}{\|}}{\text{C}}-\text{OR}'} + \text{H}_2\text{O}
$$

When dehydrated, carboxylic acids form *anhydrides:*

$$R-\overset{\overset{\displaystyle O}{\|}}{C}-OH + R-\overset{\overset{\displaystyle O}{\|}}{C}-OH \longrightarrow R-\overset{\overset{\displaystyle O}{\|}}{C}-O-\overset{\overset{\displaystyle O}{\|}}{C}-R + H_2O$$

acid anhydride

Acid anhydrides are reactive toward water, alcohols, and ammonia. With ammonia they give a class of compounds called *amides:*

$$R-\overset{\overset{\displaystyle O}{\|}}{C}-O-\overset{\overset{\displaystyle O}{\|}}{C}-R + NH_3 \longrightarrow R-\overset{\overset{\displaystyle O}{\|}}{C}-NH_2 + R-\overset{\overset{\displaystyle O}{\|}}{C}-OH$$

amide

Carboxylic acids Many carboxylic acids are isolated from natural sources.

TABLE 9.5

Some important aliphatic carboxylic acids

Name	Formula	Occurrence
formic acid	HCOOH	ants, bees, and insects
acetic acid	CH_3COOH	fermented fruit juice
butyric acid	$CH_3(CH_2)_2COOH$	butter fat
palmitic acid	$CH_3(CH_2)_{14}COOH$	animal and vegetable fats
stearic acid	$CH_3(CH_2)_{16}COOH$	animal and vegetable fats
oleic acid	$CH_3(CH_2)_7CH{=}CH(CH_2)_7COOH$	olive oil
linoleic acid	$CH_3(CH_2)_4CH{=}CHCH_2CH{=}CH(CH_2)_7COOH$	linseed oil
oxalic acid	$(COOH)_2$	rhubarb leaves
lactic acid	$CH_3CH(OH)COOH$	sour milk
tartaric acid	$HOOCCH(OH)CH(OH)COOH$	fermented fruit juice
citric acid	$HOOCCH_2C(OH)(COOH)CH_2COOH$	citrus fruits

Table 9.5 lists the more important aliphatic carboxylic acids. A careful examination of the acids listed shows that some of them are complicated molecules. Some are saturated acids, some are unsaturated, some have more than one carboxylic group, and others have besides carboxylic groups other groups (such as OH, NH_2, and so on) present as well. The acids in the last category exhibit the properties of each of the functional groups present.

Formic acid $\left(H-\overset{\overset{\displaystyle O}{\|}}{C}-OH \right)$, the simplest of the carboxylic acids, was first prepared by the destructive distillation of ants (*formica* in Latin). In fact, the swelling caused by the bites of many kinds of insects is due to formic acid in the sting venom. Formic acid is commercially prepared by reacting carbon monoxide with sodium hydroxide at 200°C, followed by HCl:

$$NaOH + CO \xrightarrow{200°C} H-\overset{\overset{\displaystyle O}{\|}}{C}-ONa \xrightarrow{HCl} H-\overset{\overset{\displaystyle O}{\|}}{C}-OH + NaCl$$

It is extensively used for dyeing in the leather and textile industries and as a coagulant for rubber latex.

From a commercial standpoint, *acetic acid* $\left(CH_3-\overset{\overset{O}{\|}}{C}-OH\right)$ is the most important member of the carboxylic acid family. It is the principal active ingredient (about 4–10%) of *vinegar* and is produced by aerobic fermentation, that is, fermentation in the presence of air, of grape juice or apple cider. Besides its use as vinegar, acetic acid is a raw material in the manufacture of some textile fibers, paint pigments, dyes, and even drugs.

Another carboxylic acid, called *butyric acid* $\left(C_3H_7-\overset{\overset{O}{\|}}{C}-OH\right)$, can be isolated from butter fat. It is one of the ingredients of "body odor" and has the smell of rancid butter. It is this smell (and perhaps other ones as well) that enables the bloodhound to track humans. Substituted butyric acid, *γ-aminobutyric acid* $\left(H_2NCH_2CH_2CH_2\overset{\overset{O}{\|}}{C}-OH\right)$, is.a powerful depressant.

Some organic acids are obtained from fats or oils. Examples of such fatty acids are *palmitic acid* $\left(C_{15}H_{31}\overset{\overset{O}{\|}}{C}-OH\right)$ and *stearic acid* $\left(C_{17}H_{35}\overset{\overset{O}{\|}}{C}-OH\right)$. Among other aliphatic acids, mention may be made of the unsaturated *acrylic acid* $\left(CH_2\!=\!CH\overset{\overset{O}{\|}}{C}-OH\right)$, which is the raw material for the plastic industry, and of the hydroxytricarboxylic *citric acid*, which is a constituent of many soft drinks.

The aromatic carboxylic acids behave much like the aliphatic ones discussed above. *Benzoic acid* $\left(C_6H_5\overset{\overset{O}{\|}}{C}-OH\right)$ is the simplest of the aromatic acids. It was first obtained from the distillation of gum benzoin, an aromatic resin produced in Asia. It is now produced by the oxidation of toluene with atmospheric oxygen in the presence of catalysts:

$$2C_6H_5CH_3 + 3O_2 \longrightarrow 2C_6H_5\overset{\overset{O}{\|}}{C}-OH + 2H_2O$$

Its sodium salt, *sodium benzoate* $\left(C_6H_5\overset{\overset{O}{\|}}{C}-ONa\right)$ is frequently added to foods as a preservative. *p-Aminobenzoic acid* is one of the B-group vitamins, and a derivative is the local anesthetic *novocaine*:

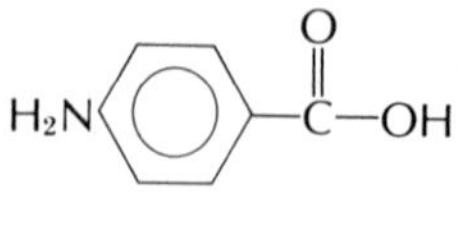

p-aminobenzoic acid 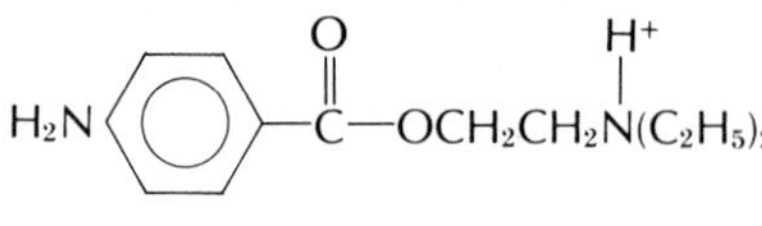   novocaine (procaine hydrochloride)

Among other aromatic acids, mention may be made of the dicarboxylic *phthalic acid*, which eliminates water easily to form *phthalic anhydride*:

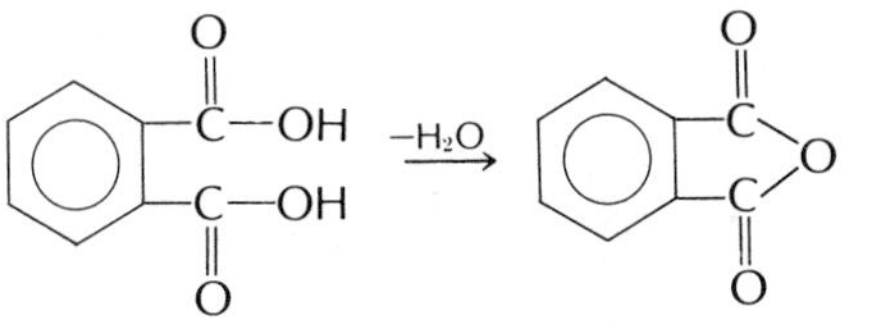

phthalic acid   phthalic anhydride

Phthalic anhydride is used in the synthesis of dyes and an important group of paint and varnish resins called the *glyptals*. *Benzene sulfonic acid* ($C_6H_5SO_3H$) is used in making dyes. If a carboxylic group is also present the compound is called *sulfobenzoic acid*. The well-known sweetening agent *saccharin* is a derivative of o-sulfobenzoic acid.

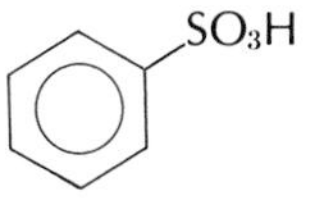

benzene   o-sulfobenzoic   saccharin
sulfonic acid   acid

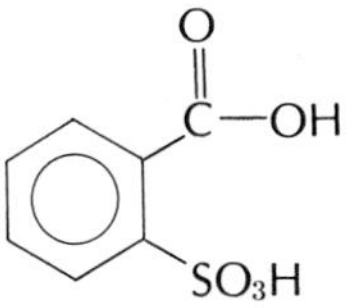

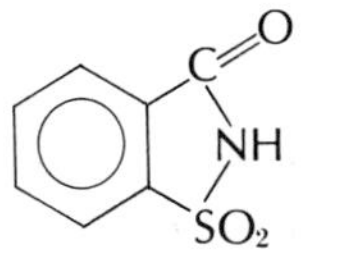

Similarly, the well-known household remedy *aspirin* is a derivative of o-hydroxybenzoic acid (salicylic acid):

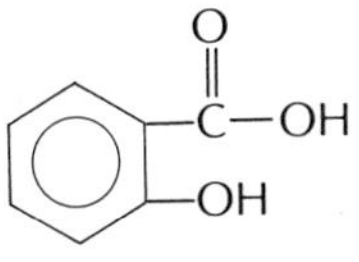

salicylic acid

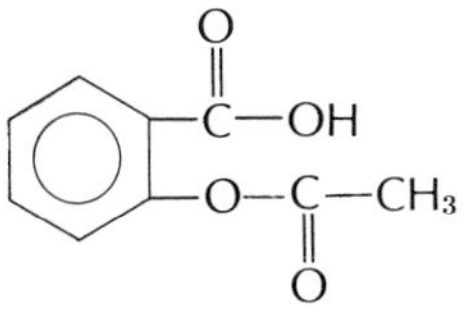

aspirin

Some important classes of carboxylic acid derivatives are the esters and the amides.

Esters   Esters have the general formula R—$\overset{\overset{\text{O}}{\|}}{\text{C}}$—OR' and are prepared by reacting a carboxylic acid with an alcohol. For example, the ester methyl butyrate is made from butyric acid and methanol:

$$C_3H_7\overset{\overset{\text{O}}{\|}}{\text{C}}\text{—OH} + CH_3OH \longrightarrow C_3H_7\overset{\overset{\text{O}}{\|}}{\text{C}}\text{—OCH}_3 + H_2O$$

Unlike the foul rancid odor of butyric acid, the ester methyl butyrate has a pleasant fruity aroma — what a difference a little change in the molecular architecture makes! In fact, many low molecular weight esters have pleasant fruity odors and are often responsible for the odors of ripe fruit, flowers, and other plant substances. Therefore, they are used in flavoring extracts, perfumes, medicines, and candies. Synthetic esters that are commonly used as food flavors are amyl acetate for banana, octyl acetate for orange, ethyl butyrate for pineapple, amyl butyrate for apricot, isobutyl formate for raspberry, and ethyl formate for rum. However, the esters used for the manufacture of perfumes are usually esters of aromatic acids.

In contrast, higher molecular weight esters often have a distinctly unpleasant odor. Common examples are the animal fats and vegetable oils sometimes called *lipids*. The term *fat* is usually reserved for solids (butter, lard, tallow, and so forth) and *oil* for liquids (castor, olive, linseed, and so forth). Both fats and oils are esters of the trihydroxy alcohol glycerol and have the formula

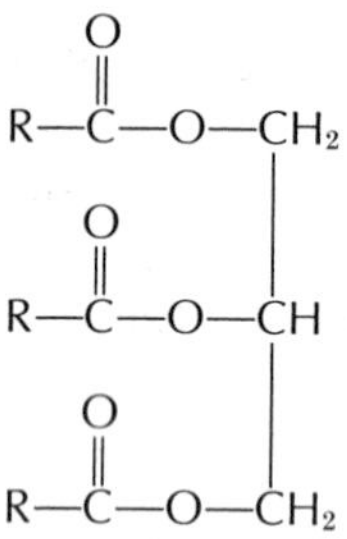

$$
\begin{array}{c}
\text{O} \\
\| \\
\text{R—C—O—CH}_2 \\
\\
\text{O} \\
\| \\
\text{R—C—O—CH} \\
\\
\text{O} \\
\| \\
\text{R—C—O—CH}_2
\end{array}
$$

In many cases the R groups are long hydrocarbon chains with 11, 13, 15, or 17 carbon atoms; they may be saturated or unsaturated. These esters are hydrolyzed by the action of basic substances such as NaOH to produce salts that are soaps. The reaction is often referred to as *saponification:*

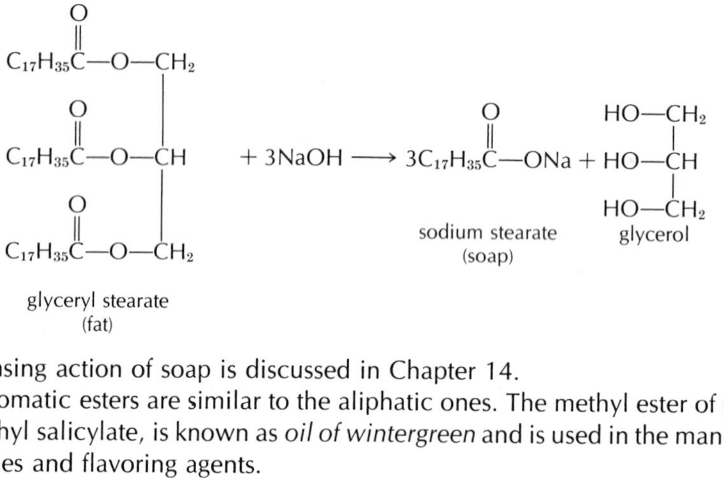

$$
\begin{array}{c}
\text{O} \\
\| \\
\text{C}_{17}\text{H}_{35}\text{C—O—CH}_2 \\
\\
\text{O} \\
\| \\
\text{C}_{17}\text{H}_{35}\text{C—O—CH} \quad + 3\text{NaOH} \longrightarrow 3\text{C}_{17}\text{H}_{35}\overset{\text{O}}{\overset{\|}{\text{C}}}\text{—ONa} + \begin{array}{c}\text{HO—CH}_2 \\ | \\ \text{HO—CH} \\ | \\ \text{HO—CH}_2\end{array} \\
\\
\text{O} \\
\| \\
\text{C}_{17}\text{H}_{35}\text{C—O—CH}_2
\end{array}
$$

glyceryl stearate
(fat)

sodium stearate
(soap)

glycerol

The cleansing action of soap is discussed in Chapter 14.

Aromatic esters are similar to the aliphatic ones. The methyl ester of salicylic acid, methyl salicylate, is known as *oil of wintergreen* and is used in the manufacture of perfumes and flavoring agents.

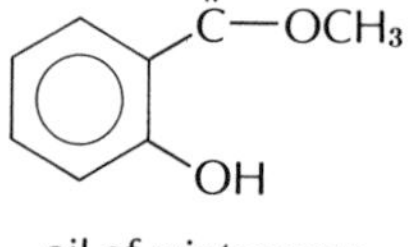

oil of wintergreen
(methyl salicylate)

Another aromatic ester, involving the phenolic OH and acetic acid, is acetyl salicylic acid or aspirin (see p. 229).

Amides   Amides are the carboxylic acid derivatives formed by replacement of the OH group with NH$_2$:

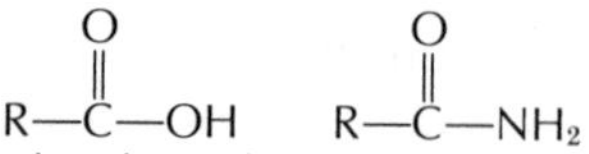

$$
\begin{array}{cc}
\overset{\text{O}}{\overset{\|}{\text{R—C—OH}}} & \overset{\text{O}}{\overset{\|}{\text{R—C—NH}_2}} \\
\text{carboxylic acid} & \text{amide}
\end{array}
$$

Acetamide $\left(CH_3-\overset{\overset{\displaystyle O}{\|}}{C}-NH_2\right)$, the simplest of the amides, is prepared by the following reactions:

$$2CH_3\overset{\overset{\displaystyle O}{\|}}{C}-OH \xrightarrow[-H_2O]{H_2SO_4} CH_3-\overset{\overset{\displaystyle O}{\|}}{C}-O-\overset{\overset{\displaystyle O}{\|}}{C}-CH_3 \xrightarrow{NH_3} CH_3\overset{\overset{\displaystyle O}{\|}}{C}-NH_2 + CH_3-\overset{\overset{\displaystyle O}{\|}}{C}-OH$$

It is used as an industrial solvent.

Urea, $\left(\overset{NH_2}{\underset{NH_2}{\diagdown\diagup}}C=O\right)$, diamide of carbonic acid, is a waste product of metabolism in the animal body and is excreted in the urine. It is manufactured as an important fertilizer and also constitutes an ingredient of a major class of plastics (see Chapter 9). In addition, urea is used in making soporifics such as Veronal and phenobarbital.

Salicylamide $\left(C_6H_4(OH)\overset{\overset{\displaystyle O}{\|}}{C}-NH_2\right)$ is an important aryl amide having many of the properties of aspirin but reportedly less toxic. Certain substituted amides are used as sedatives and hypnotics in medicine (see Chapter 12). An important polyamide, *nylon*, is discussed in Chapter 10.

QUESTIONS

9.21 What are the products of the destructive distillation of
    a. coal
    b. wood.

9.22 Name some important alcohols and their uses.

9.23 How is phenol prepared? List some of its important derivatives.

9.24 Discuss the nomenclature of doubly substituted aromatic compounds using the xylenes as an example.

9.25 What are the basic differences between aldehydes and ketones?

9.26 Write the formulas of
    a. chloral
    b. vanillin
    c. benzophenone
    d. saccharin
    e. aspirin

9.27 What are the important derivatives of carboxylic acids? How are they prepared?

9.28 What is vinegar? How is it prepared?

9.29 What is meant by saponification?

9.30 Name and give formulas for some organic compounds that are used as
    a. anesthetics
    b. refrigerants
    c. disinfectants
    d. industrial solvents
    f. food flavors
    g. drugs

SUGGESTED PROJECT 9.2
The art of fermentation (or how you too can brew)

The art of fermenting fruit juices and starchy materials to alcoholic beverages has been known and used since ancient times. Today, fermentation is a highly developed art. In fact, alcoholic beverages have become part of the diet of many people throughout the world by reason of their thirst-quenching and nutritional value, their taste, and the euphoria they create when freely used. Many households make their own beers and wines by fermenting malted cereals and grape juices, respectively. In this project, you will ferment household sugar using brewer's yeast.

Place 3 tablespoons of sugar and 1 pint of water in a quart bottle. If necessary, warm very gently to dissolve the sugar. Add 1 tablespoon of brewer's yeast and shake the mixture well. Close the bottle loosely with a stopper and leave it to stand in a warm place (20–25°C) for at least 3 days. You may test for the escaping CO_2 by introducing one end of a straw pipe into the fermenting mixture and holding a lighted match above the other end. When the CO_2 evolution ceases, decant the fermented mixture. The decantate consists of a dilute solution of ethanol in water. If you have distillation facilities available, you may distill this solution to increase the ethanol concentration to 40 percent by volume and produce the common alcoholic beverage rum, which is 80 proof.

SUGGESTED PROJECT 9.3
Types of alcoholic beverages

In preparing a list of the common alcoholic beverages available in your neighborhood liquor store, group them according to the categories: malt liquors, wines, distilled beverages, and liqueurs. Distinguish among the various types of beverages within each group and discuss the reasons for variations such as alcohol content. Because alcohol acts as a depressant of the central nervous system, refer to the Merck Index in your library to determine the lethal dose. Also, ascertain the legal limit set for intoxication in your state.

SUGGESTED PROJECT 9.4
Esters as perfumes and flavors

Many esters are used for flavoring and perfumery. Examine the odor of a number of candies, beverages, and perfumes available in your local store to find out the esters that are used as flavoring or odor concentrates. From the regulations of the U.S. Food and Drug Administration, if they are available in your library, obtain precise information on the permitted levels of use of these esters.

ORGANIC NITROGEN COMPOUNDS

A number of different families of organic nitrogen compounds are known. In the preceding section we have already encountered some organic nitrogen compounds,

the amides $\left(R-\overset{\overset{O}{\|}}{C}-NH_2 \right)$, that are related to acids by replacement of the OH group with NH_2. Some other important families are the nitriles ($R-C\equiv N$), the nitro compounds $\left(R-N\overset{\overset{O}{\diagup}}{\diagdown_O} \right)$, the amines ($R-NH_2$), and amino acids $\left(RNH_2-\overset{\overset{O}{\|}}{C}-OH \right)$.

The nitrogen atom may also be part of a ring system (see p. 236). The basic nitrogen-containing organic compounds of plant origin are known as *alkaloids*. Some of these classes of compounds are discussed in this chapter; amino acids are discussed in Chapter 11, and alkaloids in Chapter 12.

Organic *nitriles* are compounds with a cyanide ($-C\equiv N$) group linked to a hydrocarbon (R) group. Examples of nitriles are acetonitrile ($CH_3C\equiv N$), also called methyl cyanide, and benzonitrile ($C_6H_5C\equiv N$). They are conveniently prepared from alkyl and aryl halides, respectively. They react with acids or bases forming carboxylic acids. The chief use of these compounds is in organic synthesis and in the manufacture of certain types of plastics.

Organic *nitro* compounds are derivatives of nitric acid in the sense that the OH group in nitric acid may be considered to be replaced by an alkyl or aryl group:

$$HO-N\overset{\overset{O}{\diagup}}{\diagdown_O} \qquad CH_3-N\overset{\overset{O}{\diagup}}{\diagdown_O} \qquad C_6H_5-N\overset{\overset{O}{\diagup}}{\diagdown_O}$$

nitric acid   nitromethane ·   nitrobenzene

They are conveniently prepared by *nitration* of hydrocarbons:

$$CH_4 + HNO_3 \xrightarrow{400°C} CH_3NO_2 + H_2O$$

$$C_6H_6 + HNO_3 \xrightarrow{H_2SO_4} C_6H_5NO_2 + H_2O$$

An accumulation of nitro groups on one parent molecule results in explosives such as

$$\begin{array}{c} CH_2NO_2 \\ | \\ CHNO_2 \\ | \\ CH_2NO_2 \end{array}$$

nitroglycerine   trinitrophenol (picric acid)   trinitrotoluene (TNT)

When cotton is treated with nitric acid, it is converted into cellulose nitrates. "Guncotton" is a nitrocellulose containing about 13 percent nitrogen and is used in the production of smokeless powder and high explosives.

Amines may be looked upon as derivatives of ammonia in which one, two, or all three of the hydrogens are replaced by R groups:

$$H—\underset{\underset{H}{|}}{N}—H \qquad R—\underset{\underset{H}{|}}{N}—H \qquad R—\underset{\underset{H}{|}}{N}—R \qquad R—\underset{\underset{R}{|}}{N}—R$$

ammonia   primary   secondary   tertiary
               amine   amine   amine

Methylamine (CH_3NH_2), dimethylamine [$(CH_3)_2NH$], and trimethylamine [$(CH_3)_3N$] are examples of primary, secondary, and tertiary amines, respectively. They are conveniently prepared by reacting ammonia with an organic halogen compound:

$$C_2H_5Br + 2NH_3 \longrightarrow C_2H_5NH_2 + NH_4Br$$
$$C_2H_5NH_2 + C_2H_5Br \longrightarrow (C_2H_5)_2NH + HBr$$
$$(C_2H_5)_2NH + C_2H_5Br \longrightarrow (C_2H_5)_3N + HBr$$

The simpler amines are similar to ammonia in odor, basicity, and other properties. They react with water to form basic solutions as a result of the unshared electron pair on the nitrogen atom:

$$RNH_2 + H_2O \longrightarrow RNH_3^+ + OH^-$$

Just as ammonia reacts with HCl to produce NH_4Cl, amines react with HCl to give amine hydrochlorides,

$$RNH_2 + HCl \longrightarrow RNH_3^+Cl^-$$

which are ionic in character.

The simplest aromatic amine is *aniline*, $C_6H_5NH_2$. It is prepared by the reduction of nitrobenzene:

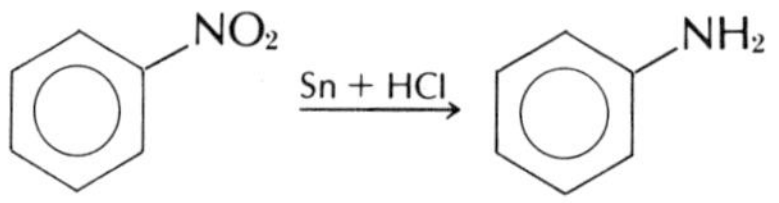

Aniline is extremely important as a starting material for the manufacture of many dyes. It is also the basis of certain medicinals including acetanilide

$$\left(C_6H_5NH—\overset{\overset{O}{\|}}{C}—CH_3 \right),$$ which is used as an antipyretic (fever-reducer), the sulfa drugs which are well known for their ability to combat infections, and an arsenical that is used for treating syphilis. Some other aromatic amines of interest are the following:

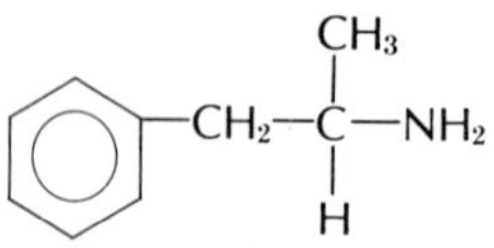

amphetamine
(speed)

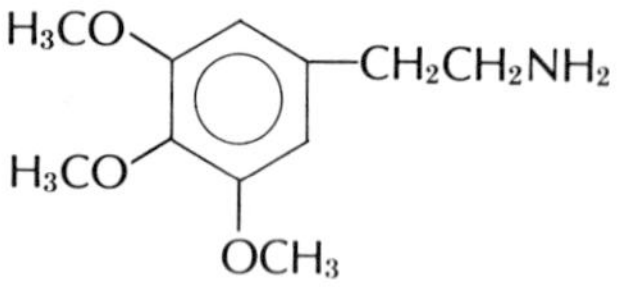

mescaline

Mescaline is one of the oldest known hallucinogens and is isolated from the peyote plant. Amphetamine, which is also known as "speed," is about 18 times more active than mescaline as a hallucinogen. Because it depresses the appetite, it is used as a treatment for obesity.

SULFA DRUGS A group of substances derived from aniline, sulfa drugs have been of great importance in fighting bacterial infection. They were discovered in 1935 by the German chemist Gerhard Domagk. The simplest of them is *sulfanila-mide* [$C_6H_4(NH_2)SO_2NH_2$] which is prepared according to the following scheme:

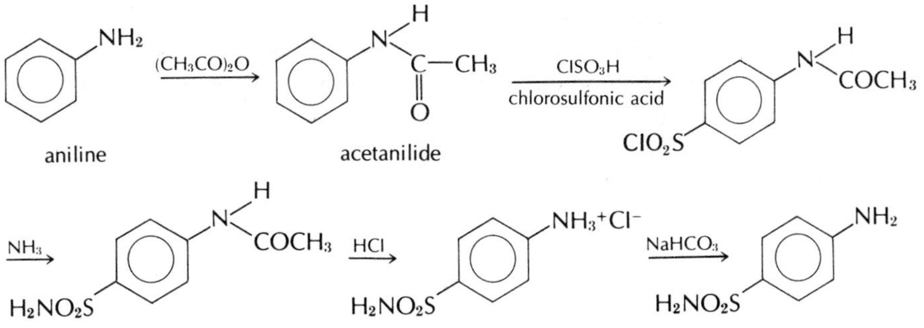

In order to overcome certain undesirable side effects of sulfanilamide, various other groups have been substituted for H atoms of the —SO_2NH_2 group.

Sulfa drugs inhibit the growth of bacteria by preventing the synthesis of a compound called folic acid from *p*-aminobenzoic acid. Actually, the close structural similarity of sulfanilamide and *p*-aminobenzoic acid,

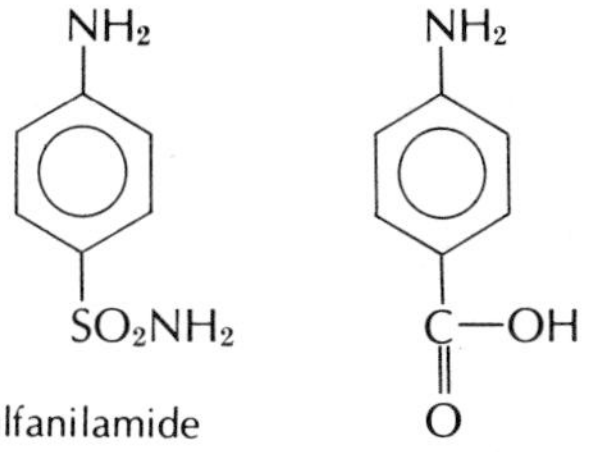

results in some of the sulfanilamide being mistakenly used instead of the *p*-aminobenzoic acid required for folic acid synthesis. This action, a case of mistaken identity, shuts off the production of the essential folic acid with the result that bacteria cease to grow. However, not all bacteria are susceptible to sulfa drugs—the most susceptible are the well-known streptococci and staphylococci bacteria. Nevertheless, the yearly production of more than 5 million lb of sulfa drugs is proof of their widespread use and general effectiveness in treating bacterial infections.

QUESTIONS

9.31   **Distinguish between amines and amides.**
9.32   **Give two examples of organic compounds containing nitrogen as part of a ring system.**
9.33   **What is meant by nitration? Describe a method for the nitration of benzene.**
9.34   **What are amino acids? Give examples.**

9.35 How is aniline produced? Discuss the use of aniline in the manufacture of sulfa drugs and organic dyes.

9.36 Discuss the mechanism by which sulfa drugs fight bacterial infections.

HETEROCYCLIC COMPOUNDS

Ring compounds that contain one or more atoms of elements such as oxygen, nitrogen, and sulfur besides carbon in the ring skeleton are generally known as *heterocyclic compounds*. Such compounds are very common in plant and animal tissues, and form the basis of much of what are called *natural products*. A majority of these compounds have five- or six-membered rings that contain either one or two elements other than carbon:

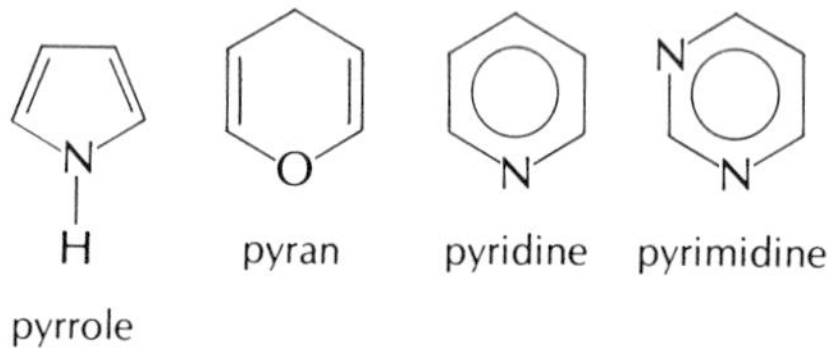

pyrrole   pyran   pyridine   pyrimidine

The *pyrrole* nucleus is an integral constituent of the *porphyrins*, which are present in many *natural pigments*. The nitrogen atoms of the porphyrin combine with metals such as iron to form the heme of hemoglobin (Fig. 9.16) and with magnesium to form chlorophyll (Fig. 9.17). The cytochromes, which are enzymes involved in biological oxidations and reductions, are also related to the porphyrins and contain groups similar to heme (see Chapter 11). The *pyran* ring is present in anthocyanin, which is responsible for the color of flowers, and in rotenone, a plant material that is used as an insecticide.

Pyridine is a liquid obtained from coal tar. It is a weak base and is used as a solvent, as a denaturant for ethanol, and as a rubber accelerator. It is also used in the

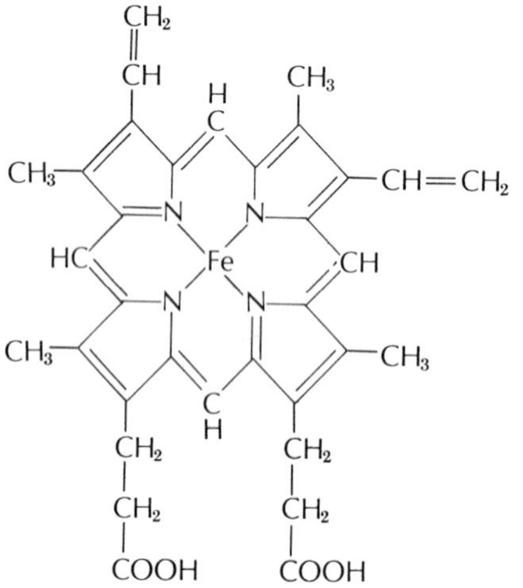

FIGURE 9.16

Molecular structure of the heme of hemoglobin.

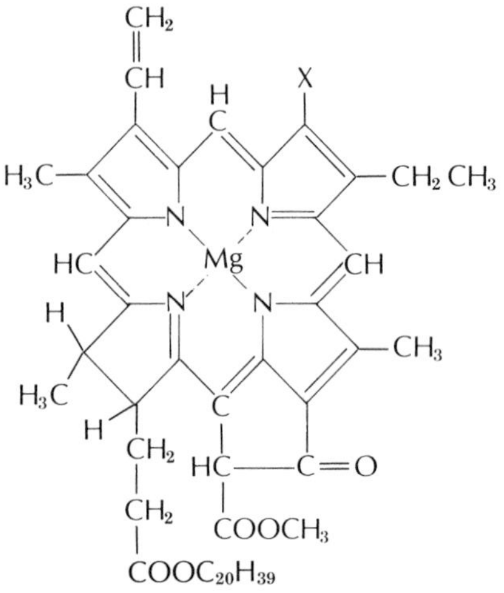

FIGURE 9.17

Molecular structure of chlorophyll. In chlorophyll a, X = —CH₃; in chlorophyll b, X = —CHO.

preparation of waterproofing agents for textiles and in the manufacture of pharmaceuticals such as sulfa drugs, antihistamines, and steroids. Its methyl derivatives,
called *picolines*, are easily oxidized to the corresponding picolinic acids. The acid
obtained from the oxidation of β-picoline (3-methylpyridine) is known as *nicotinic
acid*:

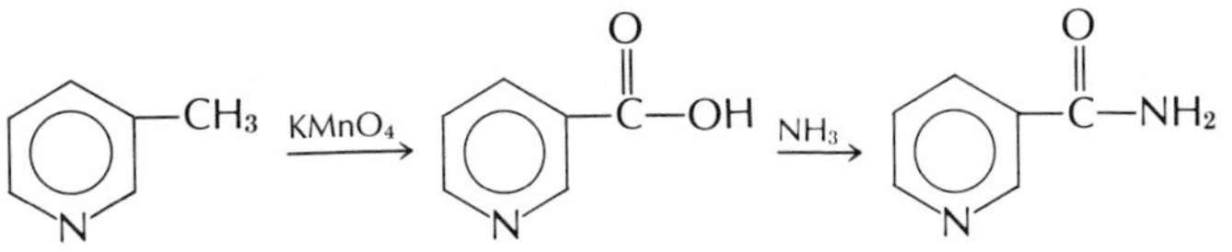

β-picoline   nicotinic acid   nicotinamide

Nicotinic acid and its amides are members of the vitamin B complex which is an
important dietary supplement. A lack of these compounds in the diet causes the
deficiency disease pellagra. Over 2 million lb of these compounds are manufactured
annually for use as additives to wheat flour or in vitamin preparations.

Pyrimidine derivatives are found in nucleoproteins, which are essential
constituents of all living cells. A class of compounds closely related to the pyrimi-
dines are the *purines* which are synthesized by animals (including man) from simple
nitrogen and carbon compounds.

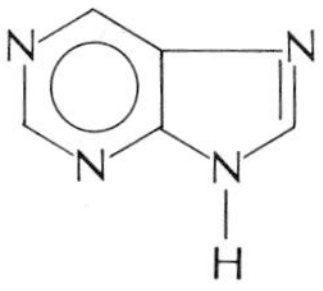

purine

Purine derivatives present in the meat contribute to the flavor. Metabolic oxidation
of purines in the body results in *uric acid* which is present in the blood and is excreted in the urine.

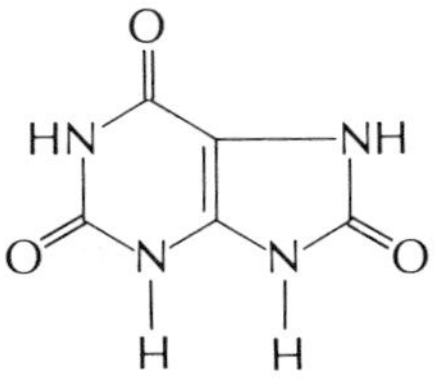

uric acid

Under abnormal conditions uric acid may form insoluble deposits, or stones, in the
kidney or bladder, or it may crystallize in the joints and cause the painful condition
known as gout.

QUESTIONS

9.37 **What are heterocyclic compounds?**
9.38 **Give the names and formulas of some heterocyclic compounds that have**
   a. five-membered rings
   b. six-membered rings

9.39 What are porphyrins? Discuss the structures of
a. heme of hemoglobin
b. chlorophyll
with respect to that of porphyrins.

9.40 What are purines? How is uric acid produced from purines?

ORGANIC MATTER AND OUR ENVIRONMENT

The present startling rate of population increase together with our increased use of organic matter for food and for fuel is posing a variety of threats to the very environment which is essential for our habitat. What are these threats? Let us consider, for example, the increased use of organic matter as fuel to supply our energy needs. When the organic matter is burned no doubt there is a slow erosion of atmospheric free oxygen due to the combustion process

$$C + O_2 \longrightarrow CO_2$$

However, through photosynthesis, most of the released carbon dioxide is recycled back to oxygen and organic matter in due course of time (Fig. 9.18). Although the composition of our atmosphere is such that at this time the slow depletion of oxygen is of little consequence, the increased level of carbon dioxide results in the *greenhouse effect*—the retention of heat on earth because of the interaction between the increasing amount of atmospheric carbon dioxide and solar radiation (see Chapter 15). In the United States the combustion of coal alone produces nearly 9 million tons of various particulates each year. The combustion of other types of organic matter produces an additional 12 million tons of particulates. These particulates contribute to respiratory illnesses and climatic changes. Often, what is worse, the

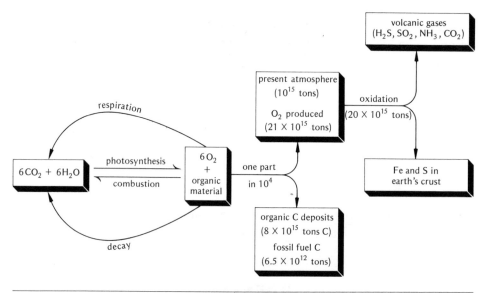

FIGURE 9.18

Combustion of organic matter and its recycling by photosynthesis. (From J. Calvin Giddings, *Chemistry, Man, and Environmental Change*, p. 197. Copyright © 1973 by J. Calvin Giddings. Redrawn by permission of Harper & Row, Inc.)

combustion process produces toxic carbon monoxide (due to insufficient oxygen and low combustion temperature):

$$2C + O_2 \longrightarrow 2CO$$
$$C + CO_2 \longrightarrow 2CO$$

It is estimated that the organic matter we burn as fuel releases about 75 million tons of carbon monoxide each year in the United States alone. The increased level of carbon monoxide is detrimental in that it reacts with the hemoglobin of the blood to form carboxyhemoglobin and thus impairs the ability of blood to transport oxygen (as oxyhemoglobin) from the lungs to the body cells and carbon dioxide from the cells to the lungs. This problem is discussed more fully in Chapter 12.

Many of our fuels consist of lower alkanes which are volatile. Their evaporation into the atmosphere by itself is not harmful to us. The alkanes, however, react with oxygen atoms and ozone to produce peroxyalkyl ($RO_2 \cdot$) radicals that react with O_2 and NO_2 to form *peroxyacylnitrates* (PAN), one of the detrimental components of *smog* (see Chapter 15).

FIGURE 9.19

Typical food chain showing the biological amplification process. The increasing concentration of DDT is shown by dots. (From J. Calvin Giddings, *Chemistry, Man, and Environmental Change*, p. 372. Copyright © 1973 by J. Calvin Giddings. Redrawn by permission of Harper & Row, Inc.)

The production of synthetic organic compounds in the United States is now estimated at 150 billion lb, which is an increase by a factor of 15 since World War II. One of the most sobering developments has been that many of these compounds, when applied for specific purposes, have had unanticipated long-term and widespread ill effects. An example is DDT, which has been used with great success in mosquito and plant pest control. However, it is slightly volatile and is not attacked easily by environmental organisms; that is, it is not readily biodegraded. In consequence, it has been distributed throughout the world by air and water currents. Concentrated in animals by the food chain* (Fig. 9.19), it is now blamed for the threatened extinction of some species. DDT is not unique in this respect; many other organic chemicals, particularly organic halogen compounds used for control of weeds and pests, have similar undesirable effects.

Finally, mention must be made of *oil pollution,* however inevitable it may seem in an oil-based technology such as ours. The use of natural fuel resources such as oil on a grand scale, without losses, is almost impossible. However, the extent of such losses has been steadily increasing. It is estimated that the total oil influx into our oceans is about 10 million tons annually. Many biological processes of importance to the survival of marine life and occupying key positions in their life processes may be affected by it. Because the food chain eventually reaches marine organisms that are harvested for human consumption, the possibility exists that not only will seafood flavor be affected but also that seafood will accumulate potential long-term poisons. The effects of oil pollution are thus by no means limited to oil-soaked birds floundering on oil-covered beaches, photographs and descriptions of which have become almost routine presentations of the visual news media.

QUESTIONS

9.41   In what respects does organic matter play a major role in our environment?

9.42   Discuss the mechanism of combustion of organic matter and its recycling by photosynthesis.

9.43   Discuss the role of hydrocarbons in the production of smog.

9.44   What are the undesirable effects of using organic halogen compounds as insecticides and pesticides?

9.45   What is meant by oil pollution? What are its consequences?

---

* A sequence in which a plant is eaten by an animal, which in turn is eaten by a predator, and so on.

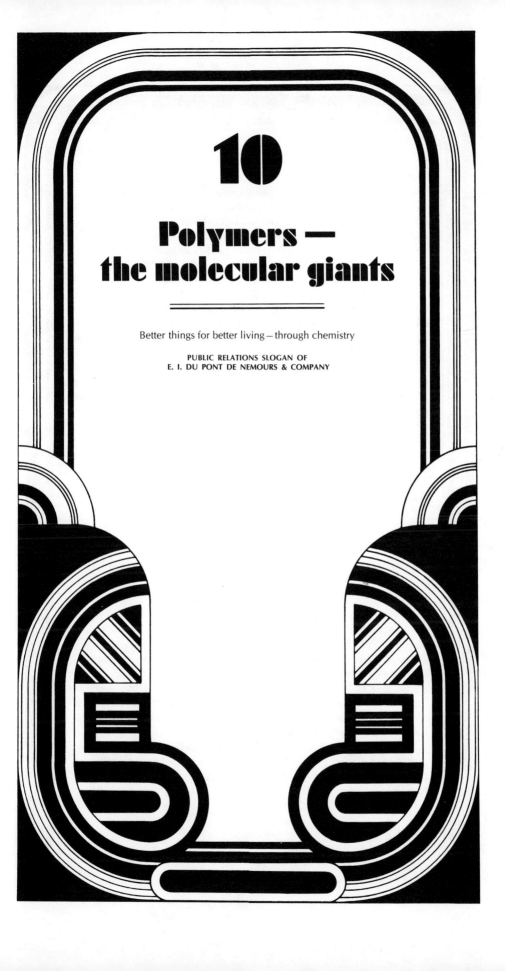

10

Polymers —
the molecular giants

Better things for better living — through chemistry

PUBLIC RELATIONS SLOGAN OF
E. I. DU PONT DE NEMOURS & COMPANY

We have already encountered some of the many carbon-containing compounds that constitute the organic matter of our environment. There exist some important organic compounds that are distinguished by the relatively big size of their molecules. In fact, many phenomena around us depend on these giant or *macromolecules* which are also called *polymers* (from the Greek *polymeres* meaning "many parts"). The celluloses that give the tree trunk rigidity with strength, the proteins that give animal muscle flexibility with strength, the viruses that infect plants and animals, the nucleic acids that store and carry information in all living systems, and the synthetic plastics and fibers that pervade our industrial society are all examples of polymeric giant molecules. Scientifically speaking, the word "plastic" refers to a substance that can be formed under pressure and softened by heat to fabricate desired shapes.

Polymers are formed from small molecules that are called *monomers*, which are linked end to end by repeating chemical bonds of the same type. An example is polythene, the material of the familiar sandwich bags and plastic wraps, which is formed from the monomer ethene, $H_2C=CH_2$. In Chapter 9, it was reported that some alkenes, such as ethene and propene, under specific conditions can react in the presence of a catalyst, so that the individual alkene molecules add to one another. Such a reaction is called *polymerization.* During polymerization, the monomer units are joined when the electrons in the double bond move to form linkages between the units:

$$n[H_2C=CH_2] \xrightarrow[\text{heat,} \atop \text{pressure}]{\text{catalyst}} \begin{array}{ccccc} H & H & H & H & H \\ | & | & | & | & | \\ H-C&-C&-C&-C&-C-\cdots \\ | & | & | & | & | \\ H & H & H & H & H \end{array}$$

ethene   polythene
(monomer)   (polymer)

Note that in an equation for a polymerization reaction, it is not possible to give the exact formula of the polymer, because its individual molecules, made up of hundreds or thousands of monomer units, vary in chain length. However, the polymer structure could be represented by a more compact notation using the repeating sequence, as for example, $[-CH_2CH_2-]_n$ in the case of polythene, where the atoms in brackets are the repeating unit of the polymer, and the subscript *n* is some large number corresponding to the number of monomer units comprising the polymer.

Besides the *addition* polymerization reaction described above, there are two other polymerization processes, namely, (1) *condensation* polymerization reactions, in which two small molecules react to form bonds to each other splitting out a smaller molecule such as water, and (2) *rearrangement* polymerization reactions, in which two different molecules form bonds with each other by rearranging their bonding patterns. These processes are discussed later in the chapter when we examine the chemistry and technology of synthetic polymers. First, we shall direct our attention to some natural polymers — the giant molecules that are found in plants and animals.

NATURAL POLYMERS

A natural polymer that is perhaps most familiar to all of us is *rubber,* which is obtained from the milky sap, or latex, of various tropical plants, especially the rubber tree (*Hevea brasiliensis*). Natural rubber, $(C_5H_8)_n$, is a light cream-to-dark amber, amorphous, elastic solid polymer of isoprene:

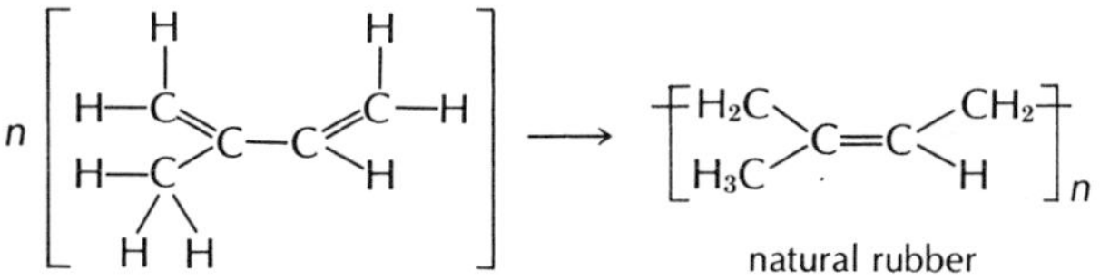

isoprene → natural rubber

A single rubber molecule can have a mass of 500,000 amu and contain as many as 7500 monomer units.

Natural rubber molecules present a type of isomerism different from the one we encountered in the previous chapter. Whereas the portions of a molecule attached by a C—C single bond are free to rotate, the planar structure that exists among compounds containing a C=C double bond anchors the attached atoms to a particular position in space:

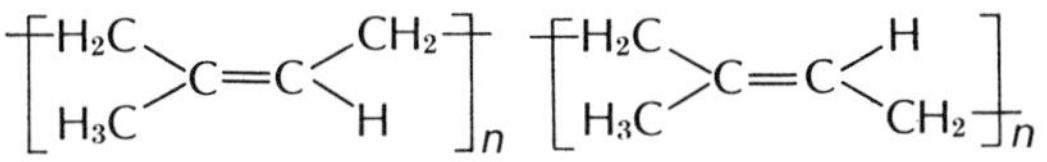

cis-polyisoprene   *trans*-polyisoprene
(natural rubber)   (gutta-percha)

Natural rubber or *cis*-polyisoprene has a structural arrangement so that the —CH₃ and —H attachments are on the same side of the C=C structure. Molecules with the other-extreme structural arrangement (*trans*-polyisoprene), do not exhibit the desirable properties of natural rubber. They constitute a rubberlike material called *gutta-percha* obtained from trees of the Sapotaceae family that grow in the East Indies.

The long-chain polyisoprene molecules have some freedom of rotation about the C—C single bonds. Consequently, they can assume many arrangements in space which lead to coiled, twisted, or intertwined chains. The straightening of such chains is responsible for the 500–1000 percent extensibility of rubber (Fig. 10.1) and its classification as an *elastomer* (elastic polymer). However, natural rubber hardens when cooled and becomes soft and tacky when heated. In 1839, Charles Goodyear demonstrated that treating raw rubber with sulfur removed these undesirable properties. This process of compounding rubber with sulfur is known as *vulcanization*. In the process some of the sulfur atoms react with the polyisoprene chains to produce *cross-linked* structures (Fig. 10.1c), which, in turn, reduce the extensibility of rubber and make it harder. Soft rubber contains about 5 percent sulfur, whereas hard rubber may contain up to 30 percent sulfur.

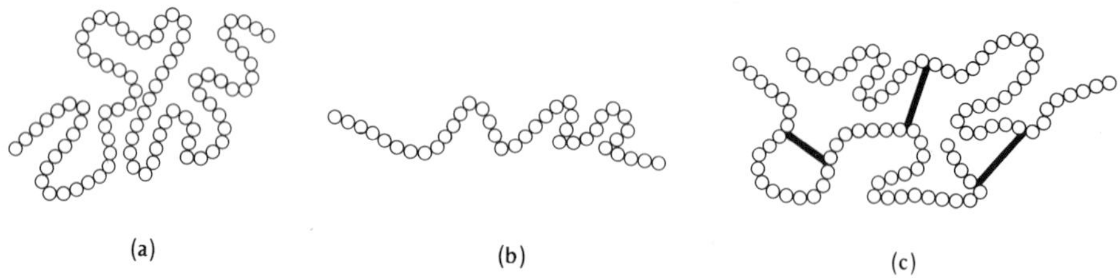

(a)   (b)   (c)

FIGURE 10.1

(a) Coiled, (b) stretched, and (c) cross-linked polymeric structures of rubber.

Most of the rubber used nowadays in the United States is *synthetic rubber*. More than 2 billion tons of synthetic rubber are produced each year using petroleum hydrocarbons as the raw materials. *Styrenebutadiene rubber,* known as SBR, is the synthetic rubber produced in the greatest amount since the beginning of World War II:

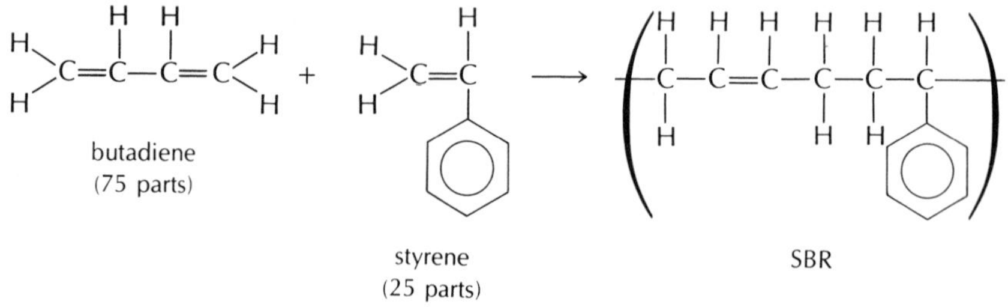

butadiene
(75 parts)

styrene
(25 parts)

SBR

The molecules can *copolymerize* as Bu—Bu—Bu—S, Bu—S—Bu—Bu, or in any other possible sequence. Other synthetic rubbers with unique properties needed to make them valuable for special uses have been developed. One of these is polychloroprene, the synthetic rubber known as *Neoprene:*

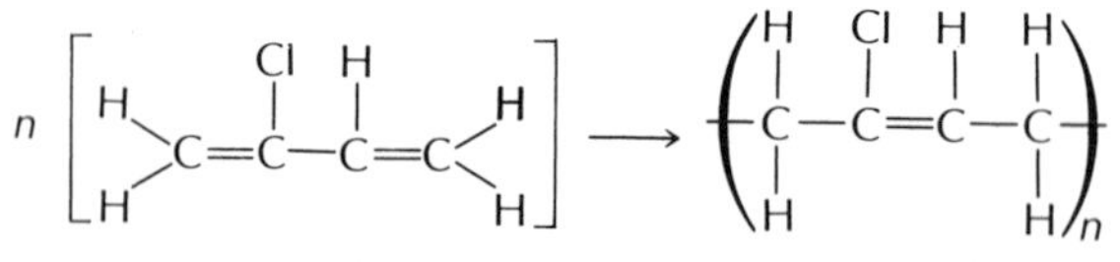

chloroprene

Neoprene

Notice that in chloroprene the chlorine atom occupies the place of the —CH_3 group of isoprene. This difference is considered responsible for the increased resistance of Neoprene to solvents such as gasoline, oil, or fats, and its chief use as an oil-resistant rubber. In 1954, both the Goodyear and Firestone companies succeeded in finding a catalyst to develop synthetic, "natural" isoprene rubber. Since that time, this synthetic "natural" rubber, prepared in the laboratory by duplicating the polymerization process originally accomplished by a living plant, is used when natural rubber is in short supply. In fact, today it is actually preferable in many ways to the actual product of the rubber tree.

Another natural polymer is *cellulose* $[(C_6H_{10}O_5)_n]$, which is found in the cell walls and wood fibers that form the skeletal structure of all plants. It is a polymer composed of 1800–3000 units of the monomer glucose $(C_6H_{12}O_6)$ united by elimination of one water molecule for each link formed (see Chapter 11):

$$nC_6H_{12}O_6 \longrightarrow (C_6H_{10}O_5)_n + nH_2O$$

Among naturally occurring cellulose materials special mention may be made of *cotton*, which contains mostly (95%) cellulose. Depending on its hardness, *wood* contains varying amounts of cellulose and cellulose-like compounds. *Rayon*, the first important synthetic fiber to be produced, is reconstituted cellulose. It is manufactured by treating wood pulp first with sodium hydroxide and then with carbon disulfide:

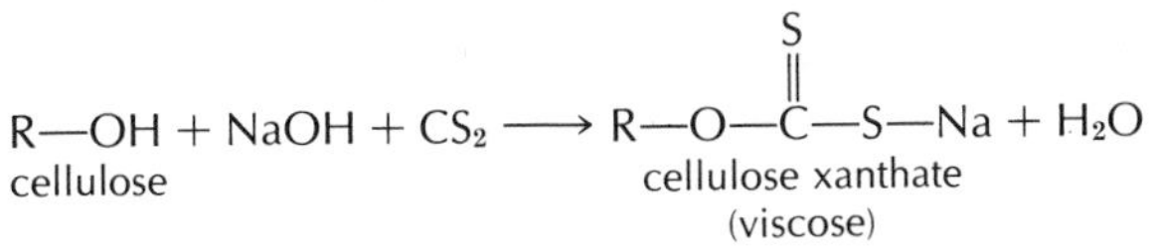

$$\underset{\text{cellulose}}{R{-}OH} + NaOH + CS_2 \longrightarrow \underset{\substack{\text{cellulose xanthate} \\ \text{(viscose)}}}{R{-}O{-}\overset{\overset{\textstyle S}{\|}}{C}{-}S{-}Na} + H_2O$$

The viscous solution of cellulose xanthate is forced through small holes into an acid solution to regenerate the cellulose in the form of long fibers:

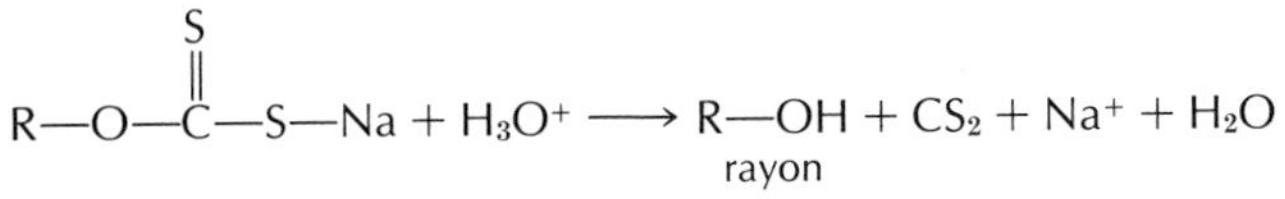

$$R{-}O{-}\overset{\overset{\textstyle S}{\|}}{C}{-}S{-}Na + H_3O^+ \longrightarrow \underset{\text{rayon}}{R{-}OH} + CS_2 + Na^+ + H_2O$$

When extruded in the form of a film, the product is known as *cellophane*. Viscose rayon is an efficient absorbent and is used extensively for all types of fabrics.

Proteins constitute a group of important polymers found in every living organism. They are solid polymers formed by condensation reactions from amino acids—compounds containing an amine and an acid functional group in the same molecule. But proteins are a different sort of polymer in that the amino acid monomer units are not the same:

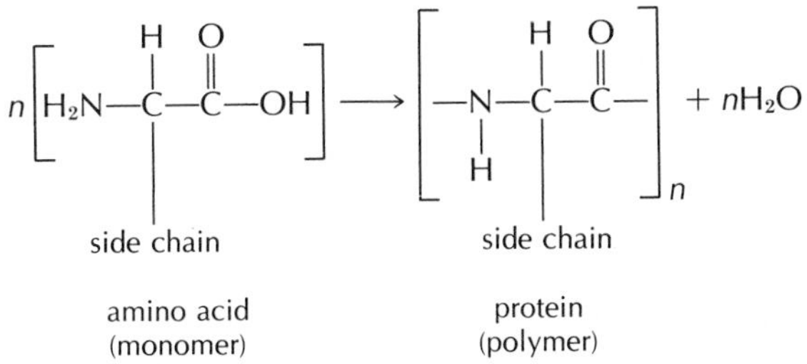

amino acid
(monomer)

protein
(polymer)

There are 20 different kinds of amino acids, and it is in their side chains that the monomer units differ from one another. Each monomer unit, as in the foregoing structure, is called an *amino acid residue*.

Protein polymers are sometimes called *polypeptides*. The amino acid residues (monomer units) in proteins are linked together by the *amide* or *peptide bonds* —bonds formed between nitrogen and carbon atoms when the amino group (of one amino acid) reacts with the carboxylic acid group (of a second amino acid):

H—N—C—C—OH + H—N—C—C—OH ⟶

amino acid   amino acid

H—N—C—C——N——C—C—OH + H₂O

peptide bonds

dipeptide

The peptide chains thus formed are, however, not free to assume random configurations because of *hydrogen bond* formation between oxygen atoms of one peptide linkage and amide hydrogen atoms of other peptide linkages (Fig. 10.2). Extensive structural investigations have shown that the amino acid residues can be held in *helical* or *spiral* configurations (see Chapter 11).

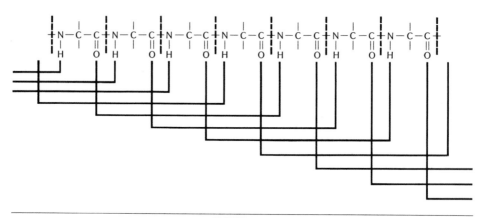

FIGURE 10.2

Diagram showing the hydrogen bonds (heavy lines) formed between peptide linkages.

In plants, proteins occur as the interior substance of cells, whereas, in animals, proteins are found in the blood and they also constitute a major portion of the body in the form of skin, muscles and other soft tissues, and hair. *Enzymes* or body catalysts and the disease-causing *viruses* are also protein molecules. Proteins have indefinite melting points and molecular masses ranging from a few thousand (insulin, 6000 amu) to many millions (tobacco mosaic virus, 4×10^7 amu) atomic mass units. The study of proteins is one of the major topics of *biochemistry,* the branch of chemistry that is concerned with living matter (see Chapter 11). Before considering the role of proteins in life processes, we shall look at some polymers synthesized in the laboratory.

QUESTIONS

10.1 Distinguish between polymer and plastic.

10.2 List the different types of polymerization processes, citing an example for each process.

10.3 Discuss the structure of natural rubber.

10.4 Why is natural rubber called cis-polyisoprene?

10.5 What is meant by a cross-linking structure? How is it effected in the case of natural rubber?

10.6 Describe the preparation of a synthetic rubber.

10.7 What do wood, paper, and cotton have in common?

10.8 How is rayon manufactured?

10.9 What are proteins?, Where are they found?, and Why are they important?

10.10 What is a peptide bond? What is the role of hydrogen bonding in protein structures? Illustrate your answers using the general equation for deriving a protein structure.

SYNTHETIC POLYMERS

Modifications of naturally occurring polymers began to be manufactured as long as 100 years ago. The first plastic to be manufactured was *celluloid,* a substitute for the scarce and expensive elephant-tusk ivory used in making billiard balls. Celluloid was prepared by modifying and reconstituting the natural polymer cellulose.

The first truly synthetic polymers were the phenol–formaldehyde resins, the manufacture of which began as early as 1900. These were patented under the trade name *Bakelite* in 1909. Unlike *thermoplastic* materials such as polythene which may be softened by heat and re-formed, the phenol–formaldehyde resins are permanently hardened when formed and are, therefore, known as *thermosetting* polymers.

The phenol–formaldehyde resins are formed by the condensation polymerization reaction:

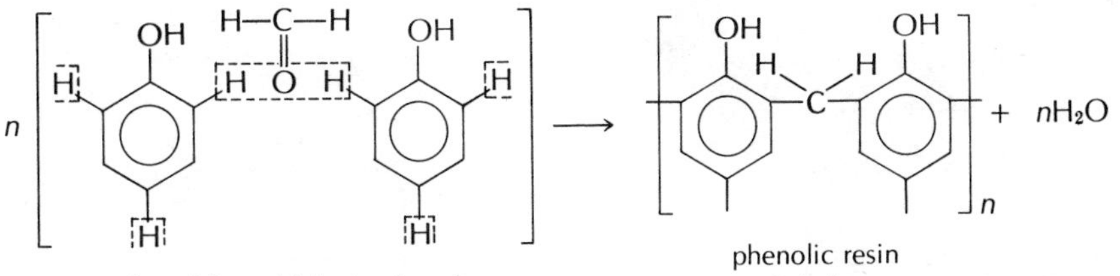

phenol formaldehyde phenol

phenolic resin
(Bakelite)

The degree of condensation and cross-linking can be controlled to give a range of different properties, including high resistance to water, heat, and most chemicals. Other condensation polymerization reactions such as those between urea and formaldehyde and between phthalic anhydride and glycerol provide the commercially important *urea plastics* and *alkyd plastics,* respectively:

FIGURE 10.3

(a) Pirns of nylon are checked on this creel by an operator at du Pont's Seaford, Delaware, plant—the world's first nylon plant built in 1939. (b) Moving down across the picture from the upper right corner, Dacron polyester fiber in ropelike tow form rushes into the drawing machine (not shown) where all fibers will be stretched to give them proper strength. This is one of many steps through which Dacron staple and tow must pass before they are ready to be shipped to textile mills. Photographs courtesy of E. I. du Pont de Nemours & Company.

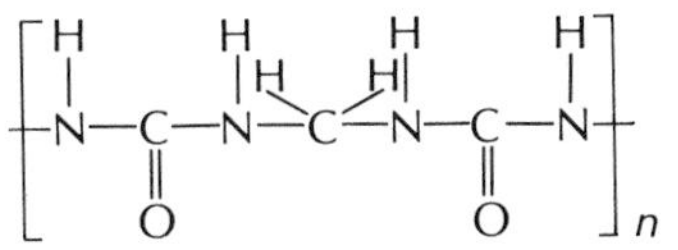

urea plastic

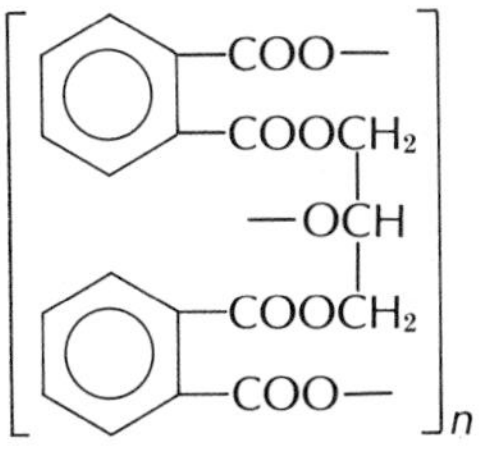

alkyd plastic
(polyester)

Phenolic plastics are excellent electrical insulators and find much application in the electrical industry. The transparent urea plastics have the advantage of taking colors and are used in the manufacture of picnicware, toys, and buttons. The alkyd resins are very hard and wear-resistant and find application in the manufacture of floor varnishes, automobile enamels, and a variety of adhesives and cements (bonding materials). Some modified unsaturated alkyds, reinforced by fiberglass, are widely used as building materials for boats, trucks, and so forth.

Condensation polymerization processes also yield the synthetic fabric materials such as nylon and Dacron. In the 1930s the du Pont Company began extensive efforts to produce a polymer that would find use as a fiber to replace natural silk and cotton. The result was *nylon,* a synthetic silk used for the production of stockings and other items of clothing (Fig. 10.3a). However, it is interesting to note that the first nylon fabrics were used for military purposes during World War II rather than for peacetime clothes. The nylon success spurred further efforts that led to the development of the useful polyesters known as Dacron in the form of fiber and Mylar in the form of film (Fig. 10.3b).

Nylon is a generic term for polyamide fibers, which are polymers that contain

the amide $\left(\begin{array}{c} O \\ \parallel \\ -C-NH_2 \end{array}\right)$ functional group. In Chapter 9 it was noted that carboxylic acids react with amines to form amides. If a diamine is reacted with a diacid, it is possible to form long-chain polymers. The diacid, adipic acid, and the diamine, hexamethylenediamine, which contain six carbon atoms each, are used to produce nylon-6,6:

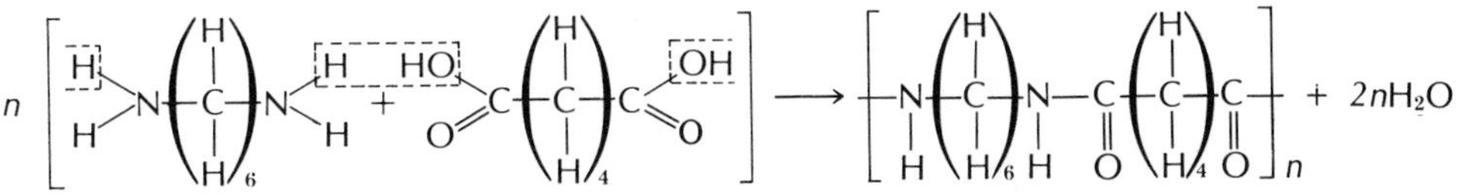

hexamethylenediamine   adipic acid   nylon-6,6
(diamine)   (diacid)   (polyamide)

Diamines and diacids with different numbers of carbon atoms have been used to produce a number of different nylon fibers with good abrasion resistance and extremely good stretch and recovery properties. An important property of nylon strands is their ability to form hydrogen bonds with similar strands lying side by side with them (Fig. 10.4), the energy of which causes the polymer chains to fall into order. Compare this nylon structure with that of natural silk (Fig. 10.5) and note the elements of similarity between the two structures.

Polymerization by *rearrangement reactions* is known to yield products structurally similar to the nylons (polyamides). For example, hexamethylene diisocyanate and butanediol react forming the *urethane linkage* by the rearrangement process:

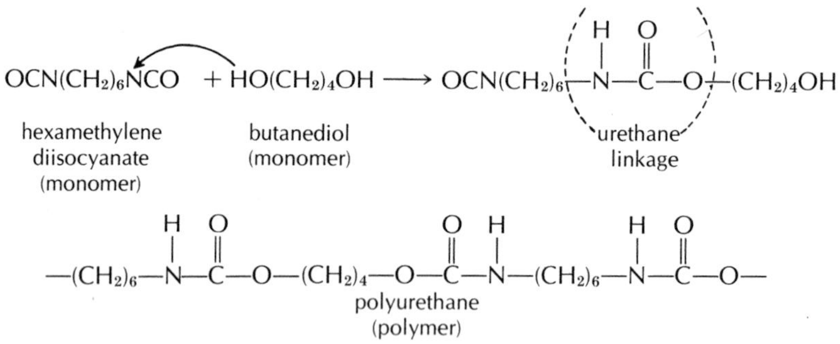

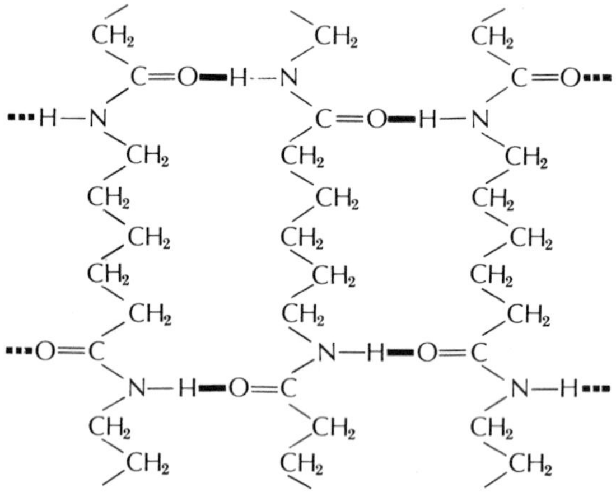

$$OCN(CH_2)_6NCO \quad + HO(CH_2)_4OH \longrightarrow OCN(CH_2)_6 \left[-N-C-O \right] (CH_2)_4OH$$

hexamethylene   butanediol   urethane
diisocyanate   (monomer)   linkage
(monomer)

$$-(CH_2)_6-N-C-O-(CH_2)_4-O-C-N-(CH_2)_6-N-C-O-$$

polyurethane
(polymer)

FIGURE 10.4

Structure and hydrogen bonding (heavy lines) in nylon 6.

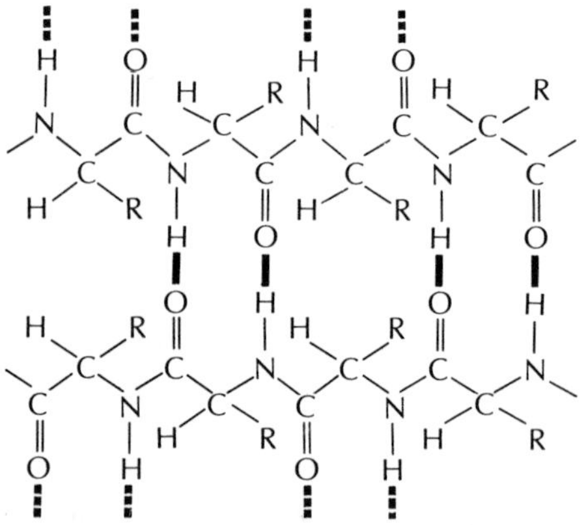

FIGURE 10.5

Two-dimensional representation of the structure and hydrogen bonding (heavy lines) in natural silk. Actually silk has a pleated sheet structure in which the amino acid side chains (indicated by R) protrude above and below the plane of the sheet.

In fact, any molecule containing the isocyanate group (—NCO) will react with almost any other molecule containing an active hydrogen atom (such as in an —OH or —NH₂ group) by a rearrangement process that, when continued, yields a variety of *polyurethanes*. Polyurethanes are used as substitutes for nylons in Europe, but mostly as "foam rubber" and "foamed plastics" which are produced by injecting a gas into a polyurethane.

Just as the diamines react with diacids so do the dialcohols react with diacids to form long-chain fibrous polymers called *polyesters*. One such polyester is *Dacron*, prepared by the condensation reaction of ethylene glycol with terephthalic acid:

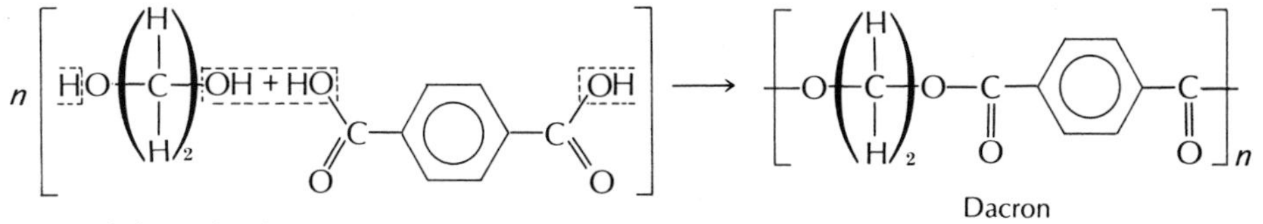

ethylene glycol   terephthalic acid   Dacron
(dialcohol)   (diacid)   (polyester)

Dacron fibers have high resilience, and because they soften on heating, they can be "permanently" pleated.

Another condensation polymerization reaction, which involves molecules such as dialcohols, is the basis of a useful and interesting group of polymers known as *silicones* (see Chapter 5):

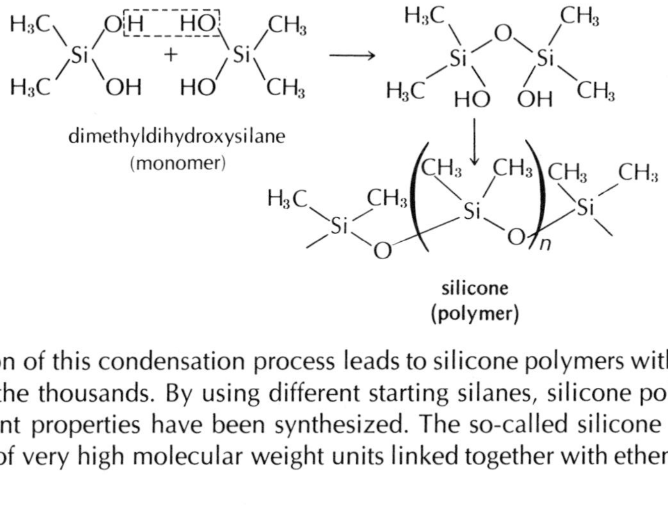

dimethyldihydroxysilane
(monomer)

silicone
(polymer)

Continuation of this condensation process leads to silicone polymers with molecular weights in the thousands. By using different starting silanes, silicone polymers with very different properties have been synthesized. The so-called silicone rubbers are composed of very high molecular weight units linked together with ethene or similar groups:

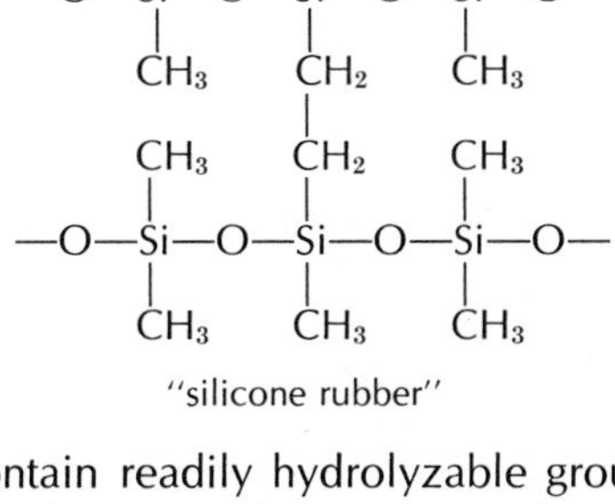

"silicone rubber"

Some silicone rubbers contain readily hydrolyzable groups such as (CH₃COO—) that cross-link at room temperature in the presence of atmospheric moisture, result-

TABLE 10.1
Polythene and its thermoplastic derivatives

Monomer	Polymer	Uses
$H_2C{=}CH_2$ ethene	$\text{---}[H_2C{-}CH_2]\text{---}_n$ polythene	packaging, bottles, toys
$H_2C{=}C{\overset{\displaystyle H}{\underset{\displaystyle CH_3}{}}}$ propene	$\left[H_2C{-}CH{-}\underset{CH_3}{|}\right]_n$ polypropene	fibers, films, carpets, laboratory and kitchenware
$H_2C{=}C{-}H$ (with benzene ring) styrene	$\left[H_2C{-}CH\right]_n$ (with benzene ring) polystyrene	Styrofoam insulations, packaging and packing material, household articles and toys
$H_2C{=}C{\overset{\displaystyle H}{\underset{\displaystyle Cl}{}}}$ vinyl chloride	$\left[H_2C{-}CH{-}\underset{Cl}{|}\right]_n$ polyvinyl chloride	floor coverings, garden hose, phonograph records, packaging
$H_2C{=}C$ with H and $O{-}C({=}O){-}CH_3$ vinyl acetate	$\left[H_2C{-}CH\right]$ with O, $O{=}C{-}CH_3$, subscript n polyvinyl acetate	latex paints
$H_2C{=}C$ with H and $C{\equiv}N$ acrylonitrile	$\left[H_2C{-}CH{-}\underset{CN}{|}\right]_n$ polyacrylonitrile (Orlon or Acrilan)	textile fibers
$H_2C{=}CCl_2$ vinylidene chloride	$\text{---}[H_2C{-}CCl_2]\text{---}_n$ polyvinylidene chloride (Saran)	self-adhering food wrapping
$H_2C{=}C$ with CH_3 and $O{=}C{-}OCH_3$ methylmethacrylate	$\left[H_2C{-}\underset{|}{\overset{CH_3}{C}}{-}\ C{=}O,\ O,\ CH_3\right]_n$ polymethylmethacrylate (Acrylic, Lucite, or Plexiglas)	unbreakable glass for windows, windshields, and outdoors; water-based latex paints
$F_2C{=}CF_2$ tetrafluoroethane	$\text{---}[F_2C{-}CF_2]\text{---}_n$ polytetrafluoroethene (Teflon)	chemically inert items, gasket materials, and frying pan coatings

ing in a "cure" which is similar to the vulcanization of rubber. Nearly 5 million lb of silicone rubber are produced each year and used for a variety of purposes that includes window gaskets, O-rings, sealants, and insulation. The soles of lunar boots worn by the Apollo astronauts were made of high-strength silicone rubber which readily withstood the extreme lunar surface temperatures.

Ethene ($H_2C\!=\!CH_2$) is the basis of the plastic material that holds first place in our markets, namely polythene. The popularity of polythene, $+CH_2\!-\!CH_2\!+_n$, may be attributed to its thermoplastic properties; that is, it can be softened by heat and re-formed into desired shapes. We have already discussed the addition polymerization mechanism by which unsaturated hydrocarbons are united to form giant molecules. The actual manufacture of polythene involves heating ethene under pressure in the presence of a peroxide initiator. The peroxide first decomposes to form free radicals* that react with the ethene monomers:

$$R\!-\!O\!-\!O\!-\!R \longrightarrow 2R\!-\!O\cdot$$

peroxide   peroxide free radicals

$$R\!-\!O\cdot + CH_2\!=\!CH_2 \longrightarrow R\!-\!O\!-\!CH_2\!-\!CH_2\cdot$$
$$R\!-\!O\!-\!CH_2\!-\!CH_2\cdot + CH_2\!=\!CH_2 \longrightarrow R\!-\!O\!-\!CH_2\!-\!CH_2\!-\!CH_2\!-\!CH_2\cdot$$
$$R\!-\!O\!-\!CH_2\!-\!CH_2\!-\!CH_2\!-\!CH_2\cdot + (n-2)CH_2\!=\!CH_2 \longrightarrow R\!-\!O\!+\!CH_2\!-\!CH_2\!+_n\cdot$$

The process repeats to cause the chain to grow very rapidly, ultimately to contain many ethene monomers. The chain terminates when two growing, free radical fragments encounter each other:

$$R\!-\!O\!+\!CH_2\!-\!CH_2\!+_n\cdot + \cdot\!+\!CH_2\!-\!CH_2\!+_{n'}O\!-\!R \longrightarrow$$
$$R\!-\!O\!+\!CH_2\!-\!CH_2\!+_n\!+\!CH_2\!-\!CH_2\!+_{n'}O\!-\!R$$

Under some conditions, such as high temperatures and pressures, the unpaired electron end of a growing free radical chain can loop back and pull a hydrogen off from the body of the chain resulting in a chain with an unpaired electron in the body rather than at the end of the chain. Further growth will result in a branched chain,

$$R\!-\!O\!+\!CH_2\!-\!CH_2\!-\!\underset{\displaystyle \overset{\displaystyle |}{\underset{\displaystyle \vdots}{\overset{\displaystyle |}{\underset{CH_2}{\overset{CH_2}{\overset{|}{C}}}}}}{\overset{\displaystyle \overset{H}{|}}{C}}\!-\!CH_2\!-\!CH_2\!-\!\cdots\!-\!CH_2\!-\!CH_3]$$

the amount of which can be expected to alter the properties of the resulting polymer.

A tremendous variety of polymers is possible if the hydrogen atoms on ethene are replaced by other atoms or groups of atoms. Some of these are listed in Table 10.1. Those polymers which have the $CH_2\!=\!CH\!-$ fragment (vinyl group) are often called *vinyl polymers*. About two decades ago, the German chemist Karl Ziegler and the Italian chemist Guilio Natta independently discovered that certain catalysts could control the spatial ordering of the substituent groups on the extended vinyl polymer chain and thus improve the qualities of many vinyl-type polymers. Let us exemplify this discovery with the three possible structures of polypropene, as shown in Fig. 10.6. In the first structure, called *isotactic*, all the $-CH_3$ groups have the same

---

* A *free radical* is an atom or a molecule that has one or more unpaired electrons.

spatial orientation along the polymer chain; in the second, called *syndiotactic*, the —CH_3 groups extend in alternate directions away from the chain; and in the third, called *atactic*, there is a random arrangement. Polypropenes of these three types have actually been prepared using novel catalysts such as aluminum or iron attached to organic molecules. These polymers were found to have properties that can be related to their structures. As a result of this discovery, we are able to choose the extent of polymerization, that is, the polymer molecular weight as well as the fine structural features of the polymer chain itself. In short, chemists today are able to prepare the so-called tailor-made or made-to-order plastics.

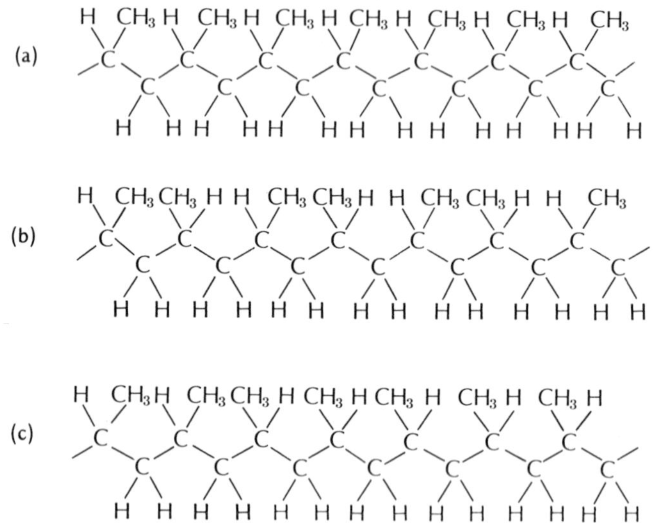

FIGURE 10.6

Contrasting spatial orientations of the methyl (—CH_3) groups give rise to three different structures of polypropene: (a) isotactic, (b) syndiotactic, and (c) atactic. Reproduced, by permission, from Mark M. Jones, John T. Netterville, David O. Johnston, and James L. Wood, *Chemistry, Man and Society*, p. 331. Copyright © 1972 by W. B. Saunders Company.

Finally, mention must be made of the fact that few plastics manufactured today find end uses without some kind of modification to achieve desirable qualities. For example, many plastics intended for outdoor use are treated with ultraviolet stabilizers that absorb sunlight and thus prevent broken bonds in the polymer structure. Plastics used as insulation are foamed by the addition of chemicals that are capable of decomposing into gaseous products and thus produce bubbles in the plastic. Where greater flexibility is required, certain compounds, called *plasticizers*, are added to the plastic to render it more flexible. The plasticizers exert a partial dissolving effect by fitting in between the polymer chain and weakening the attractions between chains, thereby increasing the flexibility.

QUESTIONS

10.11 What is the main difference between thermosetting and thermoplastic materials?
10.12 Write equations for the condensation polymerization reactions between

  a. urea and formaldehyde
  b. glycerol and phthalic anhydride

10.13 How is nylon-6,6 prepared? In what respect is polyurethane different from nylon?

10.14 What are polyesters? Give two examples.

10.15 Write the reaction for the formation of a silicone from $(C_2H_5)_2Si(OH)_2$.

10.16 How is the vulcanization of silicone rubber induced at room temperature?

10.17 Discuss the mechanism of ethene polymerization.

10.18 Suppose you want to prepare a polythene squeeze bottle that weighs 10 g. How many moles of ethene gas do you think you would need?

10.19 What are the so-called vinyl polymers? Discuss the works of Ziegler and Natta using polypropene as an example.

10.20 How are plastics modified for end uses?

10.21 The average molecular weight of the polystyrene you prepared in the laboratory is 10,400 amu. What do you think is the average number of styrene monomers in each polystyrene unit?

10.22 Write the structures of
  a. Neoprene
  b. Bakelite
  c. Dacron
  d. Orlon
  e. Saran
  f. Lucite
  g. Teflon

PLASTICS AND OUR ENVIRONMENT

What began in the 1930s as an effort to find a synthetic substitute for natural rubber to overcome a possible threat of Japanese control over natural rubber supplies has now culminated in what many people refer to as our somewhat unnatural world made of "ticky-tacky" plastics (Fig. 10.7). Approximate estimates indicate that each year over 100 billion lb of plastics are used globally. Inevitably, this enormous consumption of plastic materials has created *waste disposal* problems. It is not unusual to see plastic litter along our roadways, in our parks and recreation areas, and elsewhere. Most plastics that are presently available are not biodegradable and, in consequence, when they are disposed of in sanitary land fills (the so-called open dumps), they do not rot and disappear but just remain buried among other garbage. What if we were to burn them? This solution will only add to the pollution of our air environment. For example, automobile tires burn with sooty, stinking smoke that increases particulate matter and sulfur dioxide in the air. What is worse is that some of these plastics such as polyvinyl chloride (PVC) and polyacrylonitrile (PAN) produce hazardous combustion products such as toxic hydrogen chloride and hydrogen cyanide gases.

Recent concern over the amount of plastic polluters has led to the development of recycling processes. Unfortunately, only thermoplastic materials can be recycled, that is, melted down and remolded. Thermosetting plastics cannot be treated this way because breaking the cross-linking would cause complete molecular degradation. It will, therefore, be necessary to restrict their use to producing items that normally are not thrown away. Or, they will have to be used to serve the purpose of fuels in such a way that the combustion products could be recycled as raw materials for other chemical syntheses. Work is now in progress to study the breakdown of thermosetting plastics and foams. Ford Motor Company has reported the development of a process in which polyurethane foam is hydrolyzed to a liquid mixture of polyether glycol monomers and toluene diamine. The liquid mixture can be separated by distillation and the products used to make more polyurethane foam with an overall efficiency of 80–90 percent. Many laboratories are also developing "biodegradable plastics," but whether these will be practical remains to be seen.

FIGURE 10.7

The "ticky-tacky" world of a child. (Photograph courtesy of J. S. Venugopalan.)

In spite of the threat to our litter and solid waste disposal problems, as each year passes by, the demand for thermosetting plastics is increasing and new plastics are being synthesized for a variety of applications. Clearly, plastics have become very much a part of our lives, so much so that we could not do away with them. The question is: What to do with them when we're through with them? Throw them into our neighbor's backyard or litter our streets and highways so that our homes and cars are clean? Hopefully, there will be other solutions as more intensive research on the disposal of plastics is under way in many chemical laboratories.

QUESTIONS

10.23 What are the hazards involved when plastics such as PVC and PAN catch fire? Design experiments by which these hazards can be demonstrated.

10.24 Why is it not possible to recycle thermosetting plastics?

10.25 How do you suggest we dispose of waste plastics? Should plastics be banned?

SUGGESTED PROJECT 10.1

Plastics and synthetic fibers in your environment (or what thou shalt not burn)

With the increasing use of synthetic materials examples of synthetic fibers and plastics around us are not scarce. Our clothing and footwear, our furniture and kitchenware, our wall paints and floor covers are made either entirely of synthetics or contain synthetics to some degree. The aim of this project is to take an inventory of the plastics and other synthetic fibers in your home and to become aware of the potential hazards presented by some of these materials.

Prepare a list of the plastics and synthetic fibers in your home. Then identify those plastics that contain

a.   polyvinyl chloride

b.   polystyrene

Similarly, identify those garments that contain polyacrylonitrile (Orlon). Because, on combustion, the identified items give off hazardous products such as HCl, HCN, SO_2, and soot, you now have an opportunity to label these items to the effect that they should not be disposed of by burning.

11

Living matter

It is notoriously difficult to define the word "living." In many cases we all know whether something is alive or dead. You are alive; cats and dogs are alive; whereas a rock or a pane of glass is dead. But the word "dead" is a bad one, because it half implies that the object was once alive and is now dead. It is interesting that there is no *simple* word for something that is not alive and never has been.

FRANCIS H. C. CRICK (1916-   )
British biochemist

When chemists began to study the properties of matter, they quickly realized that the substances in living systems are quite different from those in the nonliving world. Most of the substances making up living systems were found to be very unstable compared to those making up the earth, sea, and air. For many years the substances derived from living systems were called *organic* substances and plants and animals were considered to possess some "vital force" that is responsible for producing organic substances. Once formed, and present in dead matter, organic substances could be changed into inorganic ones, but never was the reverse sequence observed until 1828, when the German chemist Friedrich Wöhler transformed ammonium cyanate, an inorganic substance, into urea, an organic substance produced in the bodies of animals and excreted in urine:

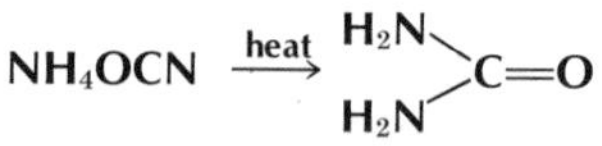

$$NH_4OCN \xrightarrow{\text{heat}} \begin{array}{c} H_2N \\ H_2N \end{array}\!\!>\!\!C{=}O$$

ammonium cyanate   urea

Soon afterward the vital force concept was abandoned. But the fact remained that organic substances form the basis for life and that the functioning of living systems is the result of chemical changes undergone by complex organic substances.

The branch of chemistry concerned with the chemical changes that take place within living systems is known as *biochemistry*. Progress in our understanding of the fundamental biochemical facts of life began with the realization that everything living, from microbe to man, ingests food, uses the chemical raw materials thus obtained to build its own structure, and in the process transforms and converts energy to its own needs. On this basis, it was possible to explore the mechanism by which living things reproduce themselves and transmit their characteristics to their descendants.

LIVING VERSUS NONLIVING MATTER

What are the distinguishing characteristics of living matter? To answer this question, we have to recognize that the living matter on earth is constituted of a huge collection of organisms that includes animals ranging from Protozoa to whales, plants ranging from algae to giant sequoias, and all the microscopic bacteria. Each of these organisms is a highly structured, highly ordered, highly integrated, but very complex self-contained entity, set off from its surroundings by some kind of boundary or limited surface through which it can interact with its environment. The interaction is called *metabolism* and involves the flow of energy and materials into and out of the organism for the maintenance of its state of organization as well as for growth to even more complex and highly organized states. Metabolism includes all of the chemical reactions that take place within the living organism and is absolutely necessary if the organism is to maintain the high degree of organization necessary for life.

The high degree of organization coupled with the metabolic interaction with the surroundings is one of the major characteristics of living systems. Another major characteristic of life is its ability to reproduce its own kind, to make replicas of itself generation after generation. If any one characteristic could be said to define life, it would probably be this property of *self-replication*. Other characteristics that we normally associate with living organisms are growth and reaction to an external stimulus, but these characteristics are really qualities that follow from structural organization, interaction with surroundings, metabolism, and reproduction.

It must, however, be realized that nonliving matter may also possess some of the above-mentioned properties of living systems. In other words, the distinction between living and nonliving matter is not always clear-cut. The modern approach is, therefore, to describe the living state in terms of its behavior over long periods of time. The apparent self-replication of living organisms will then allow for changes, for adaptations through evolution, which reproducing nonliving systems cannot achieve. These changes, or *genetic mutations,* which prove useful for the existence and stability of the organism, are themselves reproduced. Thus, a living system is any self-replicating, metabolizing system capable of mutation. *Viruses* are considered by many to be the simplest organisms that can be termed living or "nearly living." They probably do not metabolize in the ordinary sense, but they are capable of mutation and certainly reproduce themselves with the help of a host biological cell.

The *cell* is considered to be the fundamental unit of structure in living matter. It is the basic unit of life much as an atom is the basic unit of matter. With few exceptions all cells have the same intracellular organization, consisting of a *nucleus* surrounded by the nuclear membrane and imbedded in the *cytoplasm,* which, in turn, is surrounded by the cell membrane. The smaller organizations within these two major components of the cell are shown in Fig. 11.1 which is a simple representation of a highly generalized cell. It is the organization of the components of a cell that gives rise to the characteristics of life—characteristics that are different from the properties of known combinations of elements present in inanimate nonliving matter. As we shall see later in this chapter, our interest in the cell centers around its activity as a miniature but very intricate chemical factory where complex energy-laden molecules are degraded, and the energy released in the oxidative degradation process is used to synthesize other biological molecules and to perform various kinds of work.

LEFT-HANDED MOLECULES AND LIFE From a chemical point of view the cell is made of many types of organic molecules and water. Many of these organic molecules are *optically active,* that is, they rotate the plane of polarization of light. Ordinary light is composed of electric and magnetic fields that vibrate in all planes. Plane-polarized light has the electric (and magnetic) field vibrating in only one plane (Fig. 11.2) and is produced by passing a beam of light through special crystals such as a pair of joined crystals of the mineral Iceland spar (a form of $CaCO_3$).

Originally it was the French biochemist Louis Pasteur who discovered the existence of optically active molecules. He found that tartaric acid solutions, obtained as a by-product of making wine from grapes, rotated the plane of polarization

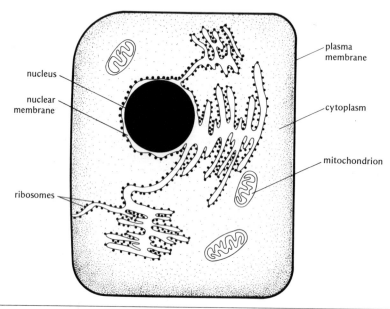

FIGURE 11.1

Schematic drawing of an animal cell. From Roger G. Gymer, *Chemistry: An Ecological Approach,* p. 574. Copyright © 1973 by Roger G. Gymer. Redrawn, by permission from Harper & Row, Inc.

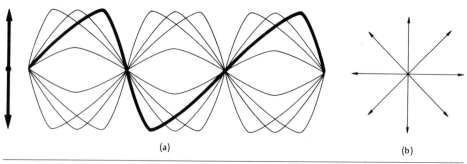

FIGURE 11.2

A representation of the rays of ordinary light and of plane-polarized light. The electric field of plane-polarized light (a) vibrates in one plane, whereas that of ordinary light (b) in many planes.

of light clockwise (to the right of the observer). He also found that tartaric acid solutions, synthesized from simpler molecules in the laboratory, were optically inactive to begin with, but rotated polarized light anticlockwise (to the left of the observer) after molds were allowed to grow in them. The two forms of tartaric acid are now called *optical isomers* or *stereoisomers* and are distinguished from one another by using the designations D (dextrorotatory) and L (levorotatory) to refer to different structural configurations. The structural origin of these isomers is seen by looking at the molecular structure of tartaric acid:

COOH   COOH
| |
H—C—OH   HO—C—H
| |
HO—C—H   H—C—OH
| |
COOH   COOH
D-tartaric acid   L-tartaric acid

Like the cis and trans geometrical isomers discussed in Chapter 10, optical isomers arise because of different arrangements of atoms in space. Generally speaking, when four different atoms or groups of atoms are bonded to a carbon atom, the carbon atom is said to be *asymmetric* and the resulting molecule can take two different forms that are mirror images of each other (Fig. 11.3). These forms are not superimposable and exhibit the same type of symmetry relationship as your right and left hands. The two forms will melt and boil at the same temperatures and will

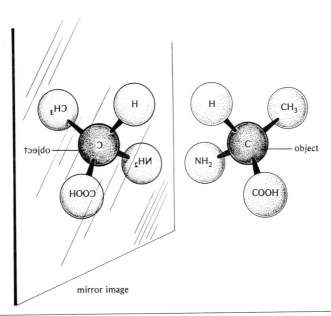

FIGURE 11.3

Spatial arrangement in optical isomers. The model shows the two possible arrangements in space of four groups around an asymmetric carbon atom. Note that the two forms have an object–mirror image relationship and that one cannot be superimposed on the other for all points to coincide. Reproduced, by permission, from Charles W. Keenan and Jesse H. Wood, *General College Chemistry*, 4th ed., p. 637. Copyright © 1957 by Harper & Row, Inc. Copyright © 1961, 1966, 1971 by Charles William Keenan and Jesse Hermon Wood.

behave alike so far as most chemical reactions are concerned. In biochemical reactions, however, a molecule with one spatial arrangement may participate freely, whereas its optical isomer will react either not at all or only extremely slowly. In laboratory synthesis of optically active compounds, there is usually a random positioning of the four groups in space so that the number of molecules of D and L configuration are equal. But, in the biological synthesis of most plant and animal materials, this randomness does not occur; rather the groups are positioned so that only the left-handed (L-configuration) molecules are present.

Clearly, somewhere in the course of the chemical evolution of life, the left-handed form was chosen, perhaps by sheer chance, and essentially all biochemical reactions and biological structures have since been restricted to this configuration. Moreover, most reactions that occur in plants and animals take place at a measurable rate only in the presence of catalytic substances and only when the molecules involved fit each other snugly at active sites. A molecule with one type of shape will not fit a molecule with the mirror-image shape or the opposite shape.

QUESTIONS

11.1 What is the "vital force" theory? What led to its downfall?

11.2 Distinguish between living matter and nonliving matter.

11.3 Explain what is meant by
   a. metabolism
   b. self-replication
   c. genetic mutation

11.4 What is the basic unit of life? Describe its structural organization.

11.5 What is meant by optical activity? Give examples. Which optical configuration is prevalent in living matter?

THE MOLECULES OF LIFE

On the molecular level, the organization and complexity of living systems is indicated by the structural complexity of the large organic polymers or macromolecules always found in these systems. Yet living systems use chiefly six of the chemical elements (see Fig. 5.1) to serve as the building blocks of a large number of macromolecules. It is the previously discussed unusual properties of the carbon atom that permit the formation of an incredible variety of complex molecules (see Chapter 9).

The macromolecules of life fall into four major classes: *carbohydrates, lipids* (fats), *proteins,* and *nucleic acids.* The first three, carbohydrates, fats, and proteins, are used as foods by living organisms. Often the body breaks down the food molecules and reconstitutes them in subtly differentiated forms. For example, the *enzymes,* the important catalysts that the body uses to accomplish virtually every chemical reaction, are specially constructed proteins. These proteins or enzymes are made from other proteins. Nucleic acid, the fourth kind of biochemical macromolecule, is built from monomers that are themselves composite molecules called *nucleotides* because of their origin in cell nuclei. The molecules known as deoxyribonucleic acid and ribonucleic acid, the carriers of inherited information, are examples of such polymers. The arrangement of the nucleotides is specific for every individual.

In this section we shall discuss the structural features of these molecules of life and comment on their functions in living systems.

CARBOHYDRATES   Carbohydrates are compounds of the elements carbon, hydrogen, and oxygen of general formula $C_n(H_2O)_m$ (hence the name carbohydrate). They are found widespread in living organisms mainly as free sugars, starches, glycogen, and cellulose. Starch occurs naturally in many plants, where it functions as a food storage system. Glycogen is a carbohydrate storage material for animals, analogous to the starch of plants. Cellulose is found mainly in the structural parts of plants and is used as food by some animals (see p. 245). All of these hydrolyze to yield simple sugars. Most simple sugars contain from three to seven carbon atoms. The six-carbon atom sugar, *glucose* ($C_6H_{12}O_6$), is the monomer constituting large carbohydrate molecules such as starch and cellulose. Such simple sugars are referred to as *monosaccharides*. Condensation polymers of monosaccharides are logically called *polysaccharides*. Starch and cellulose are common examples of polysaccharides present in living matter. The sugar most commonly encountered as a sweetening agent is a disaccharide called *sucrose* ($C_{12}H_{22}O_{11}$).

The structures of the more common carbohydrates are given in Table 11.1. A close examination of these structures will reveal the presence of asymmetric carbon atoms and hence the great variety of possible asymmetric mirror-image forms of molecules (stereoisomers). For example, the glucose structure alone allows 32 stereoisomers that will differ in the orientation of the H's and OH's attached to five of the six carbon atoms; only two isomers, however, are called glucose. The presence of so many OH groups, containing oxygen atoms with two pairs of electrons not used in bonding, allows hydrogen bonding by the hydrogen atoms in water. The consequent attractive interactions enable glucose and water molecules to intermingle freely. Because the body is about 65 percent water by weight, water-soluble glucose can readily be transported to any site for metabolism. It is the metabolic reaction of glucose, otherwise known as the "burning" or oxidation of glucose to CO_2 and H_2O,

$$C_6H_{12}O_6 + 6O_2 \longrightarrow 6CO_2 + 6H_2O$$

that gives humans and animals about half their total energy. The blood normally contains 0.08-0.1 percent glucose, but in diabetics the level may be much higher.

There are two other monosaccharides that we shall encounter later in this chapter under Nucleic Acids (p. 277). Both are five-carbon atom sugars and are important constituents of the nucleic acid group of large biological molecules. These two sugars are *ribose* and *deoxyribose*,

ribose   deoxyribose

the latter lacking one of ribose's oxygen atoms.

Many plants synthesize the disaccharide sucrose by joining together a glucose with a fructose unit:

$$C_6H_{12}O_6 + C_6H_{12}O_6 \longrightarrow C_{12}H_{22}O_{11} + H_2O$$

glucose   fructose   sucrose

TABLE 11.1
Common carbohydrates[a]

Name	Formula	Structure	Occurrence
glucose (also called dextrose or grape sugar)	$C_6H_{12}O_6$	CH₂OH ... (ring structure with H, OH, HO, OH)	fruits, honey, blood, urine of diabetics
fructose (also called levulose or fruit sugar)	$C_6H_{12}O_6$	HOCH₂ ... OH, HO, CH₂OH, OH, H (ring structure)	fruit juices, honey
sucrose (also called cane sugar)	$C_{12}H_{22}O_{11}$	CH₂OH ... CH₂OH ... (two-ring structure)	sugar cane, sugar beets, fruits and juices of many plants
lactose (also called milk sugar)	$C_{12}H_{22}O_{11}$	CH₂OH ... CH₂OH ... (two-ring structure)	milk of mammals
maltose (also called malt sugar)	$C_{12}H_{22}O_{11}$	CH₂OH ... CH₂OH ... (two-ring structure)	germinating grains
starch	$(C_6H_{10}O_5)_n$	CH₂OH ... CH₂OH ... (repeating ring structure)	plants
cellulose	$(C_6H_{10}O_5)_n$	CH₂OH ... (repeating ring structure)	plants

[a] The C's in the ring have been omitted.

The reaction is a condensation-type *dimerization* in which water is formed and removed from two OH groups, one on each unit (Table 11.1). The C—O—C linkage formed between two joined simple sugar units is called a *glycoside* linkage. The glycoside linkage of sucrose, when eaten, is hydrolyzed to yield glucose and fructose, which are then utilized by the body. Both honey and the candy syrup called "invert sugar" contain glucose and fructose. Other important disaccharides are lactose (milk sugar) and maltose (malt sugar).

Many plants also polymerize glucose by a similar condensation reaction to produce polysaccharides:

$$nC_6H_{12}O_6 \longrightarrow (C_6H_{10}O_5)_n + nH_2O$$

$$\text{glucose} \qquad \text{starch or} \atop \text{cellulose}$$

Starch, glycogen, or cellulose polymers are formed depending on the spatial arrangement of the groups around the two asymmetric carbons by which two glucose units are joined. The two forms of this linkage are called α-glycoside (in the case of starch and glycogen) and β-glycoside (in the case of cellulose) linkages. If the arrangement around these carbons in starch or glycogen is right-handed, then the arrangement around the same carbons in cellulose is left-handed (Table 11.1). Thus, starch or glycogen and cellulose are distinctly different substances with different properties. A basic difference is that humans cannot digest cellulose but can digest starch. Even though cellulose is a carbohydrate, it is not a food for humans. The digestion of starch or glycogen, obtained from such plant foods as grain or potatoes, is a *depolymerization* process that involves hydrolysis reactions, that is, reactions with water, in the presence of enzymic catalysts. Glucose is the end product and is absorbed through the intestinal walls into the blood and transferred to the cells to take part in other metabolic processes. Herbivorous (plant-eating) animals such as horses and cattle, in contrast to humans, have the ability to digest cellulose contained in food such as grass, probably due to the action of bacteria in the digestive tract of horses and cattle. Apparently, these bacteria do not exist in the human digestive system.

LIPIDS A diverse class of organic compounds, lipids are essential for all plant and animal cells. They are characterized by the presence of fatty acids or their derivatives (see Chapter 9) and by their solubility in nonpolar organic solvents. Chemically, lipids are composed of carbon, hydrogen, oxygen, and occasional

TABLE 11.2
Some classes of lipids

Class of lipid	Composition
fats and oils	esters of fatty acids with glycerol
waxes	esters of fatty acids with alcohols other than glycerol
phospholipids	compounds that contain phosphoric acid, fatty acids, glycerol, and nitrogenous compounds
glycolipids	compounds that contain a carbohydrate, fatty acid, and an amino alcohol
steroids	high molecular weight, cyclic alcohols that include sex hormones, the pill, and vitamin D

nitrogen and phosphorus. At present there is no generally accepted method for the classification of lipids due to their widely different compositions and structures, but they are known to include the classes of compounds listed in Table 11.2. The discussion given here is confined to fats and oils, phospholipids, glycolipids, and steroids. A discussion of vitamins and hormones is deferred for a later section in the chapter.

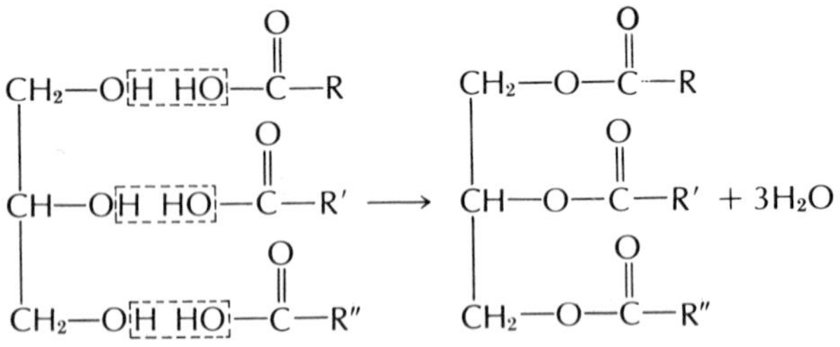

$$
\begin{array}{ccc}
\underset{|}{CH_2-O[H\ HO]}-\overset{\overset{\displaystyle O}{\|}}{C}-R & & \underset{|}{CH_2-O}-\overset{\overset{\displaystyle O}{\|}}{C}-R \\[2mm]
\underset{|}{CH-O[H\ HO]}-\overset{\overset{\displaystyle O}{\|}}{C}-R' & \longrightarrow & \underset{|}{CH-O}-\overset{\overset{\displaystyle O}{\|}}{C}-R' + 3H_2O \\[2mm]
CH_2-O[H\ HO]-\overset{\overset{\displaystyle O}{\|}}{C}-R'' & & CH_2-O-\overset{\overset{\displaystyle O}{\|}}{C}-R''
\end{array}
$$

glycerol   fatty acids   fat (solid) or oil (liquid)
(triglyceride)

TABLE 11.3
Fatty acids occurring in natural fats

Name	Formula	Occurrence
butyric acid	$CH_3(CH_2)_2COOH$	butter fat
caproic acid	$CH_3(CH_2)_4COOH$	butter fat
caprylic acid	$CH_3(CH_2)_6COOH$	butter fat
capric acid	$CH_3(CH_2)_8COOH$	butter fat, coconut oil
lauric acid	$CH_3(CH_2)_{10}COOH$	coconut oil
myristic acid	$CH_3(CH_2)_{12}COOH$	butter fat, coconut oil, nutmeg oil
palmitic acid	$CH_3(CH_2)_{14}COOH$	lard, beef fat, butter fat, cottonseed oil
stearic acid	$CH_3(CH_2)_{16}COOH$	lard, beef fat, butter fat, cottonseed oil
arachidic acid	$CH_3(CH_2)_{18}COOH$	peanut oil
cerotic acid	$CH_3(CH_2)_{24}COOH$	beeswax, wool fat
palmitoleic acid	$CH_3(CH_2)_5CH{=}CH(CH_2)_7COOH$	cod-liver oil, butter fat
oleic acid	$CH_3(CH_2)_7CH{=}CH(CH_2)_7COOH$	lard, beef fat, olive oil, peanut oil
linoleic acid	$CH_3(CH_2)_4CH{=}CHCH_2CH{=}CH(CH_2)_7COOH$	cottonseed oil, soybean oil, corn oil
linolenic acid	$CH_3CH_2CH{=}CHCH_2CH{=}CHCH_2CH{=}CH(CH_2)_7COOH$	linseed oil
ricinoleic acid	$CH_3(CH_2)_5CHOHCH_2CH{=}CH(CH_2)_7COOH$	castor oil

Fats and oils Fats and oils are the most abundant lipids in nature. They are organic esters (sometimes called *triglycerides*) formed by the reaction between glycerol (a trihydroxy alcohol) and a variety of carboxylic acids called *fatty acids* (see Chapter 9):

Because each fat or oil contains glycerol, any differences between them are probably due to differences in the fatty acid components. Many of the fatty acid components and their common sources are listed in Table 11.3. The sources listed are not exclusive. Naturally occurring fatty acids usually have an even number of carbon atoms. Solid fats, such as lard, are esters of glycerol and saturated fatty acids. Liquid fats are called oils. Structurally, these oils are glyceryl esters of unsaturated fatty acids. Corn, peanut, linseed, and olive oils are familiar liquid fats.

When esterification occurs between fatty acids (such as myristic, palmitic, and cerotic acids) and high molecular weight alcohols (such as cetyl alcohol [$CH_3(CH_2)_{15}OH$] and myrcyl alcohol [$CH_3(CH_2)_{29}OH$]), the lipids formed are called *waxes*. Lipid waxes are not to be confused with paraffin wax which is merely a mixture of hydrocarbons. Examples of common, naturally occurring waxes are beeswax (found in honeycomb), lanolin (obtained from wool), spermaceti (obtained from the sperm whale), and carnauba wax (obtained from palm leaves). Lanolin finds wide use as a base for many ointments, salves, and creams; spermaceti is used in cosmetics, pharmaceuticals, and in the manufacture of candles; and carnauba is used in floor waxes and in furniture and automobile polishes.

The reactions of fats and oils are of commercial importance. *Hydrolysis,* the reverse of the esterification reaction that produces the fat or oil, is an important reaction. The product of hydrolysis in the presence of a mineral acid, such as hydrochloric acid, is a fatty acid or a mixture of fatty acids. However, when hydrolysis is carried out in the presence of strong alkalis such as sodium hydroxide, the process is called *saponification* and the product is soap (see Chapter 9). Soaps are the sodium or potassium salts of fatty acids and find wide application as cleansing agents (see Chapter 14). Another important reaction mainly of oils is *hydrogenation* in which the degree of unsaturation is reduced by the addition of hydrogen at the double bonds of the fatty acids. The reaction is most often employed in the production of solid cooking shortenings or margarines from liquid vegetable or animal oils. A metal catalyst (finely divided nickel) is used to speed up the reaction. When fats and oils are exposed to warm moist air for a period of time they become *rancid*. Rancidity is caused by the hydrolysis of ester linkages and oxidation of carbon-chain double bonds in the fats and oils. Both these reactions produce fatty acids, many of which have disagreeable odors. The odor of butyric acid, for example, is the characteristic odor of rancid butter!

Fats and oils play an important role as sources of energy in humans. Some fats are burned as fuel, some are used to build important constituents of the body cells, and still others are stored in the body to serve as a source of energy in times of prolonged hunger. Fat also serves to insulate the body against loss of heat and to protect vital organs from injury.

Phospholipids Phospholipids are fats in which one of the fatty acid groups has been replaced by an esterified phosphate group. More specifically, they are composed of glycerol, fatty acids, phosphoric acid, and a nitrogen compound. Important phospholipids include the lecithins, cephalins, and the sphingomyelins. The structures of these compounds are such that the phosphate group contains a small polar entity such as —CH_2—CH_2—NH_3^+ or its derivatives.

Naturally occurring phospholipids may contain at least five different fatty acids. They are found largely in brain and nervous tissue and are essential constituents of the living matter (protoplasm) of cells. The cephalins are involved in the blood-clotting process, and the lecithins in the transportation of fats from one tissue to another. The toxic effects produced by some insects, spiders, and snakes are attributed to the enzymic conversion of lecithin into *lysolecithin,* a compound resulting from the hydrolytic removal of oleic acid from the central carbon atom of lecithin.

Glycolipids Lipids that contain a carbohydrate are known as glycolipids. They are also called *cerebrosides* because of their natural occurrence in brain and nervous tissue. There are four different cerebrosides, each containing a carbohydrate, an amino alcohol, but a different fatty acid. The carbohydrate in these lipids is usually galactose (an isomer of glucose), although glucose is sometimes present.

Steroids The lipid material from body tissue that is not saponifiable by alkaline hydrolysis contains compounds with the hydrogenated cyclopentanophenanthrene nucleus, commonly called the *steroid nucleus:*

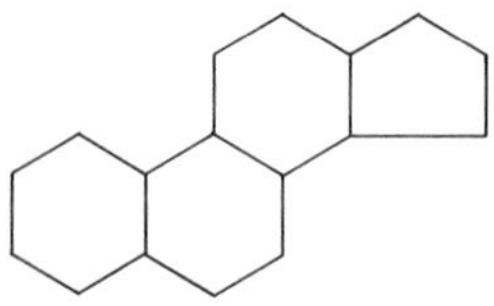

steroid nucleus

Such compounds are classified as steroids. Besides the steroid skeleton, the OH group, CO group, and double bond are often found in steroids; however, the ester linkage, so characteristic of fats, oils, and waxes is seldom present. *Cholesterol* is the most abundant steroid in the body and is found in the brain and nerve tissue as well as in gallstones. It is a sterol (steroid alcohol) and has the following structure:

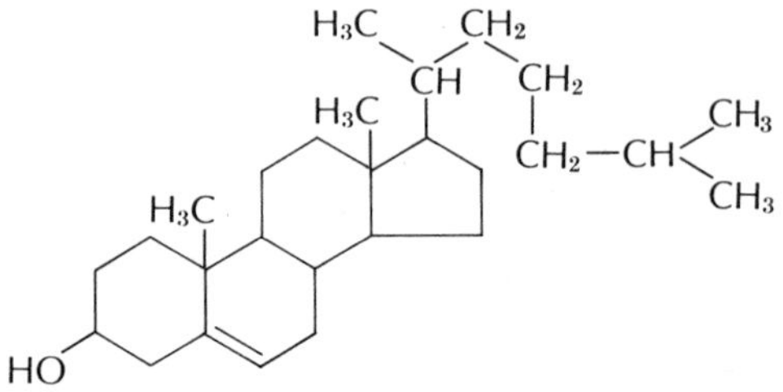

cholesterol

Several isomeric forms of cholesterol are possible because of the large number of asymmetric carbon atoms present in its structure. The importance of cholesterol arises from the fact that it is the precursor of bile acids, vitamin D, and many types of hormones, particularly the sex hormones.

Most cholesterol in the body does not come directly from foods but is synthesized in the liver from fats, carbohydrates, and proteins. All normal blood contains cholesterol. The disease known as atherosclerosis (meaning hardening of the arteries) is attributed to increasing concentrations of cholesterol in blood and the consequent accumulation of cholesterol on the artery walls. Atherosclerosis can result in blood clots and produce either coronary occlusion, coronary thrombosis, myocardial infarct, or simply a heart attack—the nation's No. 1 killer. There is considerable evidence to indicate that a diet of unsaturated vegetable fats in place of the more saturated animal fats produces a lowering of the cholesterol level in the bloodstream and a lowering of the incidence of heart disease. Why the human body is less able to synthesize cholesterol from unsaturated fats than from the saturated kind remains to be answered.

PROTEINS Proteins (from the Greek *proteios*, meaning "of prime importance") are fundamental constituents of all cells and tissues in the body and are normally present in all fluids except the bile and urine. Unlike fats and carbohydrates, proteins have a relatively high content of nitrogen. Most naturally occurring proteins contain the elements carbon (53%), hydrogen (7%), oxygen (23%), nitrogen (16%), sulfur (1%), and phosphorus (0.1%). Plant cells synthesize proteins by a process starting with photosynthesis from carbon dioxide, water, nitrates, sulfates, and phosphates. Animals can synthesize some of the building blocks (called *amino acids*) for making proteins but are dependent on plants or other animals for their source of some dietary amino acids that are essential for the synthesis of body tissue, enzymes, certain hormones, and protein components of the blood. Other elements, such as iodine and iron, may be essential constituents of certain specialized proteins. For example, iodine is a basic constituent of the protein in the thyroid gland and hemoglobin of the blood is an iron-containing protein.

The complexity of proteins is revealed by the large number of amino acids that have been isolated by hydrolyzing protein samples. It was previously pointed out that amino acids are bifunctional molecules containing both a carboxylic acid (COOH) and an amine (NH_2) functional group and that they are linked together by peptide (amide) bonds to form proteins (see Chapter 10). Table 11.4 lists the names and structures of 20 amino acids that are commonly found, but not necessarily in each protein sample. Of these, eight are considered to be "essential" for humans inasmuch as the body is unable to synthesize them from other molecules and, therefore, they must be provided in the proteins of the diet. All amino acids except glycine contain an asymmetric carbon atom in their formulas. For this reason they may exist in the D or L form. Naturally occurring amino acids from plant and animal sources have the L configuration. However, alanine and glutamic acid, which have been obtained from the cell walls of microorganisms, have the D configuration.

Although the composition of a protein can be established by separating and identifying the amino acids produced when the protein molecule is hydrolyzed, the sequence in which these amino acids are arranged in the protein molecule can only be determined by the use of difficult and laborious techniques. In consequence, the amino acid sequence has been investigated only for a small number of proteins. The protein molecules investigated include *insulin* and *ribonuclease*.

Insulin is a protein formed in the pancreas. It promotes the utilization of sugar in the body. We all know that diabetes is a disease in humans resulting from insulin deficiency. The complete molecular sequence of insulin was determined by the British biochemist Frederick Sanger (and later confirmed by synthesis) to consist of 51 amino acid units arranged in two chains connected by disulfide bridges of cystine molecules as shown in Fig. 11.4. Ribonuclease, a more complicated protein found in the pancreas, has 124 amino acids linked together in a continuous chain that is intraconnected by means of four disulfide bridges (Fig. 11.5). Its total synthesis from 124 units of 19 different amino acids was accomplished in 1968 by two independent teams, one at Rockefeller University and one at the Merck, Sharpe, and Dohme laboratories.

TABLE 11.4
Common amino acids found in proteins

Name	Abbreviated name	Structure
glycine	Gly	CH₂—COOH | NH₂
alanine	Ala	CH₃—CH—COOH | NH₂
valine[a]	Val	CH₃ | CH₃—CH—CH—COOH | NH₂
leucine[a]	Leu	CH₃ | CH₃—CH—CH₂—CH—COOH | NH₂
isoleucine[a]	Ile	CH₃ | CH₃—CH₂—CH—CH—COOH | NH₂
phenylalanine[a]	Phe	⬡—CH₂—CH—COOH | NH₂
tyrosine	Tyr	HO—⬡—CH₂—CH—COOH | NH₂
tryptophan[a]	Try	(indole ring)—CH₂—CH—COOH | NH₂
serine	Ser	OH | CH₂—CH—COOH | NH₂
threonine[a]	Thr	OH | CH₃—CH—CH—COOH | NH₂

[a] Amino acids essential for humans — must be provided in the proteins of the diet.
[b] Basic amino acids.
[c] Acidic amino acids.
[d] Sulfur-containing amino acids.

Because proteins are such large polymers, interactions between parts of the polymer chain are inevitable. One type of such interactions (see Chapter 10) results in hydrogen bonding and a helical or coiled structure for proteins. In such a structure the repeating —C=O and —NH units along the chain are twisted into just the

	TABLE 11.4 (Cont.)	
	Common amino acids found in proteins	
Name	Abbreviated name	Structure
proline	Pro	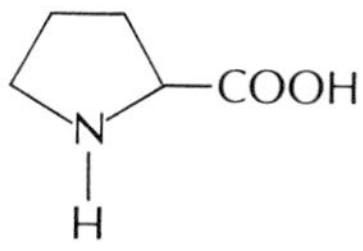
hydroxyproline	Hypro	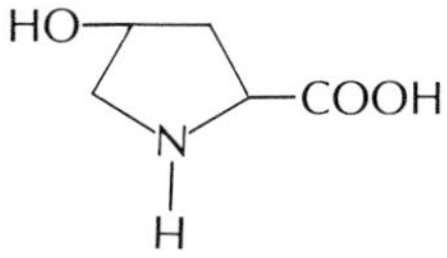
histidine[b]	His	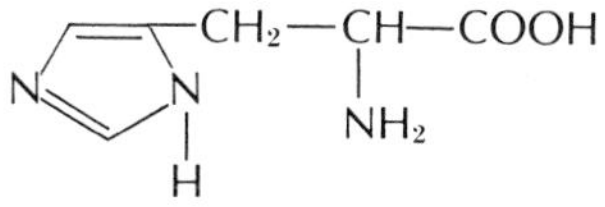 $-CH_2-CH-COOH$, NH_2
lysine[a,b]	Lys	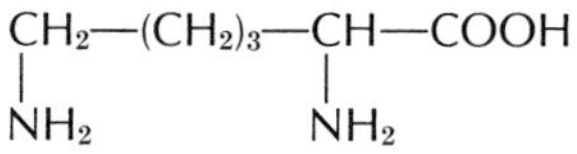 $CH_2-(CH_2)_3-CH-COOH$, NH_2, NH_2
arginine[b]	Arg	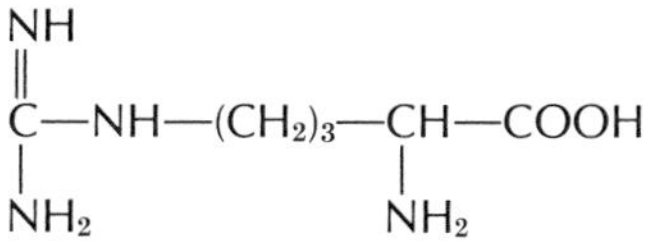 NH, $C-NH-(CH_2)_3-CH-COOH$, NH_2, NH_2
aspartic acid[c]	Asp	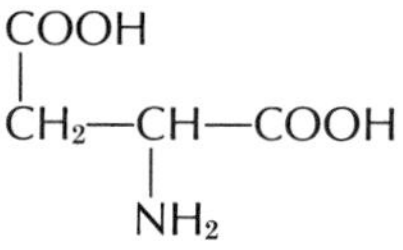 $COOH$, $CH_2-CH-COOH$, NH_2
glutamic acid[c]	Glu	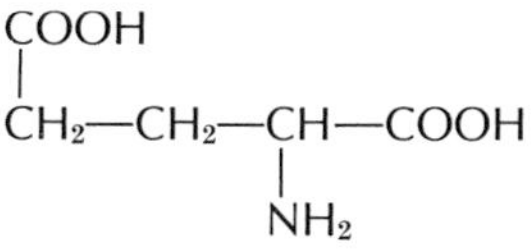 $COOH$, $CH_2-CH_2-CH-COOH$, NH_2
methionine[a,d]	Met	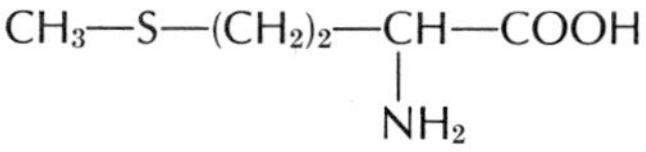 $CH_3-S-(CH_2)_2-CH-COOH$, NH_2
cystine[d]	CyS–SCy	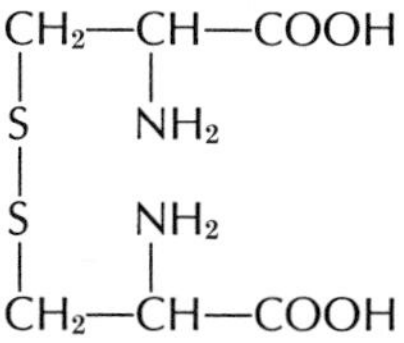 $CH_2-CH-COOH$, S, NH_2, S, NH_2, $CH_2-CH-COOH$
cysteine[d]	CySH	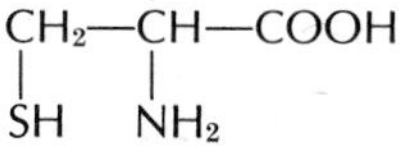 $CH_2-CH-COOH$, SH, NH_2

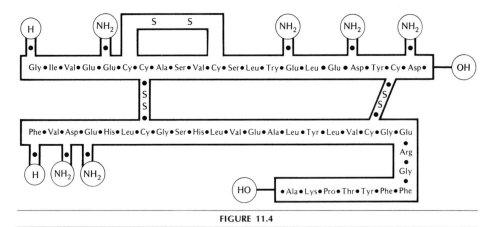

FIGURE 11.4

Structure of the bovine insulin protein molecule.

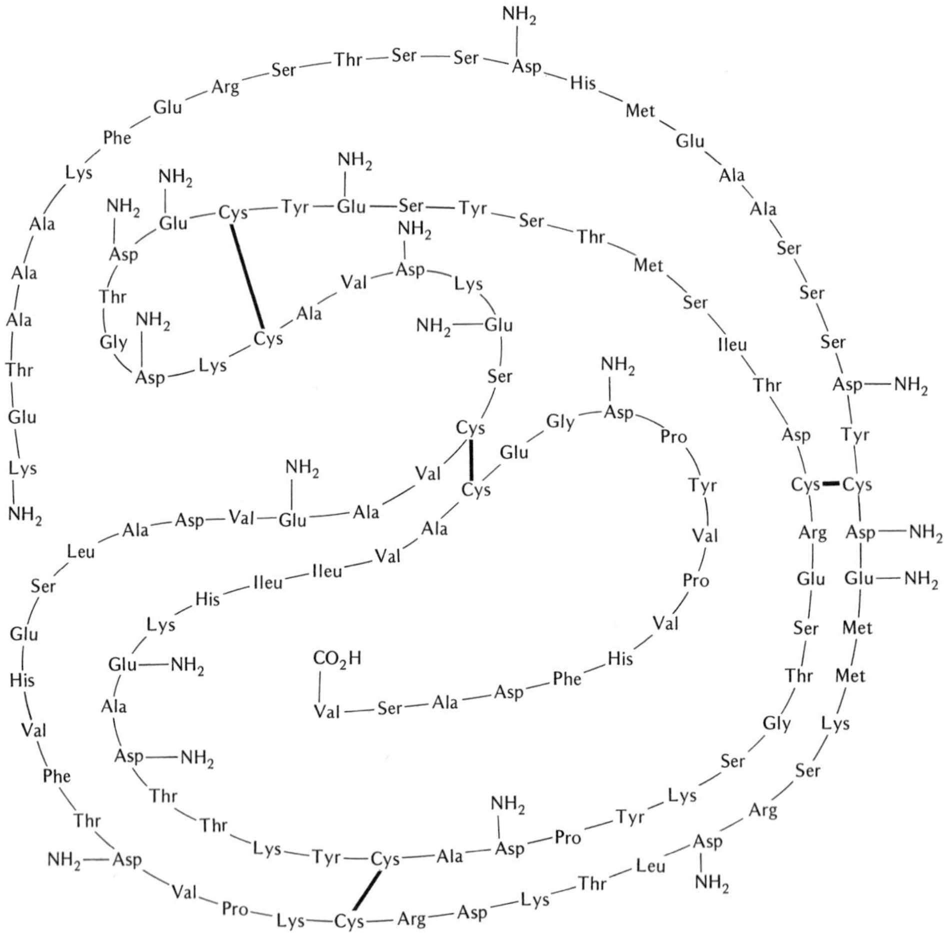

FIGURE 11.5

Structure of the ribonuclease molecule. From Glenn H. Miller, *Chemistry*, p. 373. Copyright © 1969 by Glenn H. Miller. Reproduced by permission from Harper & Row, Inc.

right spatial orientation to have the H on the nitrogen atom attracted to the electrons of the oxygen on the chain at a previous loop of the helix (Fig. 11.6). Another type of interaction results in disulfide bonding (—S—S—) between the same amino acid units, such as cysteine, having the —S—H group (Fig. 11.4). Because the major constituent of hair is a protein that contains numerous disulfide linkages, home permanents involve processes in which these sulfur-to-sulfur linkages are broken, the polypeptide chain curled, and the disulfide linkages re-formed (Fig. 11.7).

On hydrolysis, some complex proteins yield other substances in addition to amino acids. Such proteins are called *conjugated proteins*. A very important class of

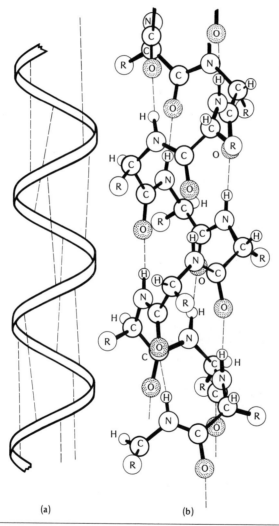

(a) (b)

FIGURE 11.6

Helical structure of proteins (the α helix). (a) Helical arrangement of the peptide chain showing the hydrogen bonds (dashed lines) formed between peptide linkages. (b) The molecular arrangement in which amino acid side chains are represented by R and hydrogen bonds by dashed lines. From Pauling, Corey, and Branson, *Proceedings of the National Academy of Sciences*, **37**, 207(1951). Redrawn, by permission from Professor Pauling and the National Academy of Sciences, Washington, D.C.

conjugated proteins, the *nucleoproteins,* are found in the chromosomes, cell nuclei, viruses, and the heads of sperm cells. Mild hydrolysis of a nucleoprotein yields a protein and a nucleic acid. Because nucleic acids are responsible for the transmission of hereditary characteristics, we shall focus our immediate attention on the chemistry of nucleic acids.

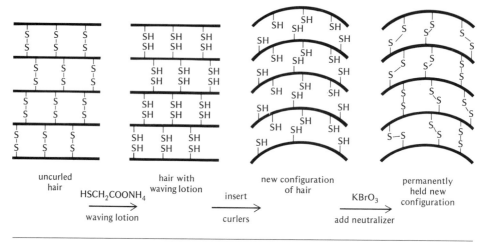

uncurled hair $\xrightarrow[\text{waving lotion}]{HSCH_2COONH_4}$ hair with waving lotion $\xrightarrow[\text{curlers}]{\text{insert}}$ new configuration of hair $\xrightarrow[\text{add neutralizer}]{KBrO_3}$ permanently held new configuration

FIGURE 11.7

Permanent wave process for hair. From Spencer L. Seager and H. Stephen Stoker, *Chemistry: A Science for Today,* p. 401. Copyright © 1973 by Scott, Foresman and Company. Redrawn with permission of the publisher.

QUESTIONS

11.6 List the major classes of molecules of life and point out their sources.

11.7 What are carbohydrates? Give three common examples.

11.8 Describe the chief metabolic reaction of glucose. In what form is glucose stored in the body?

11.9 Distinguish between starch and cellulose.

11.10 What are lipids? Classify them into structural groups.

11.11 How do you distinguish fats, oils, and waxes from each other?

11.12 List the important reactions of fats and oils.

11.13 What is the difference between a phospholipid and a glycolipid?

11.14 Write the structure of cholesterol and point out its importance. What are the consequences of a high cholesterol content in humans?

11.15 In what ways are proteins different from fats and carbohydrates?

11.16 Name the amino acids that are essential to humans. What are their structural differences?

11.17 Describe the amino acid sequence in insulin. Why is this molecule important to humans?

11.18 What interactions cause the helical structure of proteins? How are these interactions applied in the use of home permanents?

11.19 What are nucleoproteins? Give examples and indicate the important products of their hydrolysis.

SUGGESTED PROJECT 11.1

Your food consumption (or how bulky a feeder you really are)

List the foods that you eat according to the approximate amounts of each consumed over a 24-hour period. Exclude food additives (food colorings, common
seasonings such as salt, pepper, and other spices) but include beverages and snacks.
Classify them, in part or whole, as carbohydrates, fats, and proteins.

Next, write down the food items and the approximate amounts of each that
are discarded over a 7-day period in your household. This is best done by a garbage
analysis, that is, by itemizing the contents of your garbage cans before their weekly
disposal. Again, classify the wasted foods as carbohydrates, fats, and proteins.

Based on this information, estimate the approximate annual tonnage of
carbohydrates, fats, and proteins consumed and wasted in your household. Assuming your household to be a typical American household, calculate the approximate
annual tonnages of carbohydrates, fats, and proteins sold by the food industries to
consumers in the United States.

NUCLEIC ACIDS AND THE CHEMISTRY OF LIFE PROCESSES Nucleic
acids are the substances present in the central body or nucleus of a cell. They are
responsible for the action of heredity. All evidence available today indicates that
deoxyribonucleic acid (abbreviated DNA) is the compound of which *genes,* the
basic units of heredity are composed. It is a polymeric molecule and is composed of
monomeric units called *nucleotides.* Nucleotides contain one molecule of phosphoric acid (H_3PO_4), one molecule of a five-carbon atom sugar (ribose or deoxyribose), and one molecule of a nitrogen-containing substance called purine base
(adenine or guanine) or pyrimidine base (thymine or cytosine). Phosphoric acid is
usually obtained from minerals in the diet; the deoxyribose is manufactured by the
body from glucose; the nitrogen-containing purine and pyrimidine bases are made
by the body from nitrogen-containing proteins. The formation of a nucleotide from
these components is accompanied by the removal of two molecules of water:

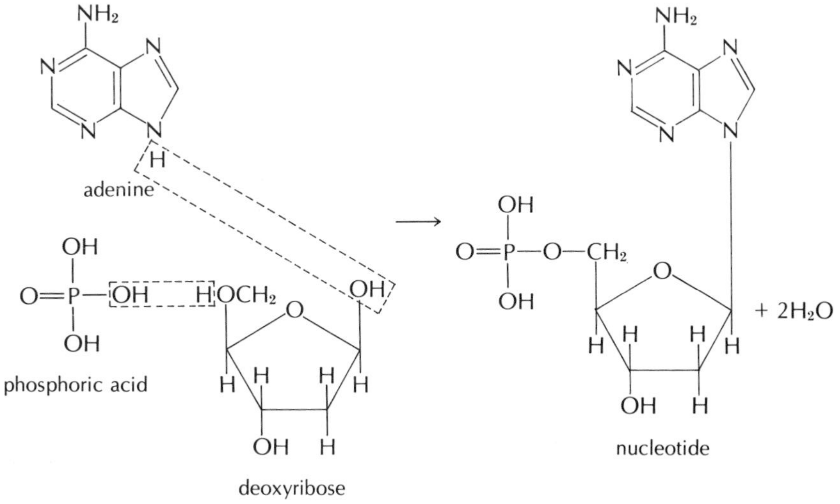

adenine

phosphoric acid

deoxyribose

nucleotide

$+ 2H_2O$

The reverse reaction, hydrolysis, generates the individual components and is used to analyze the composition of nucleotides.

Each nucleotide has the potential to form ester linkages, the OH groups of both the sugar and the phosphoric acid being available for such water-removing reactions. Bonds involving these groups are formed when nucleotides polymerize to form the nucleic acid DNA. Figure 11.8 shows the polymeric structure of DNA as well as a condensed structure used to simplify future representations. Note that the

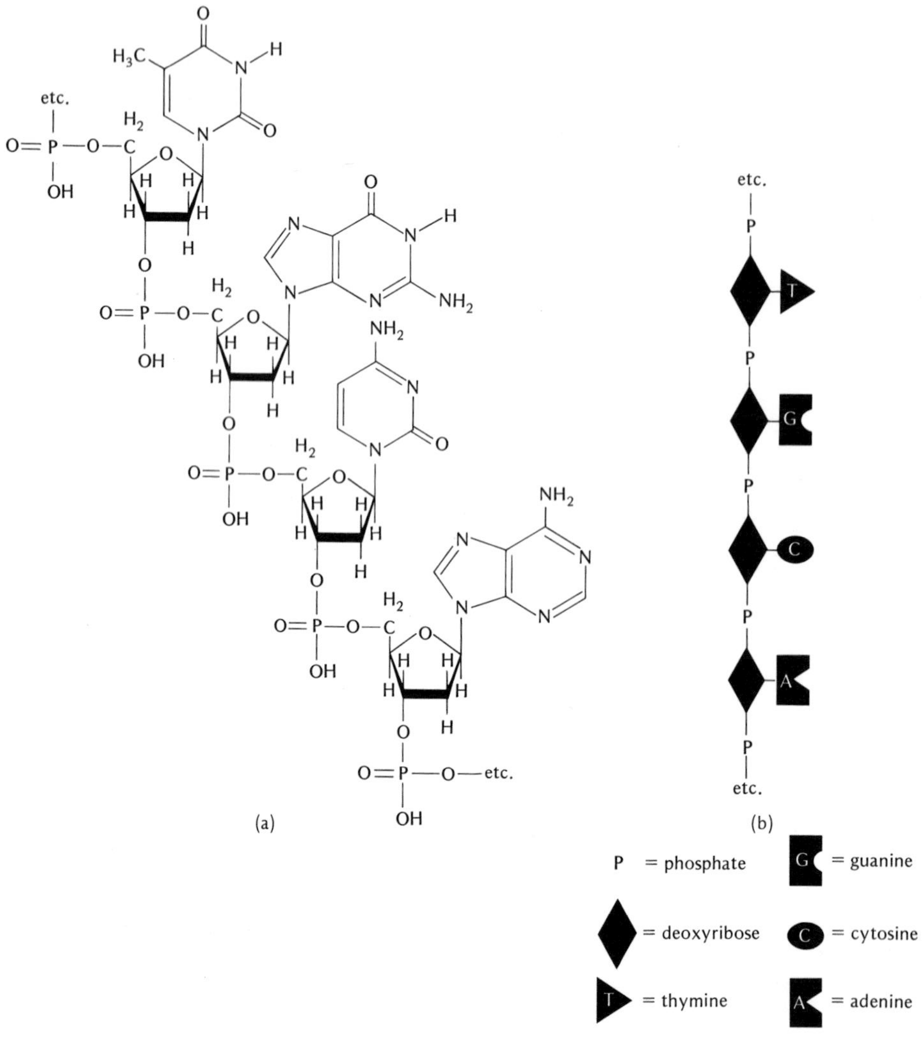

FIGURE 11.8

Polymeric structure of DNA: (a) complete formula; (b) condensed formula. From Spencer L. Seager and H. Stephen Stoker, *Chemistry: A Science for Today*, p. 422. Copyright © 1973 by Scott, Foresman and Company. Redrawn with permission of the publisher.

backbone of the structure consists of alternating phosphate and sugar units. In 1953, the molecular biologists Francis Crick and James Watson proposed that the DNA molecule consisted of two nucleic acid chains, each in the form of a helix coiled around the same axis, and that the helical configuration was the result of hydrogen bonding between base units of the chains. They further proposed that only certain

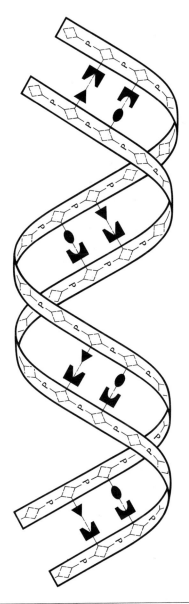

FIGURE 11.9

The double helix structure of DNA, showing the hydrogen bonding between paired bases. From Spencer L. Seager and H. Stephen Stoker, *Chemistry: A Science for Today*, p. 423. Copyright © 1973 by Scott, Foresman and Company. Reproduced by permission of the publisher.

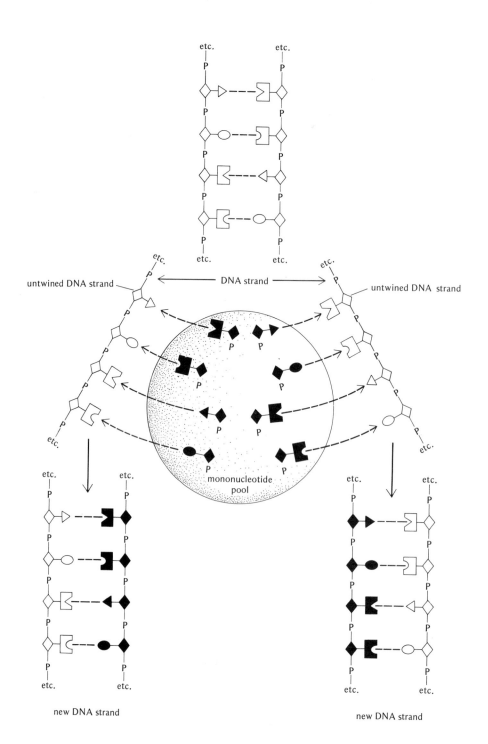

FIGURE 11.10

DNA replication. From Spencer L. Seager and H. Stephen Stoker, *Chemistry: A Science for Today*, p. 424. Copyright © 1973 by Scott, Foresman and Company. Reproduced by permission of the publisher.

pairs of bases had the correct size and structural properties to enable hydrogen bonds to form between chains. The pairs of bases able to hydrogen-bond with each other were adenine (a purine) with thymine (a pyrimidine) and guanine (a purine) with cytosine (a pyrimidine). Therefore, if adenine were in one chain, it would be hydrogen-bonded to thymine of the other chain, as illustrated in Fig. 11.9. This new concept gave the first satisfactory explanation for the process by which DNA could replicate (duplicate) in a dividing cell—truly a breakthrough step in understanding the fundamental chemistry of living matter.

DNA replication   Replication of DNA involves the unwinding of the helix, with each separated strand of nucleic acid then serving as a pattern on which a complementary strand is synthesized. The process produces two double helices of DNA where only one existed before. The new strand is identical to the former partner strand by virtue of the matching properties of the bases. The nucleotides for the new strand are obtained from the pool of liquid medium surrounding the original DNA in the cell nucleus, as shown in Fig. 11.10. Experimental evidence indicates that replication takes place during the unwinding of the strands rather than after the strands have completely separated.

Ribonucleic acid and protein synthesis   We have already noted that the chromosomal DNA has the primary function of replicating to supply new cells with exact copies for its chromosomes. How is it then possible that DNA can also direct such complicated processes as protein synthesis? The fact is that protein synthesis does not take place in the nucleus but rather on the surface of ribosomes (small granular particles in the cytoplasm of the cell; see Fig. 11.1). The nucleic acid that directs protein synthesis is *ribonucleic acid* (abbreviated RNA)—very similar to DNA, but differing in that the sugar portion of the nucleotide is *ribose* instead of deoxyribose and the base portion has *uracil* instead of the thymine of DNA. The polymer structure of RNA is shown in Fig. 11.11.

Three main types of ribonucleic acids are known: (1) ribosomal RNA (abbreviated rRNA), (2) messenger RNA (abbreviated mRNA), and (3) transfer RNA (abbreviated tRNA). Each of them is synthesized from nucleotides under the direction of DNA, that is, by the formation of complementary strands against a DNA pattern. Each has a distinctive function in protein synthesis. For example, the rRNA stabilizes some of the molecules involved in protein synthesis; mRNA carries the code or directions for protein synthesis from the DNA of the cell nucleus out to the ribosomes; tRNA delivers individual amino acids to the site of protein synthesis from an amino acid pool in the sequence called for by the code of the mRNA. At least 20 different molecules of tRNA exist, one for each of the at least 20 different amino acids used in protein synthesis.

The synthesis of a part of a protein chain is diagrammatically represented in Fig. 11.12. First, a section of a DNA helix unwinds to expose a sequence of bases that dictates the sequence of amino acids in a protein (stage I). This section serves as a pattern (template) for the synthesis of a complementary strand of mRNA. Mononucleotides from the pool of liquid medium in the cell nucleus (see Fig. 11.10) move into proper sequence as determined by the base-pairing possibilities with the exposed DNA strand (stage II). The newly formed mRNA migrates to the cytoplasm and attaches to a ribosome (stage III), thus initiating the process in which the information encoded in mRNA is expressed as a precise sequence of amino acids in a polypeptide molecule. These amino acids are then transferred to the mRNA in this precise sequence by properly coded tRNA (stage IV). Finally, the amino acids react with each other to form peptide linkages and grow into a polypeptide chain (stage V).

The order of the bases along the DNA strand is the *genetic information*. Instructions to supply such information for the sequence of amino acids to be assembled into a particular protein are written in a language that is referred to as the *genetic code*. It is a language in which only four "letters" are used, namely, the bases A, U, G, and C, to make three-letter "words" called *codons*. For example, the protein-synthesis machinery may be ordered to use a molecule of glycine amino acid by the codon sequences GGA, GGG, GGC, and GGU. Note that more than one codon can specify a particular amino acid, but no single codon specifies more than one amino acid.

Both DNA and RNA occur in the cells of all living organisms ranging from

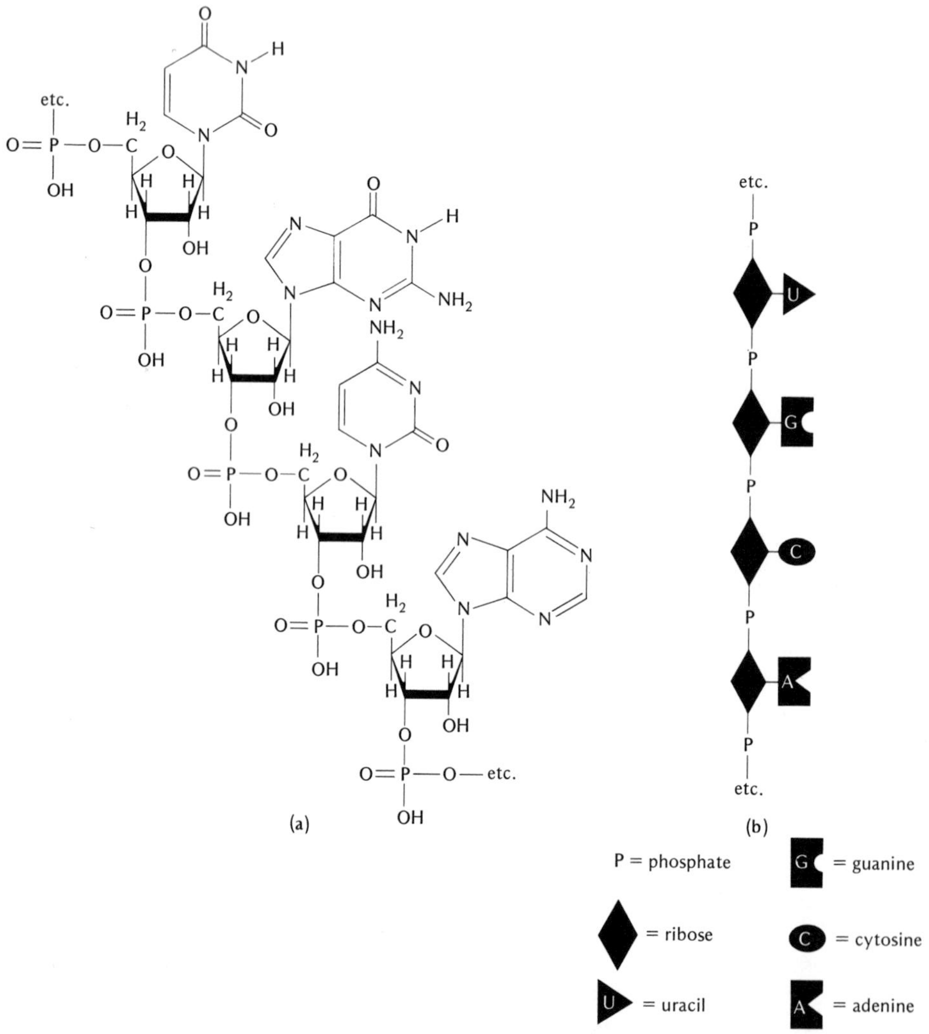

(a)

(b)

P = phosphate   G = guanine

= ribose   C = cytosine

U = uracil   A = adenine

FIGURE 11.11

Polymeric structure of RNA: (a) complete formula; (b) condensed formula. From Spencer L. Seager and H. Stephen Stoker, *Chemistry: A Science for Today*, p. 422. Copyright © 1973 by Scott, Foresman and Company. Redrawn with permission of the publisher.

bacteria to humans. Because no living organism could function without them, the nucleic acids are sometimes referred to as the "key to life." Following the discovery by Watson and Crick of the *double helix,* not only have nucleic acids been synthesized in the laboratory, but also a true *gene*— an entire double-stranded informational unit of DNA. The first total synthesis of a gene from nucleotide starting materials was achieved in 1970 by the Indian biochemist Gobind Khorana while he was working at the University of Wisconsin. Whether a synthetic gene can function in a cell or not is a topic of intensive study at the present time. At any rate, the creation of genes has opened up possibilities for manipulation and alteration of genetic heritage in living matter.

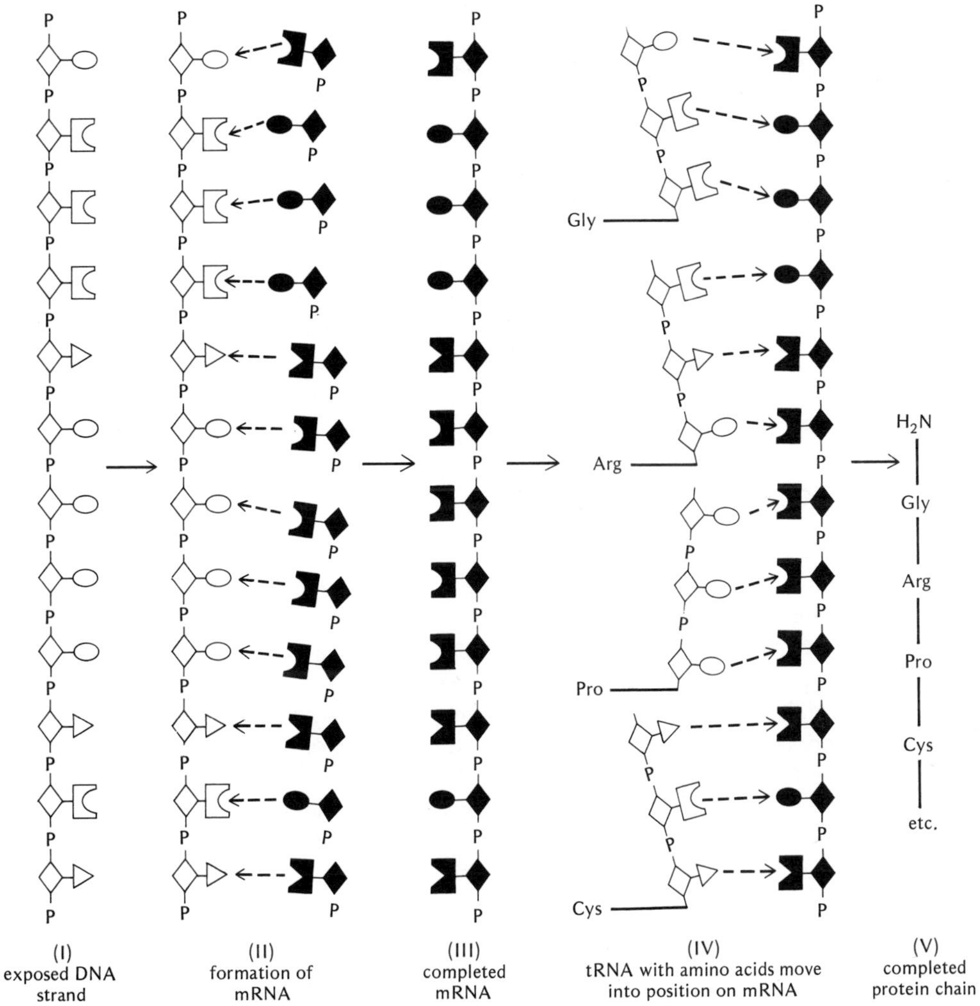

(I)	(II)	(III)	(IV)	(V)
exposed DNA strand	formation of mRNA	completed mRNA	tRNA with amino acids move into position on mRNA	completed protein chain

FIGURE 11.12

Mechanism of protein synthesis. From Spencer L. Seager and H. Stephen Stoker, *Chemistry: A Science for Today,* p. 427. Copyright © 1973 by Scott, Foresman and Company. Reproduced by permission of the publisher.

Viruses A discussion of the chemistry of proteins and nucleic acids would be incomplete without a reference to the class of nucleoproteins known as viruses. Viruses are macromolecules made up of strands of nucleic acids, usually RNA, surrounded by a protective sheath of protein (Fig. 11.13). Ordinarily they are crystalline compounds that do not appear to possess the qualities of life. But when injected into a living organism they can ingest nourishment from their surroundings and reproduce themselves like true living organisms. Apparently, viruses lack the chemical machinery to carry out these functions on their own and require the services of a living cell to perform them for them. For these reasons viruses are considered to occupy a borderline position between living and nonliving matter.

FIGURE 11.13

A representation of a part of a tobacco mosaic virus. The core is a single helical polymer of RNA; the protein coat consists of lobe-shaped globular proteins, which are bonded to one another and are wound helically around the RNA. There are 2200 protein monomers (each containing 158 amino acids) in one virus which has a molecular weight of 40 million.

Much is known about viruses because of their parasitic role in animals and plants. Viruses are responsible for diseases such as smallpox, typhus, chickenpox, measles, yellow fever, poliomyelitis, mumps, influenza, and even the common cold. When a virus attaches to a living cell, the genetic material (DNA or RNA) is injected through the cell wall; once inside the cell, the message on the viral genetic material directs the cell's metabolic machinery to other than its normal functions. Because viruses cannot be killed as can bacteria or other living organisms, the only way to avoid this is by helping the cells to resist them. This is normally accomplished by providing the cells with some means of ignoring or misreading the instructions on the viral genetic material. Such means are the antibodies generated by the cell in response to immunizing agents. The common cold is caused by a wide variety of viruses and, therefore, finding effective immunizing agents has been particularly difficult.

QUESTIONS

11.20 What are nucleic acids? Why are they important?

11.21 Describe the composition of a nucleotide. How do nucleotides form nucleic acids?

11.22 What are the important discoveries of Watson and Crick regarding the DNA molecule?

11.23 **How does DNA replicate?**
11.24 **Explain what is meant by genetic code. What is its importance?**
11.25 **Distinguish between DNA and RNA. What is the importance of RNA?**
11.26 **Discuss briefly the role of RNA in protein synthesis.**
11.27 **Why are the nucleic acids referred to as the key to life?**
11.28 **Has the total synthesis of a gene been achieved? Can a synthetic gene function in a cell?**
11.29 **What are viruses? Discuss their role in relation to animals and plants.**
11.30 **List some of the diseases caused by viruses.**

VITAMINS Vitamins are specific organic compounds that are required in small amounts in the diet to prevent specific diseases and sustain healthy growth. Their biochemical function in many cases is to aid enzymes. As long ago as 1747, a British Navy captain showed that the disease called scurvy (characterized by spongy and bleeding gums, bleeding under the skin, and extreme weakness) could be prevented and cured by the inclusion of fresh fruits and vegetables in the diet. It is now known that the compound called *ascorbic acid* or *vitamin C,* which is present in

ascorbic acid (vitamin C)

citrus fruits and tomatoes, is responsible for preventing scurvy. More recently, the same compound has been credited by the American chemist Linus Pauling to have the ability to prevent common colds or greatly lessen their symptoms.

Over a period of years many vitamins have been discovered. They are totally unrelated to one another chemically. They can arbitrarily be divided into water-soluble and fat-soluble groups. Water-soluble vitamins include vitamin C and a number of vitamins known together as the B complex. Table 11.5 lists the more important vitamins of the B group, some of the foods in which they are found, and the diseases that result if they are not present in the diet. All the B group vitamins and their derivatives occur in nature coupled with protein molecules and function in several important reactions in metabolism.

The fat-soluble vitamins include vitamins A, D, E, and K; their sources and deficiency symptoms are summarized in Table 11.6.

HORMONES Hormones, like vitamins, are organic compounds needed in very small amounts in the body. However, unlike vitamins, they are synthesized in the body by the endocrine (derived from the Greek *endo* and *krinein,* meaning "to secrete within") glands, the locations of which in the human body are shown in Fig. 11.14. The hormones these glands secrete are listed in Table 11.7 together with their principal physiological effects. Although hormone-producing glands are found throughout the body, the principal hormones secreted in humans fall into three classes which are based on structural features. These are the amino-acid-derivative hormones, protein hormones, and steroid hormones. Since the discovery of the structural features of naturally occurring hormones, chemists have synthesized many of them in the laboratory. In addition, structurally related compounds, called *mimetic hormones*, have been synthesized and found useful for the treatment of many illnesses and as birth-control agents.

Hormones generally differ from vitamins in their action. Most vitamins initiate chemical activity and continue to participate as long as the reaction occurs. Hormones, however, function only as "chemical messengers" — they initiate chemical

TABLE 11.5
Water-soluble vitamins (B complex)

Vitamin	Sources	Deficiency effects
B$_1$ (thiamine)	whole cereal grains, legumes, nuts, milk, pork	beriberi, diseases of nervous system
B$_2$ (riboflavin)	milk, lean meat, liver, fish, egg whites	dermatitis, glossitis, inflammation of mouth and tongue
B$_3$ (pantothenic acid)	yeast, egg, liver, kidney, milk	retarded growth, emotional instability, gastrointestinal discomfort
B$_4$ (choline)	egg yolk, fats	hepatosis, alcoholic cirrhosis of liver, dermatosis, anemia
B$_5$ (niacine; nicotinic acid)	liver, coffee, lean meats	pellagra; stunted growth
B$_6$ (pyridoxine)	meat, fish, egg yolk, whole cereal grains	convulsion in infants, retarded growth
B$_7$ (biotin)	yeast, liver	dermatitis, scaly and greasy skin
B$_9$ (folic acid)	yeast, soybean, wheat, liver, kidney, eggs	mitosis, inhibition of cell division
B$_{12}$ (cyanocobalamin)	liver, meats, eggs, seafood	pernicious anemia

TABLE 11.6
Fat-soluble vitamins

Vitamin	Sources	Deficiency effects
A	fish, liver, eggs, butter, cheese, precursor in carrots and other vegetables	night blindness; excessive light sensitivity
D (calciferol)	irradiated milk; produced in body by ultraviolet irradiation of skin	abnormal growth of bones and teeth, rickets
E (α-tocopherol)	vegetable oils, leafy vegetables, wheat germ oil	sterility
K	green leaves, alfalfa leaves	hemorrhage, slow clotting of blood

activity and are then inactive while the action takes place. In general, hormones appear to function by changing the permeability of cell walls and by activation processes.

How hormones exert their effect in changing the permeability of cell walls is best illustrated by the action of the protein hormone called *insulin* (see Fig. 11.4). Although little is known about the actual mechanism of insulin action, it is well known that glucose can enter the cell only if insulin is present. In the absence of insulin, glucose level rises in the bloodstream and if it continues diabetes ensues. A currently held theory is that insulin attaches itself to pores in the cell membrane to form a protein-to-protein unit and there serves as a specific escort for glucose through the membrane and into the cytoplasm of the cell. Other protein hormones are also thought to alter the permeability of the cell wall in a manner similar to the insulin–glucose mechanism.

We know more about the action of hormones that activate enzymes and genes. For example, a tiny amount of the amino-acid-derivative hormone called *epinephrine* (commonly called *adrenaline*) is sufficient to cause a great increase in blood pressure. When one is under stress or is frightened, the flow of adrenaline prepares the body for fight or flight by releasing glucose molecules that supply quick energy. In the absence of such hormone action, frustration and mental depression may result. In fact, one widely held molecular theory for mental illness involves an imbalance of the hormones *norepinephrine* and *serotonin* in the brain:

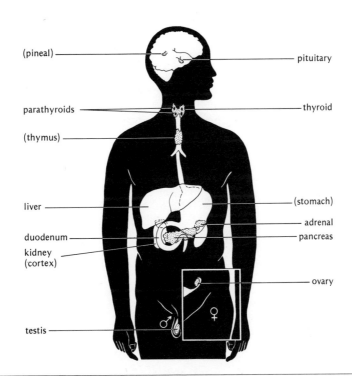

FIGURE 11.14

The approximate locations of the endocrine glands in humans. Although the pineal body, thymus, and stomach are shown, they are not definitely known to secrete hormones. Reproduced, by permission, from David O. Johnston, John T. Netterville, James L. Wood, and Mark M. Jones, *Chemistry and the Environment*, p. 209. Copyright © 1973 by W. B. Saunders Company.

TABLE 11.7
Principal hormones of humans and their physiological effects

Endocrine gland	Hormone	Classification	Physiological effects
adrenal cortex (outer layer)	aldosterone	sterol	regulates metabolism of Na and K
	cortisol	sterol	stimulates conversion of proteins to carbohydrates
	dehydroepiandrosterone	sterol	stimulates development of male characteristics
adrenal medulla (inner layer)	epinephrine	amino acid derivative	reinforces action of sympathetic nerves; stimulates breakdown of liver and muscle glycogen
	norepinephrine	amino acid derivative	constricts blood vessels
ovary (corpus luteum)	progesterone	sterol	acts to regulate menstrual cycle
ovary (follicle)	estradiol	sterol	stimulates development and maintenance of female sex characteristics
pancreas	insulin	protein	increases glucose utilization; decreases blood sugar concentration; increases glycogen storage and metabolism of glucose
	glucagon	protein	stimulates conversion of liver glycogen to blood glucose
parathyroid	parathormone	protein	regulates Ca and P metabolism
pituitary anterior	adrenocorticotropin (ACTH)	protein	stimulates adrenal cortex to grow and produce cortical hormones
	follicle-stimulating hormone (FSH)	protein	stimulates growth of graafian follicles in ovary and of seminiferous tubules in testis
	human growth hormone (HGH)	protein	controls bone growth and general body growth; affects protein, fat, and carbohydrate metabolism
	luteinizing hormone (LH)	protein	controls production and release of estrogens and progesterone by ovary and of testosterone by testis
	thyrotropin (TSH)	protein	stimulates growth of thyroid and production of thyroxin

TABLE 11.7 (Cont.)
Principal hormones of humans and their physiological effects

Endocrine gland	Hormone	Classification	Physiological effects
	prolactin	protein	maintains secretion of estrogens and progesterone by ovary; stimulates milk production by breast; controls "maternal instinct"
pituitary posterior (hypothalamus)	oxytocin	protein	stimulates contraction of uterine muscles and secretion of milk
	vasopressin	protein	stimulates contraction of smooth muscles; antidiuretic action on kidney tubules
testis	testosterone	sterol	stimulates development and maintenance of male sex characteristics
thyroid	thyroxin	amino acid derivative	increases metabolic rate

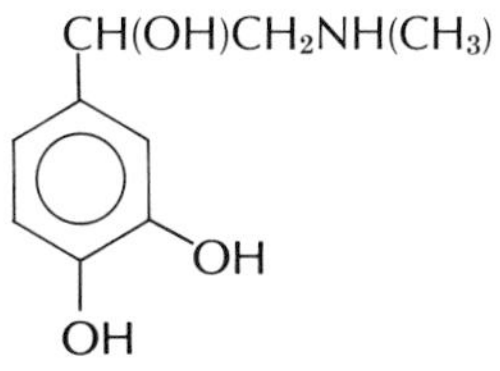

epinephrine 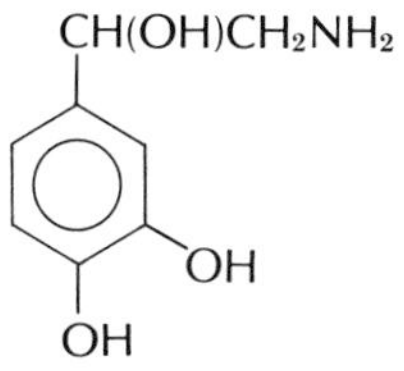   norepinephrine 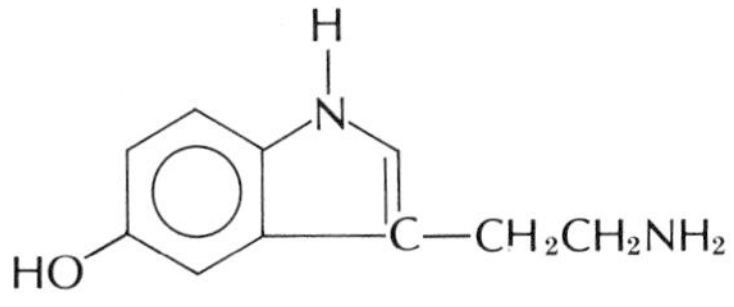   serotonin

Whereas norepinephrine, like adrenaline, is a powerful stimulant, serotonin is a depressant. When an excess of norepinephrine is formed the person is in an elated, perhaps manic state. If, on the other hand, serotonin is formed in excess, the person is in a depressed state. Clearly, a normal mental condition involves a very delicate balance of the two compounds in the brain.

Another important group of 28 hormones is produced by the outer part of the adrenal gland. An important member of this group is *cortisone:*

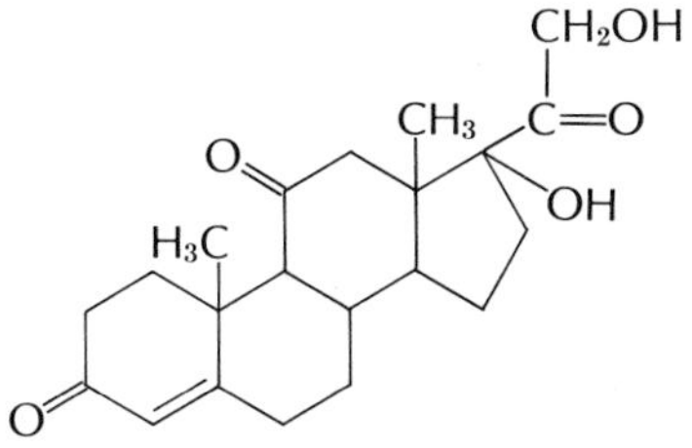

cortisone

These hormones affect the metabolism of food taken into the body, maintain the proper balance of ions in the various body fluids, and help control inflammation and allergic reactions in the body. Cortisone has been found particularly effective in relieving symptoms of rheumatoid arthritis.

Many hormones with steroid structure are popularly known as *sex hormones* because they control the development of secondary male and female sexual characteristics at puberty and are important in the normal reproductive processes. Their structures are given in Table 11.8. The testes of the male produce *androgens (andro-*

TABLE 11.8
Natural sex hormones

Sex	Hormone Name	Structure
male	androsterone	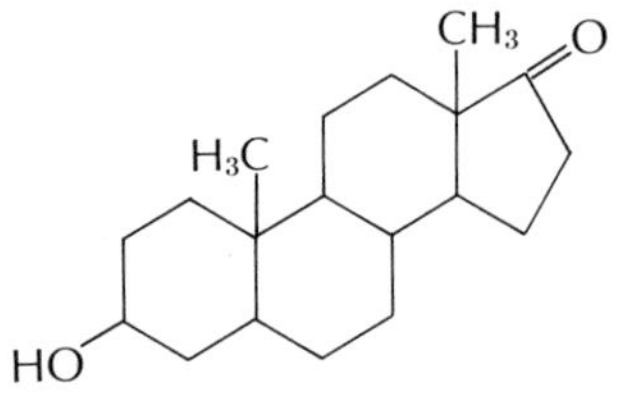
	testosterone	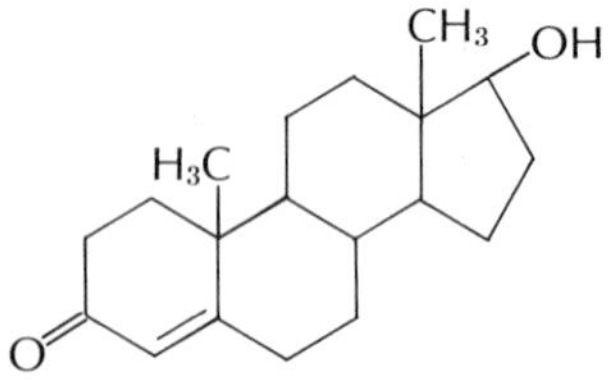
female	estradiol	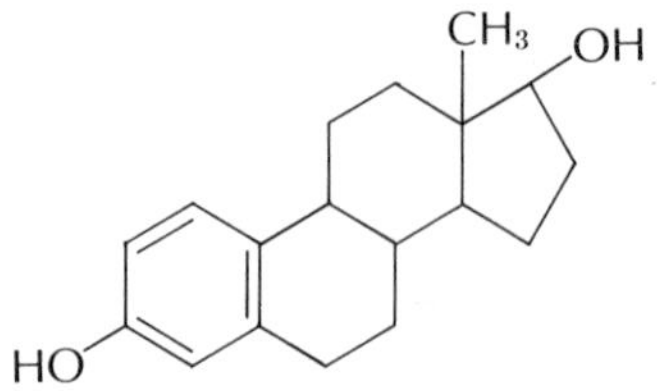
	estrone	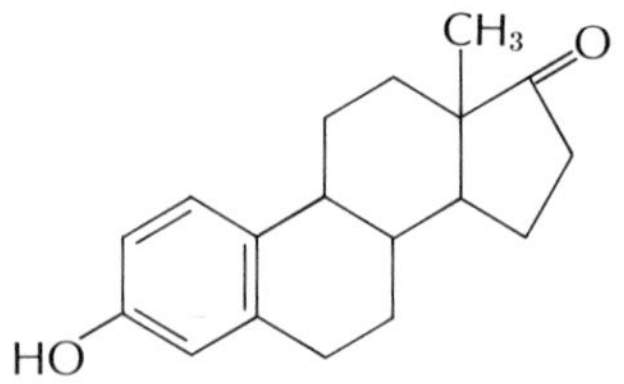
	progesterone	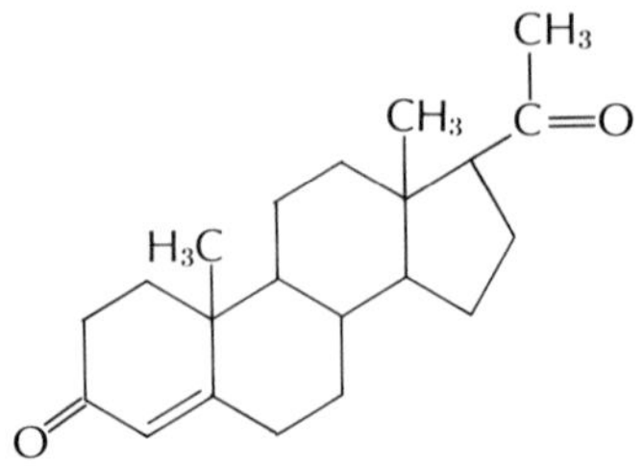

sterone and *testosterone)* both of which are responsible for the development of the sex organs and for secondary sex characteristics such as voice and hair distribution. The ovaries of the female produce *estrogens* which are responsible for the development of breasts and other secondary sexual characteristics that occur at puberty, regulate the *estrus* or *menstrual* cycle, and function in pregnancy. (Figure 11.15 shows the sequence of events in the menstrual cycle.) The liquid within the follicle contains at least two female hormones, known as *estrone* and *estradiol;* the liquid within the corpus luteum contains *progesterone.* Estrone is involved in the development of secondary female characteristics, estradiol in the control of the ovulation cycle in the female, and progesterone in the preparation of the inner wall of the uterus for implantation of a fertilized ovum. Recent experiments support the possibility that the estrogen hormones act chemically by activating genes to produce proteins. When pregnancy occurs, progesterone also prevents further ovulation, retains the embryo in the uterus, and develops the mammary glands prior to lactation. During pregnancy these three hormones are excreted in the urine in increased amounts.

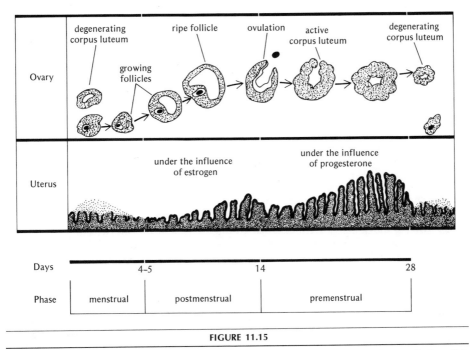

FIGURE 11.15

The sequence of events in the menstrual cycle. Redrawn, by permission, from Joseph I. Routh, Darrell P. Eyman, and Donald J. Burton, *Essentials of General, Organic and Biochemistry,* p. 527. Copyright © 1969 by W. B. Saunders Company.

Synthetic derivatives of the female sex hormones were developed following the realization that progesterone injections are effective in birth control. The oral contraceptive known as "the pill" was first developed in 1951 largely by the American chemist Carl Djerassi who removed a methyl (—CH₃) group and introduced the ethynyl group (—C≡CH) into the progesterone molecule:

"the pill"

Apparently the removal of a methyl group made progesterone more effective in simulating pregnancy; the introduction of the ethynyl group allowed its oral administration. Since then, many variations in the basic molecular architecture of progesterone have been tested and successfully marketed.

All birth-control pills contain a progestin and an estrogen, such as estradiol. When taken regularly at appropriate intervals, the pill can function very effectively in preventing ovulation and pregnancy and in regulating the menstrual cycle. In fact, millions of women around the world now take the pill to prevent ovulation and pregnancy. While the gravity of the population problem may justify such wide use of the pill, the ultimate effects of the chemical alteration of a female sex cycle are yet unknown. In other words, the biochemistry of the pill is largely unknown. Indeed, a number of undesirable side effects have been noted. Some of the side effects can be changes in carbohydrate metabolism, hypertension, acne, hirsutism, amenorrhea, mental disturbances, breast cancer, and the suppression of adrenocortical pituitary functions. These side effects are, however, quite rare and, in fact, on the average, less dangerous than some of the complications of pregnancy and delivery.

QUESTIONS

11.31 What are vitamins? Why are they important?

11.32 Write the structure of vitamin C. Discuss its ability to prevent common colds.

11.33 Discuss the deficiency effects arising from the lack of specific vitamins.

11.34 What are hormones? How are they different from vitamins?

11.35 How do hormones function in the body? Describe the function of the hormone insulin.

11.36 How are hormones possibly involved in mental illness?

11.37 Discuss the role of cortisone hormones in humans.

11.38 What are sex hormones? Give examples and describe their individual functions.

11.39 List some of the synthetic derivatives of the female sex hormones that are used as oral contraceptives.

11.40 Discuss your views on the use of "the pill."

SUGGESTED PROJECT 11.2

The nutritional value of food (or Is it really worth your eating?)

The value of the food you are eating depends on its nutrient content, in particular on the vitamins and minerals. The United States Department of Agriculture (USDA) Handbook 8, Composition of Foods, gives information on the nutrient content of foods and the daily dietary requirements for the maintenance of good nutrition. Any nutrition textbook will also give the same information.

If you need to know the value of the foods you are eating, record everything you consume and their approximate amounts for seven consecutive days. Use Handbook 8 or some other nutrition textbook to determine how much of each of the following nutrients was provided by each of your food items for each day:

a.   **vitamins (A, B_1, B_2, B_5, C, D, and E)**
b.   **minerals (Ca, Fe, I, Mg, and P)**

Compare these values with the amounts given in a Recommended Daily Dietary Allowances table and make adjustments in your diet to improve its nutritional quality.

SOME BIOCHEMICAL PROCESSES AND THEIR ENERGETICS

A living organism is a collection of chemicals that is largely water, carbohydrates, fats, proteins, minerals, and salts. The food it ingests is largely carbohydrates ($\sim65\%$), fats ($\sim10\%$), and proteins ($\sim20\%$) as well. In this section we shall examine how these chemicals are changed into other chemicals and how some of these chemical reactions also provide energy for the living organism.

In all living organisms the important process of *digestion* changes the complex molecules of food into relatively simple and smaller molecules that can be readily absorbed into the body and used as sources of energy and as raw building materials. The most common reaction of digestion is the hydrolysis of fats, proteins, and polysaccharides. When the hydrolysis reaction is carried out in a laboratory, it takes place at a temperature near the boiling point of water, 100°C. Yet the same hydrolysis reaction takes place in the living body at the relatively low temperature of 37°C. What is it that makes reactions which ordinarily occur at temperatures higher than body temperature possible at body temperature? In Chapter 7 it was noted that the chemical reactions in most living organisms are speeded up by catalysts called *enzymes*. It is the presence and activity of enzymes that make such reactions as the hydrolysis of fats, proteins, and polysaccharides possible at the relatively low body temperature. Before we examine the biochemical processes themselves we shall study the chemical nature of enzymes and their mode of action.

ENZYMES   Most enzymes consist of a protein coupled with some other substance called a *coenzyme*. The protein by itself shows no enzyme activity until it is combined with a proper coenzyme. In some enzymes the coenzyme is a nonprotein organic molecule; in some others it is a simple divalent metal ion such as Mg^{2+}, Ca^{2+}, Mn^{2+}, or Zn^{2+}; and in still some others both types of coenzymes are necessary. Many vitamins, particularly vitamins of the B group (see Table 11.5), have been found to be coenzymes or their precursors.

All catalytic behavior is accounted for on the basis of an intermediate compound formed between the catalyst and a reacting substrate molecule. In biochemical terms, the substrate and enzyme form an intermediate compound which then breaks apart yielding the products and the enzyme. To illustrate the mode of enzyme action for a specific process, let us consider the enzyme-catalyzed hydrolysis of cane sugar (sucrose):

$$\text{sucrose} + \text{sucrase} \longrightarrow \text{intermediate compound} \xrightarrow{H_2O} \text{glucose} + \text{fructose} + \text{sucrase}$$

    (substrate)   (enzyme)   (substrate-enzyme)   (products)   (enzyme)

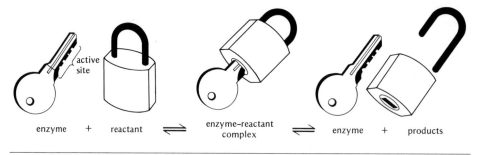

FIGURE 11.16

Lock-and-key theory for enzymic catalysis. Redrawn, by permission, from Mark M. Jones, John T. Netterville, David O. Johnston, and James L. Wood, *Chemistry, Man and Society*, p. 381. Copyright © 1972 by W. B. Saunders Company.

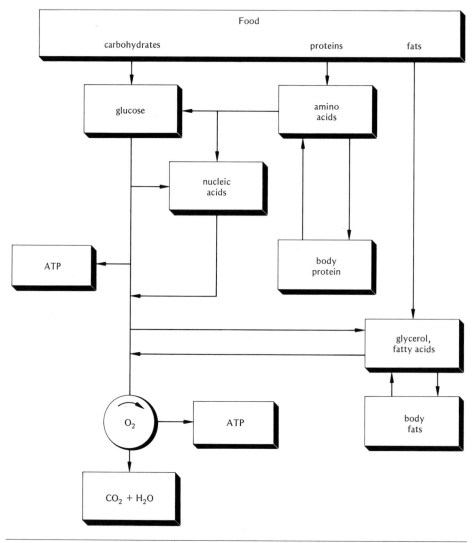

FIGURE 11.17

Major metabolic pathways. Adapted, by permission, from T. P. Bennett and E. Frieden, *Modern Topics in Biochemistry*, p. 118. Copyright © 1966 by the Macmillan Company. Redrawn from William F. Kieffer, *Chemistry: A Cultural Approach*, p. 329, Figure 13-4. Copyright © 1971 by William F. Kieffer. Reprinted by permission of Harper & Row, Inc.

It has been verified that the formation of intermediates occurs at specific locations called *active sites* on the enzyme surface. Clearly, the active site must have a conformation that fits only a particular substrate in a manner analogous to a *lock and key* (Fig. 11.16). This fact accounts for the specificity of catalysts and enzymes in particular.

Notice that the name of the enzyme ends in -*ase* (in fact, the names of most enzymes end in -ase). Prefixes to the ending indicate either the type of reaction catalyzed or the type of substrate upon which the enzyme acts. In the above example, sucrase is the enzyme that catalyzes the hydrolysis of sucrose. Because sucrase is active in a hydrolysis reaction, it would also be called *hydrolase*. Other group names of enzymes include *oxidoreductase* that catalyzes oxidation–reduction reactions and *transferase* that catalyzes group transfer reactions. However, some common enzymes are exceptions to these rules. For example, the enzymes of the digestive system are known by their common names—pepsin, trypsin, and chymotrypsin.

BIOCHEMICAL ENERGETICS   The essential energy-providing reactions of the living organism involve the enzyme-catalyzed combustion or oxidation of carbohydrates and fats at the body temperature of 37°C. The greater bulk of the energy thus generated is not instantly consumed but stored in molecules capable of storing potential energy in what are called *high-energy* bonds. The key energy-rich compound most frequently found in cells is an organic phosphate called *adenosine triphosphate* (abbreviated ATP). In aqueous solution the molecule exists primarily in the ionized form:

adenine
(a purine base)

ribose
(a sugar)

3 phosphate groups
attached by ester linkages

The symbol $\sim$ is used to designate the high-energy bonds in the molecule. On the addition of water to one of the high-energy ester linkages in the line of three phosphates, a molecule of *adenosine diphosphate* (abbreviated ADP) and phosphoric acid result:

$$A(PO_3)_3^{4-} + H_2O \longrightarrow A(PO_3)_2^{3-} + H_2PO_4^- + energy \quad (8 \text{ kcal})$$
$$\text{ATP} \qquad\qquad\qquad\qquad \text{ADP}$$

where A = adenosine. Further hydrolysis results in a molecule of adenosine monophosphate (abbreviated AMP) and so on. The body uses the energy available when the high-energy bond is hydrolyzed to accomplish simultaneously other desirable reactions in which energy must be supplied to drive the reaction. The process is called a *coupled reaction*. A number of specifically identified and understood biochemical reactions such as the synthesis of glycogen from glucose, proteins from amino acids, and so on occur in this fashion. As the body uses ATP to accomplish reactions that are energetically difficult without it, more and more ATP is resynthesized and stored.

The major metabolic pathways and the production of the energy-rich ATP are outlined in Fig. 11.17. Note that part of the ATP is produced by processes that do not require the direct addition of oxygen. The circular pathway shown at the lower left corner is the chief energy-converting cycle of the metabolic scheme and represents the reactions by which oxygen is used and CO_2 and H_2O are produced. It is shown as a cycle because some of the compounds in the complex mixture that results are needed to effect reactions with compounds entering the cycle from the prior steps in the metabolism of glucose and fats. Many reactions, all catalyzed by enzymes, are involved in these sequences. The overall process provides the organism with energy in the form of the high-energy bonds of ATP and gives CO_2 and H_2O as waste products.

There exists a complementary relationship between *respiration* in living systems (animals and plants) and *photosynthesis* in plants. Respiration in animals results in the oxidation of carbohydrates into CO_2 and H_2O; photosynthesis in plants results in the formation of carbohydrates and oxygen from CO_2 and H_2O:

$$C_6H_{12}O_6 + 6O_2 \underset{\text{photosynthesis}}{\overset{\text{respiration}}{\rightleftarrows}} 6CO_2 + 6H_2O + \text{energy} \quad (688 \text{ kcal})$$

In both processes ATP, which is present in both plants and animals, plays important roles. We have already pointed out that ATP is the essential converter of the chemical potential energy into biological energy in animals. In plants, ATP helps to convert the light energy that chlorophyll has absorbed from the sun into the potential energy ultimately represented by the synthesis of carbohydrates. Clearly, all the energy of life on earth can be traced back to that source of all energy, the sun.

QUESTIONS

11.41 List two important biochemical processes that occur in a living organism.

11.42 What are enzymes? What is their role in biochemical processes?

11.43 Describe the mode of enzyme action. Give examples.

11.44 What are high-energy bonds? Write the structure of a molecule that contains such bonds.

11.45 Name the major metabolic pathways that produce the energy-rich ATP molecule.

11.46 Discuss the complementary relationship between respiration in animals and photosynthesis in plants. What is the role of ATP in these processes?

SUGGESTED PROJECT 11.3

Simulation of a biochemical process (or readers digest)

The process of digestion is a biochemical process essential for the utilization of food ingested by all living organisms. It involves the enzyme-catalyzed breakdown of carbohydrates, fats, and proteins into simpler molecules which can be transported in the blood to various parts of the body where they provide materials and energy for building bones and tissues. The process of breaking down carbohydrates such as starch into sugar molecules is catalyzed by the enzyme ptyalin that is present in saliva.

To simulate the digestion of starch, collect a teaspoon of your saliva in a small cup. (The secretion of saliva can be stimulated by chewing some paraffin or rubber.) Heat a pinch of household starch with a tablespoon of water until a solution is obtained. Mix this solution with the saliva to obtain a digestion mixture. Note the time of mixing. At 1-min intervals, test a drop of the digestion mixture using a strip of commercially available iodine paper or a drop of tincture iodine. Because the undigested starch left in the digestion mixture reacts with iodine to give a blue color, the completion of the digestive reaction will be indicated by the absence of a blue color. Record the time required for the completion of the digestion at room temperature.

Repeat the experiment at normal body temperature by placing the cup containing the digestion mixture in a pan of warm water at 98.6°F, as recorded by a clinical thermometer. Record the time required for the completion of digestion and compare it with the value at room temperature (see Chapter 7).

If the commercial test paper commonly used by diabetic patients is available you may test for the formation of glucose in the digestion mixture.

12

The biosphere

. . . There is an ecology of the world within our bodies. In this unseen world minute causes produce mighty effects; the effect, moreover is often seemingly unrelated to the cause, . . . A change at one point, in one molecule even, may reverberate throughout the entire system . . .

RACHEL CARSON
Author of Silent Spring

That portion of the earth inhabited by life constitutes what is called the *biosphere.* It includes all the living matter of the world and its physical environment (air, water, and soil); but is largely restricted to areas where solar energy is available. Of the three main areas where there is life—land, freshwater, and ocean water—the oceans cover about three-fourths of the surface of the earth and provide three hundred times as much inhabitable space as the other two areas combined. The great bulk of life in the ocean water consists of small organisms—phytoplankton (minute, floating aquatic plants) and zooplankton (floating, often microscopic aquatic animals). Yet, most of our planet's oxygen and organic matter are produced in the oceans. The land, on the other hand, is on the average less than one-half as productive as the ocean is, and an increasing human population is looking at the oceans as a source of food more than ever before.

Human beings are a part of the biosphere and exert a significant, perhaps polluting, effect on the other inhabitants of the biosphere. At the same time the functioning of the biosphere is essential for humans. In this chapter we shall examine the origin and functioning of the biosphere, the significance of the biosphere to humans, and the impact of humans on their cohabitants (other humans and other living organisms) in the biosphere.

ORIGIN AND FUNCTIONING OF THE BIOSPHERE

From time immemorial the question of how the biosphere originated has fascinated human beings (and perhaps other living beings too!). It is a question that embodies other questions such as how the earth was formed and how life came into existence and evolved on the earth. The earlier attempts to answer these questions speak in philosophical or religious terms. As late as the middle of the nineteenth century, there was still a widespread belief in the concept of a vital or supernatural force and the spontaneous generation of life. Speculation concerning the chemical processes whereby atoms became living molecules has developed only recently as a meaningful pattern of scientific thought.

ORIGIN AND EVOLUTION OF THE EARTH Many conflicting theories have been advanced to explain how our physical world attained its present condition. According to one well-received hypothesis developed by the American physicist George Gamow, the earth and the rest of our universe had a common origin billions of years ago in highly compressed primordial matter consisting perhaps only of neutrons. A violent expansion of this primordial matter, often called the "big bang," and the consequent radiation of energy resulted in some element formation. Condensation of these elements around centers of turbulence in the expanding material gave rise to clusters of matter that became stars, planets, and so on. It is possible that some element building did occur before differentiation into stars and planets occurred. Our solar system, consisting of the sun, the earth and its moon, other planets and their satellites, the asteroids, the comets and the meteors, is part of a galaxy of billions of stars (suns). It is only one of a large number of galaxies thus formed and scattered in space at tremendous distances from one another. Two entirely different calculations* indicate that about 5 billion years had elapsed since the solar system formed and the big bang occurred.

Following its accretion the earth underwent a process of evolution. Figure 12.1 gives an estimated evolutionary history of the earth. At the beginning of the formative stage, the temperature was relatively low, but owing to the heat released by gravitational compression and radioactive changes, the earth became molten and remained molten until most of the original radioactive nuclides† present had changed to stable nuclides. As the earth's rocky crust began to accumulate, volatile substances, such as H_2, CH_4, NH_3, H_2O, and the rare gases that had been trapped beneath the early crust began to escape to form the proto-atmosphere and the proto-ocean. As concentrations of these simple molecules built up in the atmosphere and oceans, more complex molecules were formed by chemical reactions largely aided by sunlight. As time passed, some of the lighter molecules such as hydrogen escaped into outer space and the ultraviolet radiation from the sun decomposed water vapor in the upper atmosphere to form oxygen and ozone. It is now known that the oxygen–ozone layer provides an effective barrier against incoming solar radiation of shorter wavelengths, which has a tendency to decompose complex

---

* One, on the age of the solar system, calculated from isotopic ratios obtained by radioactive dating; the other, on the time since the hypothetical chaotic explosion sent the galaxies apart from a common origin (primordial matter) to their present positions calculated from the recession velocities as determined by the spectral red shifts.
† The term *nuclides* is used to refer to all the different kinds of atoms of all the elements without regard to their atomic numbers or mass numbers. It is a broader and more inclusive term than *isotopes* which refer to the atoms of a given element that have different masses.

organic molecules. Therefore, it is possible that the appearance of that layer in the upper atmosphere contributed to the conditions necessary for the development of life on earth.

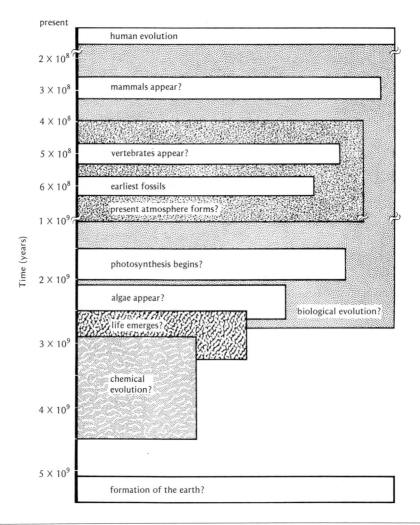

FIGURE 12.1

Estimated evolutionary history of the earth. Adapted, by permission, from Melvin Calvin, *Chemical Evolution*, p. 13. Copyright © 1969 by the Oxford University Press.

It is believed that about 3.5 billion years ago the earth's surface began taking on the appearance that we are familiar with today (Fig. 12.2). Rocks as old as perhaps 3.4 billion years that contain organic molecules have been found, and rocks formed about 3.1 billion years ago contain what appear to have been bacteria and simple algae. According to one theory, these organisms totally changed the surface chemistry of the earth through energy transformation processes, particularly photosynthesis, and laid the foundations for the eventual appearance of life as we know it. Yet, the oceans began teeming with life only 650 million years ago. Man himself evolved only a couple of million years ago. If the earth's nearly 5 billion-year-old history were to be arbitrarily compressed into a 24-hr day, beginning with midnight, it can be postulated that life arose around noon and evolved into man 24 sec before the following midnight (present time). According to this scheme, as shown in Fig. 12.3, modern man appeared only about 1 sec before midnight, and the whole span of recorded human history takes place in the last 0.1 sec. Surely, the evolution of life was linked to the processes shaping the physical appearance of the earth itself. How then did life evolve and what was the driving force behind it?

FIGURE 12.2

This Apollo 17 view of the earth, photographed by Astronaut Ronald E. Evans, extends from the Mediterranean Sea area to the Antarctica ice cap. (Photograph courtesy of the National Aeronautics and Space Administration.)

ORIGIN AND EVOLUTION OF LIFE ON EARTH   Before we proceed to discuss theories about how life evolved let us first attempt to trace the origin of life. Following the French biochemist Louis Pasteur's discovery in the 1860s that living organisms could not originate spontaneously, the concept that life started as a result of a sequence of many chemical reactions was given important consideration (Fig. 12.4). The origin of life on earth is thought to have been intimately bound up

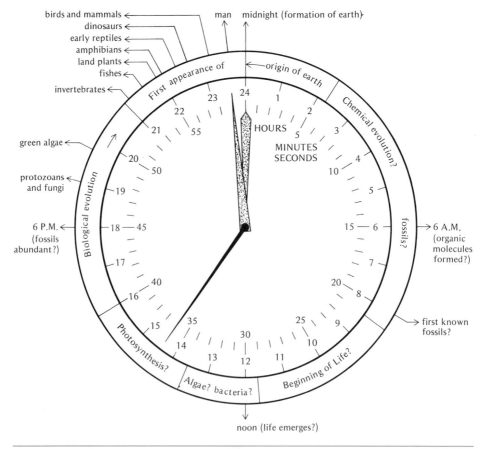

FIGURE 12.3

A geological clock—the history of earth on a 24-hr scale. The clock's hands show 23 hr, 59 min, and 36 sec, the rather late hour of man's appearance in earth history.

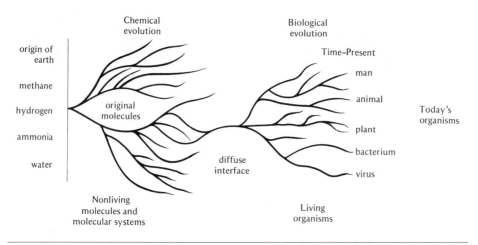

FIGURE 12.4

The origin of life from a nonliving milieu. Adapted, by permission, from Melvin Calvin, *Chemical Evolution*, p. 4. Copyright © 1969 by the Oxford University Press.

with the composition of its early atmosphere. Instead of containing the oxygen and nitrogen that make up the bulk of our present atmosphere, the primordial atmosphere is believed to have consisted of a mixture of methane, ammonia, hydrogen, and helium together with some water vapor. Ultraviolet radiation from the sun, lightning discharges in thunderstorms, or even shock waves produced by falling meteors could cause these gases to react with each other and form compounds such as amino acids, fatty acids, and carbohydrates which are, from a chemical standpoint, the materials of which living organisms are composed. Experimental evidence in support of this theory was obtained in the early 1950s by the American chemist Stanley Miller who subjected a mixture of the supposedly early atmospheric gases (CH_4, NH_3, H_2, and H_2O vapor) to an electric discharge in the apparatus shown in Fig. 12.5. The products that he obtained included several amino acids, fatty acids, and simple sugars. Similar results, using ultraviolet radiation instead of electric discharges demonstrated that solar radiation may have been chiefly responsible for promoting reaction among the gases of the early atmosphere.

However, Gustav Arrhenius of the University of California, San Diego, has very recently questioned whether bio-organic substances of any complexity could have survived the lightning storms that crackled across the primitive earth. An

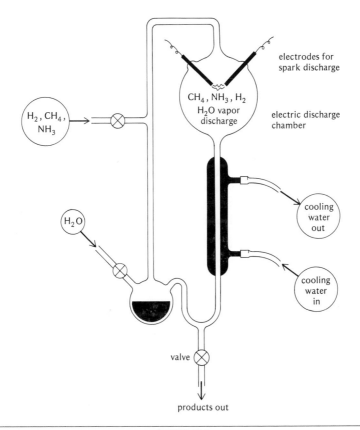

FIGURE 12.5

Diagram of an electric discharge apparatus used by Professor Stanley Miller for the synthesis of the molecules of life from the components of a postulated primeval atmosphere.

alternative theory, proposed by Arrhenius in 1973, is that organic ingredients of terrestrial life existed in the solar system even before the earth was formed. These organic materials could have been formed on the surface of tiny bits of star dust that later agglomerated into bodies as large as the earth and other planets. The theory is supported by the reported presence of several compounds of biological interest in interstellar clouds, by the results of recent examinations of moon dust and meteorites (carbonaceous chondrites), and by the discovery that complex organic molecules may be formed under conditions that supposedly duplicate those existing in the solar nebula.

The chemical evolution of all the various molecules for specific functions in life is much less clear. Experiments have shown that proteins, polysaccharides, and other naturally occurring macromolecules essential to life can be produced from simple amino acids and sugars under conditions that are thought to have prevailed in the primitive oceans. It has, therefore, been theorized that organic compounds were dissolved in the warm water of tidal pools created by the early oceans to produce a "nutrient soup" and proceeded to combine with each other to form the more complicated molecules of life. Life, however, did not really develop in the early seas until the appearance of the nucleic acids and the formation of the well-organized structures of proteins, fats, and carbohydrates. In all likelihood, the nucleic acids permitted organisms consisting of aggregates of proteins and other materials to take on nourishment by absorbing organic molecules and transforming them chemically into material needed by the organisms for growth. The nucleic acids also made it possible for organisms to reproduce themselves and pass on their traits to descendants.

Because cells are the simplest living systems, their formation with boundary membranes is considered to be the final step in the prebiological evolution of life. It is believed that the first cells were formed by physical processes such as the formation of frothy bubbles in the primeval nutrient soup. Modern experiments have, indeed, revealed that the membranes of minute bubbles formed from dilute solutions of proteins and fatlike molecules have structures similar to those of the membranes in living cells. Yet, how life emerged from the essential molecules in the form of a living cell is still a mystery. If the molecules of life can be formed abiologically, why is it that it has not yet been possible to observe the actual process of life emergence. The answer lies in the fact that the prelife environment was quite different from what now exists. It was devoid of oxygen gas that combines with and modifies many of the complicated molecules which are essential for life. It was equally devoid of bacteria and other microorganisms that, together with oxygen, are constantly at work in the present environment, changing its composition by removing some constituents and adding others. Under present conditions, living organisms selectively remove molecules for their food from any mixture of organic molecules. In the absence of life in the prebiological environment, once formed, the organic molecules could have remained unchanged during the millennia that may have passed before the next step in the complicated process of life synthesis could be taken. It is believed that more than 2 billion years passed before unicellular life developed into multicellular life. It is also believed that during the same time period, algaelike single-cell organisms not only evolved the molecules and mechanisms that made cell differentiation possible, but they also brought about many changes in the environment.

In all probability the complex evolutionary process began when mutations occurred in the primitive cells. Mutations were perhaps called for by the depletion of the specific nutrients that the cells ingested from the aqueous environment of the primeval ponds. Thus new cells, capable of using simpler molecules as nutrients, replaced the original cells. Gradually, the cells themselves took over some of the synthetic processes that previously had been accomplished abiologically and evolved metabolic cycles, ultimately resembling those universally found in all contemporary living matter.

The synthetic routes that were developed generated the enzymes to catalyze the energy-releasing and energy-utilizing processes and the information-storage and information-transmittal systems of nucleic acid reactions. Eventually some type of organism resembling the present *anaerobic* bacteria, which is capable of metabolizing food in the absence of oxygen, appeared. The reactions accomplished by yeast cells in the process of *fermentation* are typical nonoxidative metabolic reactions of anaerobic organisms:

$$C_6H_{12}O_6 \xrightarrow{\text{yeast}} 2C_2H_5OH + 2CO_2$$

As the primitive anaerobic organisms multiplied, the nutrients were depleted and waste products accumulated. Some cells, which then evolved a different means of obtaining food and energy, eventually emerged as the aerobic organism that first accomplished *photosynthesis* with the aid of chlorophyll. The carbon dioxide accumulated in the atmosphere and oceans by anaerobic and volcanic activity was converted photosynthetically to carbohydrates and gaseous oxygen:

$$6CO_2 + 6H_2O \longrightarrow C_6H_{12}O_6 + 6O_2$$

Subsequently, organisms adapted to *respiration* as a more efficient means of utilizing the energy potential in carbohydrates than that presented by fermentation. Because glucose undergoing metabolism to CO_2 and H_2O makes several times (~35) more energy available than glucose fermenting to C_2H_5OH and CO_2, organisms capable of respiring would have evolutionary advantage over those capable only of fermentation. As the oxygen content of the atmosphere began to build up, organisms capable of respiring multiplied. As a consequence, the present population of the earth is dominated by plants and animals that use respiration as their device for unlocking food energy.

Our understanding of how life has evolved into its present forms is also based on information gained from the study of *fossils*. Fossils are the remains of ancient plants and animals that have been preserved in rock and in other media. The earliest fossils are probably more than 2 billion years old, but the fossils that are abundant are only about 600 million years old. All of the earlier fossils were forms that lived in water, showing that life probably originated in the oceans. According to the available fossil records the various types of animals came into existence in about the following order: Protozoa, jellyfish, shellfish, bony fish, amphibians, reptiles, and mammals. It was not until about 400 million years ago that land-dwelling plants and animals appear in the fossil record.

LIFE ELSEWHERE IN THE UNIVERSE? The abiotic origin of life need not have been confined to the earth or even to our solar system. In fact, a number of molecules that are known to play important roles in chemical evolution have already been detected in the vast gas and dust clouds of outer space. The estimate that there are billions of planets makes it likely that life must exist somewhere else in the universe, even though no definite proof of this exists at the present time. Of the other planets in our solar system, Mars, based on the evidence presently available to us, appears to be the most likely to harbor some form of life. Both the inner planets

Mercury and Venus have surface temperatures that are much too high for the existence of living organisms. On the other hand, the outer planets (Jupiter, Saturn, Uranus, and Neptune) receive too little energy from the sun. However, there is plenty of methane, ammonia, and water in their atmospheres so that, given the proper conditions, there is no reason why these molecules could not combine to form more complex organic molecules and perhaps evolve into life itself, as theorized in the case of planet earth.

In order for life as we know it to be able to exist on a planet certain conditions should prevail there. First of all, a suitable energy source is a necessary condition for the existence of life. Such a source must provide the food energy for living organisms and must also maintain moderate temperatures that are high enough to initiate the chemical changes necessary for life but too low to cause thermal degradation of the chemical materials comprising living tissues. On earth, this source is originally the sun; in fact, sunlight is probably the major energy source on most planets supporting life. Solar radiation here on earth provides directly the energy needed for photosynthesis in plants which, in turn, are used as a source of food energy by animals. However, it is not impossible that energy could be derived from heat originating within the planet itself, knowing as we do that there is a steady temperature increase in the earth's crust as we bore more deeply into it. Second, the existence of life depends on the presence of a solvent, such as liquid water on earth, in which virtually all biochemical functions would take place. Third, an atmosphere of some sort is probably required for living processes because it screens out harmful radiation from the sun and because it maintains water (and other solvents) in the liquid state. (In a vacuum, water cannot exist as a liquid and sublimes directly from a solid to a gas at any temperature.) However, oxygen gas is not necessarily a prerequisite for life, because, even on earth, there are certain living organisms (the so-called anaerobic bacteria) that thrive in its absence.

WERE WE PLANTED ON EARTH? If in fact life as we know it existed, or does now exist, elsewhere in the universe, then were we planted on earth? As early as 1908, the Swedish chemist Svante Arrhenius suggested the *panspermia* theory that living cells floated haphazardly through the universe, bringing life to desolate planets. More recently, the British biochemist Francis Crick and the American biochemist Leslie Orgel have jointly theorized that life on earth may have sprung from tiny organisms from a distant planet sent here by spaceship by a civilization that was perhaps only slightly more advanced than man is now. Such a "directed" panspermia project could certainly explain how life arose spontaneously on earth.

Besides the spontaneous origin of terrestrial life, a deliberate act of seeding accounts for the existence of a single genetic code for terrestrial life. It also accounts for the fact that the element molybdenum, which is rare on earth, plays a key role in many enzymic reactions that are important to life. Surely, the chemical composition of organisms must reflect to some extent the composition of the environment in which they evolved.

If life began on a distant planet, why would the beings of that planet indulge in panspermia? Was it a demonstration of their technological capability and some form of "missionary" zeal? Or, was it the fear of a cataclysmic end to life on their own planet and the need to propagate life somewhere? Indeed, on earth there are predictions of apocalyptic times. Technological predictions promise that within a few decades earthly beings will have the capability to conduct panspermia of their own. Why would man ever launch a panspermia project? To solve the mystery of the origin of life?

SIGNIFICANCE OF THE BIOSPHERE TO MAN   The biosphere is significant to human beings for a number of reasons. Processes of the biosphere are the source of all of man's food energy and nutritional requirements, the movement of energy through food chains being a basic function. Moreover, these processes play a significant role in removing air pollution, in cleansing water, and in the prevention of soil erosion. We have previously seen that plants, the cohabitants of man, are capable of photosynthesis, the process of fixing solar energy as food (carbohydrates) for animals. During photosynthesis carbon dioxide moves into the leaves of the plants together with any gaseous pollutants that may be present in the air, such as sulfur dioxide, nitrogen oxides, and ozone. In the leaves, by mechanisms as yet unknown, molecules of the air pollutants remain. Thus, plants act as natural air filters. But they have a limited tolerance, and overloading by pollution will impair their growth and even kill them. In addition to the role of food producers and air cleaners, plants are also nutrient suppliers because of their ability to absorb nutrients from the soil and make them available as part of the molecular architecture of food. Besides absorbing nutrients from the soil, plant roots hold the soil in place and thus prevent erosion of nutrients from the upper soil layers. Although soil erosion is often neglected in today's concern with pollution, it is nevertheless a major problem.

There are other cohabitants of the biosphere that are also important to humans. Some microorganisms act as decomposers of organic wastes. Without such decomposers all bodies of water would become completely polluted with organic wastes, and dead organic material would accumulate on and in the soil. Some other microorganisms produce *antibiotics,* that is, chemical compounds that are capable of killing other microorganisms. Perhaps the most important in terms of human health is penicillin, a powerful antibiotic derived from a green fungus (mold) called *Penicillium.* Antibiotics such as streptomycin and aureomycin are also natural chemicals obtained from funguslike organisms. However, some members of the biosphere are the cause of several of man's most serious diseases, including malaria, typhoid fever, plague, gonorrhea, and syphilis. Bacteria are probably the most important group of organisms pathogenic to humans, but fungi and some of the single-cell and other simple organisms are also implicated in disease.

QUESTIONS

12.1   What is the biosphere?

12.2   Which biospheric region is the most productive? Why?

12.3   How did the earth originate? Give supporting evidence.

12.4   List the various steps in the evolution of the earth.

12.5   Explain the significance of the oxygen–ozone layer in the upper atmosphere.

12.6   What is the presumed origin of life on earth?

12.7   Describe an experiment that relates the origin of life on earth to the composition of its early atmosphere.

12.8   What role did the primeval oceans play in the chemical evolution of life?

12.9   What physical process is considered responsible for the formation of the first cells? Give experimental evidence, if any.

12.10 Give examples of oxidative and nonoxidative metabolic reactions.

12.11 What are fossils? How are they formed?

12.12 How does a study of fossils support the evolution of life on earth?

12.13 What conditions must prevail on a planet to support life?

12.14 Discuss the possible existence of life elsewhere in the universe.
12.15 What is the significance of the biosphere to human beings?

MAN VERSUS OTHER LIVING ORGANISMS

Every human being is intimately involved with the rest of the biosphere. Nothing in the biosphere is more important to humans than food and shelter. Like the other components of the biosphere, human beings have genetically fixed minimum demands for food and shelter. Early in our history foods were eaten and shelters were used as they were found. As time passed, man learned to preserve his food and protect himself and his shelter from other living organisms in the biosphere. Gradually the human population began to increase as well (Fig. 12.6). But during the last 100 years, the human population has been increasing at a continually accelerating rate. Concomitantly, there has been increasing competition between man and other living organisms for the fixed resources — limited supply of freshwater, fertile lands, minerals, and fuels — in the biosphere.

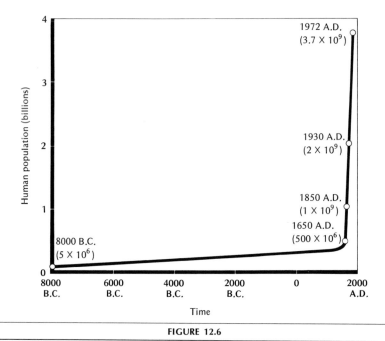

FIGURE 12.6

The growth of human population. Adapted, by permission, from Joseph M. Moran, Michael D. Morgan, and James H. Wiersma, *An Introduction to Environmental Sciences*, p. 258. Copyright © 1973 by Little, Brown and Company.

The competition between man and other species is very severe in terms of man's destruction of natural habitats of other species. It is estimated that each year an additional 1 million acres of land in the United States are intensely developed by man. The effects of massive habitat changes are considered responsible for nearly 100 species of endangered wildlife in the United States. On a worldwide scale, nearly 300 species of mammals and 360 species of birds are in danger of becoming extinct. Each species represents a unique combination of genetic characteristics (gene pool) that enables it to adapt to certain environmental conditions, and there is always the possibility that we may need these adaptations for our own survival. We have seen that members of the biosphere are important in producing food, medicine, and drugs. Clearly, it is in our self-interest to prevent the accelerated extinction of species so that we may preserve a vast gene pool to provide us with solutions to future problems.

History gives us accounts of more severe lethal forms of competition between man and other living organisms. Such competition increasingly involves man-made chemicals. We shall discuss next the lethal chemicals used by man to fight famine and parasites and their impact upon the other components of the biosphere.

FAMINE AND PESTICIDES From time immemorial man has been plagued by insect pests and has continuously been competing with rodents and insects for food. Man has long tried to control insect pests and avoid famine by using a variety of chemicals, called pesticides. Many of these chemicals initially were compounds of arsenic. A few, such as *pyrethrum* and *nicotine sulfate,* are obtained from plant

TABLE 12.1
Classification of major biocides

Class	Chemical group or action	Examples
fungicides	mercurials	$HgCl_2$
	quinones	Phygon
	dithiocarbamates	nabam, ziram
	others	captan, thiram
herbicides	contact toxicity	sodium arsenite, oils, "dinitro" compounds
	translocated (particularly hormone types)	2,4-D; 2,4,5-T
	soil sterilants	borates, chlorates
	soil fumigants	methyl bromide, Vapam
insecticides		
inorganic	arsenicals	lead arsenate
	copper-bearing	copper sulfate
organic, naturally occurring	nicotine alkaloids	nicotine sulfate
	pyrethroids (also synthetic)	pyrethrum
	rotenoids	rotenone
organic, synthetic	chlorinated hydrocarbon compounds	aldrin, dieldrin, lindane, chlordane, DDT, endrin, heptachlor, methoxychlor, toxaphene
	organic phosphorus compounds	Chlorthion, Diazinon, Malathion, parathion, TEPP
	carbamates	Sevin
rodenticides (mammal poisons)	anticoagulants	warfarin, Pival
	immediate action	strychnine, sodium fluoroacetate (1080), thallium, endrin

TABLE 12.2

The three major insecticide groups and the structures of some important insecticides

Group name	Chemical name	Structure	Toxicity to mammals
organochlorine insecticides	DDT		fairly low
	lindane		fairly low
	aldrin		fairly high
	dieldrin		fairly high
	methoxychlor		fairly low
organophosphorus insecticides	dichlorous		—
	parathion		high
	Malathion		—
carbamate insecticides	carbaryl		low
	Baygon		—

sources. Besides being extremely poisonous to animals as well as to insects, most of these chemicals are extremely poisonous and even fatal to man so that they are more appropriately called *biocides*. A classification of the major biocides is given in Table 12.1.

Since the realization of the pesticidal activity of DDT, (see Chapter 9) during World War II, competition between man and his food competitors has changed dra-

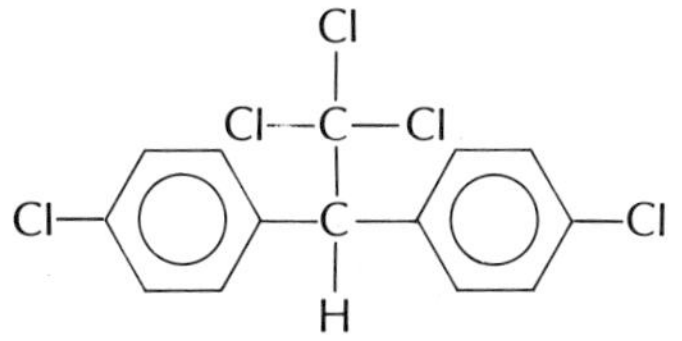

(DDT)

matically. Man now uses chemical warfare to aid in his competition for food energy produced in the world. Today, a complex arsenal of synthetic compounds of carbon, hydrogen, oxygen and at least one other element such as chlorine, arsenic, mercury, sulfur, or phosphorus are available for insect control. The chemical structures of the three major insecticide groups are given in Table 12.2. In addition to the controlling of insects, chemicals are also used to kill fungi on crop seeds and weeds in crop fields. Crop seeds are treated with chemicals containing mercury or arsenic (*fungicides*) to prevent their decomposition while in the ground prior to germination. Because weed plants compete with crop plants for nutrients, water, and light energy, chemicals (*herbicides*) are used to decrease competition and promote high crop yield. The early herbicides were solutions of copper salts, sulfuric acid, and sodium chlorate. Two herbicides now in restricted use are 2,4-dichlorophenoxyacetic acid (*2,4-D*) and 2,4,5-trichlorophenoxyacetic acid (*2,4,5-T*):

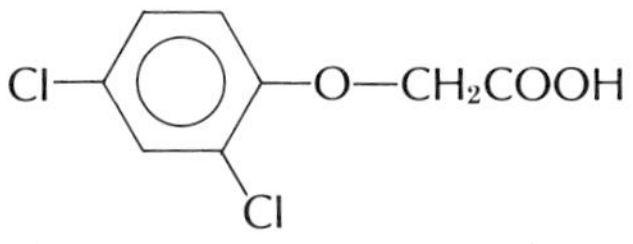

2,4-D 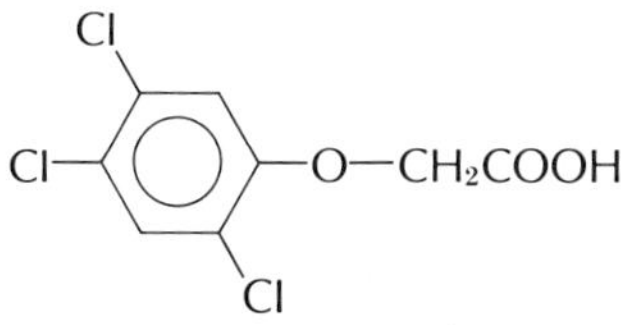   2,4,5-T

These are particularly effective against newly emerged, rapidly growing broad-leaved plants and have increased the yield of agricultural crops.

In addition to their role as killers of insects, weeds, and fungi, chemicals also find wide use as defoliants and rodenticides. Chemicals such as calcium cyanamide (CaNCN) have been used as defoliants to facilitate the harvesting of crops. They cause plants to lose their leaves, thus making possible the use of mechanical harvesters. A final type of pesticide is rodenticides, such as strychnine and sodium fluoroacetate ("1080"), which is used to kill the rats and mice that damage stored grain.

The use of pesticides does, indeed, protect plant crops from the ravage of insects and increase food production. It is estimated that pesticides have saved one-third to one-half of the world's annual food harvest that would have been lost to pests. However, a number of deleterious effects have also been observed. Broad-spectrum pesticides lack specificity and kill not only the target species but also other organisms that are present. With continued application some insects have become resistant to certain pesticides. Over 250 different species of insects have already increased their resistance, so that man is now searching for more toxic pesticides to control "superpests."

A number of environmental problems arise from the molecular stability of some pesticides. Most pesticides are synthetic compounds that are foreign to the biosphere and, therefore, many organisms lack enzyme systems that can decompose pesticide molecules by chemical and/or biological action. Hence some pesticides tend to persist in living organisms. When dead organisms containing the persistent compound are acted upon by decomposers, the pesticide is released to the abiotic environment where it can be taken up by another organism. Finally, persistent pesticides, such as the chlorinated hydrocarbons, may be transported all over the biosphere (Fig. 12.7). The Federal Food and Drug Administration has reported that chlorinated hydrocarbons such as DDT are commonly found in many foods. The dynamics of food chains (see Fig. 9.19) result in higher concentrations of pesticides in individuals at the upper levels of the chains. This phenomenon, known as *biological amplification,* is considered responsible for impairing the reproductive capacity and endangering the population levels of birds that prey upon contaminated fish (Fig. 12.7).

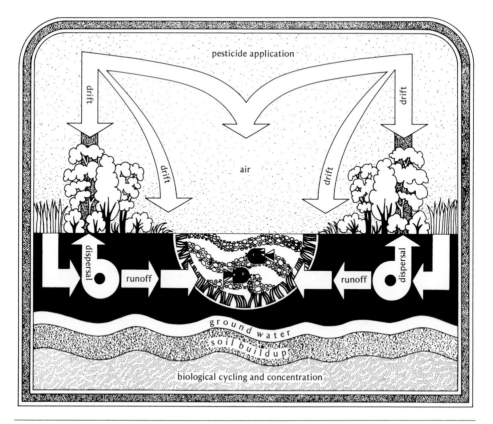

FIGURE 12.7

Pathways for the movement of pesticides in the biosphere. Activity in the center of the pond suggests that pesticides can reach the upper levels of food chains. Modified, by permission, from Robert L. Rudd, "Pesticides," in William W. Murdoch, Ed., *Environment: Resources, Pollution and Society,* p. 286. Copyright © 1971 by Sinauer Associates, Inc., Stamford, Connecticut.

Because man is at the top of food chains, there has been considerable interest in studying the persistence of pesticides, particularly DDT, in the human body and the consequential health effects. DDT is fat soluble and, therefore, it accumulates in fatty tissue in animals and concentrates as it passes up the food chain to man. The U.S. Public Health Service studies show that everybody, from infancy till death, retains DDT in his fatty tissues. Worldwide averages vary from 4 ppm in Alaskan Eskimos to 26 ppm in the people of India; the average for U.S. citizens is about 12 ppm. Biochemically, DDT is a nerve poison, but investigations focused on nervous system disorders have resulted in the observation of "no clinical effects." However, it is known to interfere with calcium metabolism essential to the formation of healthy bones and teeth. Although no harm has ever been conclusively demonstrated in man, the disruption of calcium metabolism in birds has been disastrous for some species. No humans are reported to have died from long-term, low-level exposure to DDT and even the more toxic pesticides. However, many accidental deaths from improper handling of large quantities of pesticides have been reported.

The near extinction of several species of birds, massive fish kills, and eradication of beneficial insects, such as bees, have resulted in the ban of DDT and other chlorinated hydrocarbons in a number of areas around the globe. However, in such areas the less persistent organophosphorus and carbamate pesticides are now used. *Parathion* (Table 12.2) is an insecticide that has largely replaced DDT. Unlike DDT, parathion is extremely toxic to mammals and several human deaths have been attributed to careless use of this poison. However, it does not remain long in the environment and decomposes to form inorganic phosphates, which might possibly contribute to phosphate pollution of water (see Chapter 14).

In spite of the increasing concern over the effects of pesticides their production and application have increased over the years and will continue to do so in the absence of official prohibitions. If synthetic organic pesticides are to be banned what other methods can be used to control our competitors and avoid famine? One alternative is the use of chemicals occurring naturally in the biosphere. Such chemicals could include natural poisons from certain plants, juvenile hormones of many insect species, and sex attractant chemicals. For example, the compound *pyrethrum,* which occurs sparingly in the flowers of a daisylike plant, is a naturally occurring pesticide that is biodegradable and nonresistant. Insect species that produce their own unique hormones, called *juvenile hormones,* in decreasing concentrations to allow for their growth can be made to remain in a particular stage of their life cycle by applying the juvenile hormone in the proper concentration at the right time or can be made to interrupt their life cycle by continued use of the hormone over a growing season. Because many female insects secrete chemicals that are called *sex pheromones* to attract the males for mating, insect traps with the natural or synthetically prepared sex attractants could be used for controlling certain pest populations. If large numbers of males are attracted and caught by the so-called sex traps, the breeding population is reduced. The method has shown real promise for controlling the gypsy moth which has caused considerable damage to trees in the eastern part of the United States.

An alternative to the use of chemicals is *biological control,* that is, the use of natural enemies such as the pest's own predators and parasites to control its population. These include bacteria and viruses which are often specific for a given species of pest. When some insect pests are found, they are treated with bacteria or virus which they carry back with them and then infect other insects. Bacteria have been used successfully to combat termites; viruses have been successfully tested against cotton bollworm. However, production of bacterial and viral pesticides is time-consuming and expensive. Nevertheless, this approach to insect control is advantageous because viral agents seem to be highly specific, nontoxic to man and other animals, and completely biodegradable. Other biological approaches include the sterilization of male insects and the breeding of insect- and fungus-resistant plants. Sterilization of large numbers of males through the use of radiation, chemicals, or crossbreeding has virtually eliminated many tropical fruit flies and the screwworm fly, once a serious cattle pest in the southern United States. The breeding of resistant wheat varieties has contributed immeasurably to the "Green Revolution," that is, increased crop yields from land already in production.

The foregoing discussion on pesticides clearly illustrates the wide range of effects that man's activities have on his cohabitants of the biosphere. If our population continues to grow, our competition with other consumer species for the limited food supply will become even more intense than it is today. The search will go on for effective ways to control insects, thus protecting man's food supply.

PARASITISM AND DRUGS Populations of two different species that depend on the same environmental factors may interact to determine their respective populations. *Parasitism*, an interaction in which one species may obtain nourishment from another species, has been quite important in controlling the human population. Epidemics of bubonic plague caused by a type of bacteria that is transmitted to man by fleas carried by rats and other rodents devastated pre-Renaissance Europe. With the development of chemical drugs to kill organisms that transmit parasites to man, human deaths have decreased dramatically. The use of DDT to kill the malaria-carrying mosquito is an example. However, the population-control potential of disease is still effective and even today flu epidemics caused by viruses occur rather frequently, even in developed nations. Urbanization and worldwide transportation systems are considered responsible for the spread of many diseases.

It was earlier pointed out that some members of the biosphere, particularly some types of bacterial organisms, are the cause of several of man's most serious diseases. At the same time some members of the biosphere are important in producing chemical substances that kill certain other microorganisms. These "opposed" actions form the basis for the concept of *biological antagonist*. The chemicals produced are called antibiotics and include penicillin, streptomycin, aureomycin, terramycin, and many others that have been isolated from bacteria and other species of fungi. By introducing antibiotics into the human body, the lives of millions of human members of the biosphere have been saved. Prior to the discovery of these antibiotics, man had depended exclusively on a group of synthetic chemicals called *sulfa drugs* to control bacterial infections (see Chapter 9).

The first of the antibiotics discovered was *penicillin*, the structure of which is

penicillin G

It was discovered in 1928 by the British bacteriologist Alexander Fleming in association with a mold growing on cultures of *Staphylococcus aureus*, a germ that causes boils and some other types of infections. However, it was developed as a practical drug only in the 1940s largely as a result of the work of the British bacteriologists Howard Flory and Ernest Chain. Today penicillin G is the type most widely used in medicine, although many different penicillins exist, differing in the structure of the R group only. In its mode of action, penicillin interferes with the synthesis of the mucoprotein* cell wall of pathogenic bacteria, causing the bacteria to die. The fact that penicillin destroys bacterial cells and not human cells is the basis of its success. However, penicillin is toxic to some people because of an immunity reaction and must be used with caution.

The development of a series of antibiotics followed soon after the success of penicillin. In 1937, *streptomycin* was isolated from a soil organism *Streptomyces griseus*, and in 1945, *aureomycin*, the first of the tetracycline compounds, was isolated from *Streptomyces aureofaciens*. Subsequent tests on numerous soil samples led to the isolation of *terramycin* and a number of other antibiotics having tetracyclic structures:

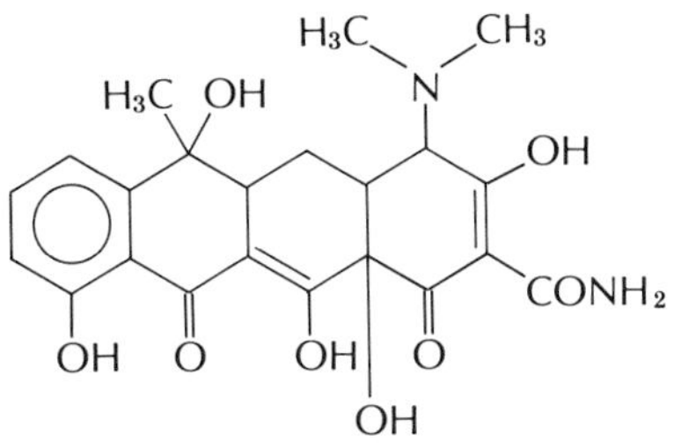

tetracycline

However only a very few tetracyclines have been found to be relatively nontoxic to humans. Streptomycin, which was quite successful in controlling certain types of bacteria, was on the market for more than 10 years before it was withdrawn owing to its bad side effects. The search continues for antibiotics that are more effective on more kinds of bacteria, but with lower human toxicity.

Besides bacteria and other species of fungi, there are other cohabitants of the biosphere that are important in terms of human health. Some species of higher plants produce drugs, such as morphine, codeine, quinine, cinchonine, and cocaine, all of which are classified under the general name of *alkaloids* (meaning "alkali-like"). Alkaloids are basic nitrogen-containing compounds of plant origin that manifest significant physiological activity. Most of them are white crystalline solids with a bitter taste and usually have rather complex molecular structures with a basic nitrogen atom incorporated into a heterocyclic ring:

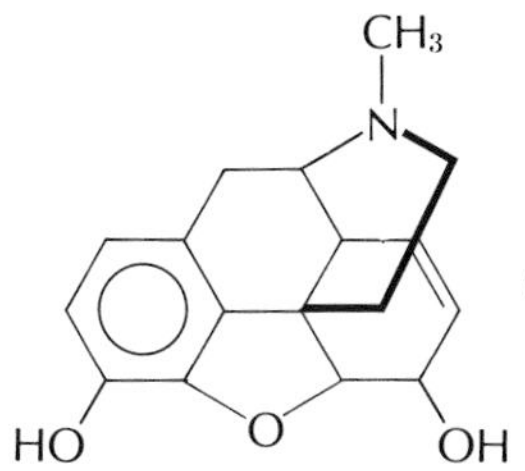

morphine 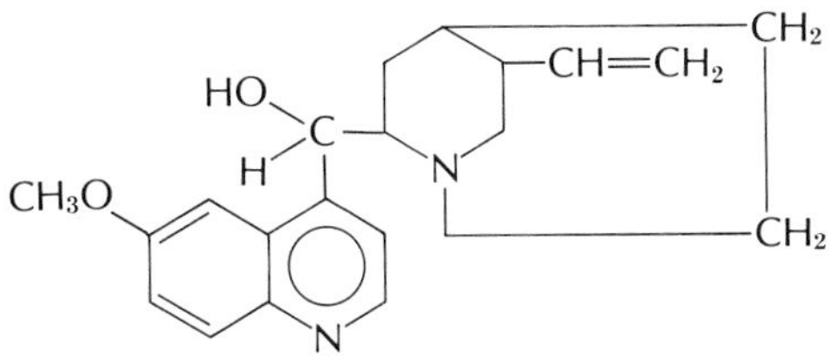   quinine 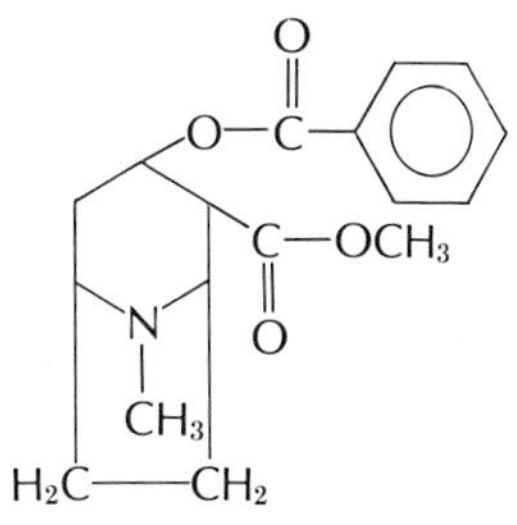   cocaine

---

* A combination of proteins and mucopolysaccharides, in which some of the glucose or galactose units contain a —NHCOCH₃ group substituted for an OH group.

At this time over 2000 alkaloids have been isolated from plants. Some important alkaloids, their plant sources, and applications are listed in Table 12.3.

Following the discovery of the physiological activity of alkaloids, some alkaloids have been mass-produced synthetically. These include morphine, codeine, quinine, and cocaine. All of these alkaloid drugs are normally used by prescription. They are strongly addictive and fatal if administered in amounts greater than the prescribed doses or for a period longer than the prescribed time.

Morphine depresses the central nervous system to produce an analgesic action that reduces pain, but in larger doses it induces sleep. *Codeine,* the methyl ester of morphine, has a weaker analgesic action and is less toxic than morphine. The mechanism of their action on human beings is not completely understood. Both morphine and codeine are habit-forming narcotics, and their continued use results in addiction. A substitute for morphine is the synthetic *Demerol,* which is not as habit-forming and as potent a narcotic as morphine. *Heroin,* the diacetate ester of morphine, is obtained by reacting morphine with acetic anhydride

$$\left(\begin{array}{c} O \quad\quad O \\ \| \quad\quad \| \\ CH_3-C-O-C-CH_3 \end{array} \right)$$ in such a manner that both OH groups become trans-

formed into $CH_3-\overset{\overset{\displaystyle O}{\|}}{C}-O-$ groups. It is much more habit-forming than morphine, and its manufacture in and importation into the United States is forbidden.

Quinine is an antipyretic (fever-reducing) and antimalarial agent that is able to poison the parasite that causes malaria without harming the host. Another alkaloid with an antimalarial action is *atabrine.* Both quinine and atabrine were extensively used as antimalarials until the widespread application of DDT annihilated most of the malaria-carrying mosquitoes.

Cocaine has been used as a local anesthetic, particularly in dentistry. It has a stimulating action on the central nervous system and increases physical endurance. The mechanism of action on humans is believed to involve the blockage of the nerves that transmit pain. Cocaine, however, exhibits toxic effects and its continued use is habit-forming, eventually producing mental deterioration. However, the local anesthetic known as *procaine* or *novocaine* (see Chapter 9) has none of the toxic effects of cocaine.

TABLE 12.3
Some alkaloids, their plant sources, and applications

Name of alkaloid	Plant source	Application
atropine	belladonna root	dilate pupils of the eye
caffeine	coffee, tea	stimulant
cinchonine	cinchona bark	antipyretic (fever reducer)
cocaine	cocoa leaf	local anesthetic
codeine	opium poppy	analgesic (pain reducer)
conine	poison hemlock	respiratory sedative
LSD	ergot fungus	produce delusions
morphine	opium poppy	analgesic
nicotine	tobacco leaf	insecticide, stimulant
quinine	cinchona bark	antipyretic
reserpine	rauwolfia	antihypertensive

The disease-preventing activities of only a few chemicals found in nature are now known. Human lives have been prolonged and saved by their use. Certainly, identifying new chemicals in the biosphere that can be used to improve the health of man is a fertile field of research.

QUESTIONS

12.16 In what respects does man compete with other living organisms? Why?

12.17 What are pesticides? Give examples.

12.18 Distinguish between a biocide and pesticide.

12.19 List the major classes of biocides, giving an example for each class.

12.20 Name the three major insecticide groups, giving one example for each group.

12.21 What are fungicides, herbicides, and defoliants? Give examples of their application.

12.22 Discuss the merits and demerits of using chemical pesticides.

12.23 What is meant by biological amplification? How does it occur?

12.24 What is DDT? What are its uses? What do you know about the results of such uses?

12.25 List possible ways by which pests can be controlled without the use of
   a. synthetic organic pesticides
   b. chemical pesticides

12.26 What is a sex trap? How does it work in the case of insects?

12.27 What are the difficulties encountered in using bacterial and viral pesticides?

12.28 What is meant by parasitism? Give examples.

12.29 What drugs are produced by the different members of the biosphere?

12.30 How was penicillin discovered? How does it work?

12.31 Name some tetracycline compounds and give their structural formulas.

12.32 What are alkaloids? Describe their sources and medicinal properties.

SUGGESTED PROJECT 12.1
Pesticides and their toxicities (or DDT for fun and profit)

This project is intended to familiarize the student with some pesticides and their human toxicities. Obtain from your garden or farm supply store the names of a dozen different pesticides, their active ingredients, and inactive ingredients, if any. Use the Merck Index to find out the formula, use, and toxicity for each of the ingredients. Tabulate your results.

MAN VERSUS CHEMICALS

Human beings are unique among living organisms in that they pollute not only their surrounding biosphere but also themselves. They pollute their own bodies with foreign substances which were not intended to be present in the collage of chemicals that constitutes a normal human body. What are these foreign substances? They are the chemical additives we find in our foods; the chemicals we come in contact with in our households; or they may be the chemical compounds that come from air, water, and soil pollution. This section is devoted to food additives and household chemicals some of which contribute to better living, some others to miserable or worse living, and still some others to no living at all. Pollution of the air, water, and soil, which constitute our physical environment, is discussed in the succeeding chapters.

FOOD ADDITIVES With the industrial revolution and the increase in urbanization human beings came to depend on foods prepared, shipped, and stored by others. Such foodstuffs are treated by the addition of *chemical preservatives* to retard spoilage resulting from the growth of molds (fungi) or bacteria and by the action of oxygen (oxidation). Salt is added to preserve meat and fish; propionic acid, benzoic acid, sorbic acid and their salts are added to bread and cheese as mold and yeast inhibitors; sulfur dioxide or salts of sulfurous acid are added to dried fruits, fruit preserves such as jams and jellies, and wine. Two substituted phenols, butylated hydroxytoluene (BHT) and butylated hydroxyanisole (BHA) are often added as *antioxidants:*

BHT   BHA

The use of sodium nitrite ($NaNO_2$) and sodium nitrate ($NaNO_3$) as inhibitors has become a matter of public concern recently. When fed to experimental animals in large amounts and in combination with secondary amines (see Chapter 9), sodium nitrite forms nitrosamines which are carcinogenic. Although nitrate does not react with amines to form nitrosamines, a problem may arise if nitrate is converted to nitrite in the stomach. Such a conversion can occur under some conditions.

In recent years, a wide variety of chemicals are added to food not only to preserve it but also to improve nutrition, to enhance flavor, and to impart color. The use of such additives is especially important for convenience foods. Nutrient additives such as the vitamins have helped to prevent deficiency diseases such as beriberi and rickets. The restoration of the B complex vitamins and addition of iron (as $FeCO_3$) to flour has served to eliminate the disease called pellagra, but the nutritional value of such "enriched" flour has been questioned by nutrition experts. Other food additives (Table 12.4) have replaced natural spice and fruit flavors in foods. Today, the consumer foodstuff depends for its flavor almost totally on synthetic organic chemicals, particularly the esters of some organic acids.

Monosodium glutamate (MSG),

$$HOOC-CH_2-CH_2-\underset{\underset{NH_2}{|}}{CH}-COONa$$

MSG

is the best known and most widely used flavor enhancer. A few years ago it was reported that brain damage resulted when high doses of MSG were injected under the skin of 10-day-old mice. Because of the public alarm about MSG in baby foods, the producers of these foods voluntarily stopped using the additive. The use of MSG, in reasonable concentrations, in any food has never been banned. However, a national investigative committee has recommended that it not be added to foods for infants because infants do not seem to appreciate enhanced flavor. The committee found no evidence that reasonable use of MSG in foods is hazardous to older children or to adults, except to those persons unusually sensitive to the additive. The mysterious malady referred to as "Chinese restaurant syndrome," experienced by some people after eating MSG-containing foods is merely a hypersensitivity response of a small percentage of the population.

Sugar has been used for ages to make certain foods sweeter and, thus, more palatable. In spite of the fact that there is no evidence that artificial sweeteners such as saccharin (see Chapter 9) and the cyclamates are worthwhile in controlling obesity, their use as food additives boomed until 1970. The cyclamates were banned at about that time because they were found in concentrated amounts to induce cancer in test animals, presumably by reacting in the body to form cyclohexylamine, a known carcinogen:

calcium cyclamate   cyclohexylamine

However, some carcinogens occur naturally in food. Spices such as cinnamon and nutmeg contain safrole, a carcinogen that has been banned as a flavoring in root beer. Most charcoal-broiled food contains 3,4-benzopyrene (see Chapter 9), a carcinogen also found in cigarette smoke and automobile exhaust fumes. Further, some artificial dyes (colors) that were once used to impart color to certain foods have later been shown to be harmful and even carcinogenic.

To protect the food consumer, most countries have enacted laws and set up agencies to administer these laws. In the United States these laws are called Food, Drug and Cosmetic (FDC) Acts and are administered by the Federal Food and Drug

TABLE 12.4
A partial list of food additives

Classification	Commonly used food additives
food preservatives (a) spoilage inhibitors	benzoic acid, sodium benzoate, propionic acid, sodium and calcium propionates, sorbic acid and its salts, sulfur dioxide and sulfites
(b) antioxidants	ascorbic acid and its salts, butylated hydroxytoluene, butylated hydroxyanisole, lecithin, sulfur dioxide and sulfites
food colors	annatto (yellow), carbon black, carotene (yellow–orange), cochineal (red), FD & C Red Nos. 2 and 3, Blue No. 1, and titanium dioxide (white).
food flavors	amyl acetate (banana), amyl butyrate (pear), bornyl acetate (camphor), carvone (spearmint), cinnamaldehyde (cinnamon), citral (lemon), ethyl formate (rum), ethyl propionate (fruity), ethyl vanillin (vanilla), eucalyptus oil (bittersweet), eugenol (spice, clove), geraniol (rose), geranyl acetate (geranium), ginger oil (ginger), menthol (peppermint), methyl anthranilate (grape), methyl salicylate (wintergreen), orange oil (orange), vanillin (vanilla)

Administration (FDA). At this time the FDA lists about 600 chemical substances "Generally Recognized As Safe" (GRAS) for their intended use. A part of this GRAS list is given in Table 12.4, which includes mostly the common food additives. It must be emphasized that even recognized additives are safe only if they are used in the amounts and foods specified. Although the effect of an individual additive can be easily tested, it is difficult to check its *synergistic effect,* that is, its effect in the presence of other substances in the foodstuff. However, this is being done in as many cases as practical. Today, any intentional additive that is found to induce cancer when ingested by man or animal is withdrawn and banned.

 Whereas the controlled use of intentional additives has caused relatively little danger to public health, incidental additives in some phase of food production, processing, packaging, storage, or transportation have had some deleterious effects on human health. Incidental additives include chemical contaminants, particularly the pesticides used in the agricultural industry and biological contaminants. You may recall the widely publicized contamination of cranberries by the weed-killer aminotriazole and of fishes by DDT and methyl mercury. You may also recall reports of food poisoning by staphylococci, *Salmonella* bacilli, and *Clostridium botulinum.* However, today, modern food processing has cut down on the instances of biological poisoning and chemical contamination as well.

QUESTIONS

12.33 What are food additives? Give examples.
12.34 Why is it necessary for human beings to depend on processed foods?
12.35 Name some of the food preservatives that are familiar to you.
12.36 Give two examples of nutrient additives that have helped to prevent deficiency diseases.
12.37 List five synthetic organic chemicals that are used as common food flavors.
12.38 Why were the cyclamates banned?
12.39 What is meant by synergistic effect?
12.40 Discuss the advantages and disadvantages of intentional additives as distinct from incidental additives (contamination) in food.

SUGGESTED PROJECT 12.2

Additives in your food (or some ingratiating ingredients)

The project is intended to familiarize the student with the numerous additives that are used by the food-processing industries. Check the labels on foods such as breakfast cereals, bread, canned meats, canned soups, frozen dinners, ketchup, and snack foods. Record the ingredients that are not themselves natural foods and describe their functions. Use the Merck Index to determine their toxicities. Tabulate your results.

 COMMON HOUSEHOLD CHEMICALS All of us, every day, use at least several household chemical products that are available through supermarkets, department stores, and drugstores. Detergents, bleaches, disinfectants, waxes, deodorants, insecticides, paints, and over-the-counter drugs are just a few of the hundreds of chemical products in use in the home today. Some of these have been in use for years, but many are of recent development. Some are fairly harmless to human beings and animals, and others are extremely hazardous. Of all the household chemical products, the largest volume is probably made up by the cleansing agents, followed by beauty agents and medicines.

Cleansing agents   Cleansing agents include a variety of chemicals that go under the name of laundry products, dishwashing compounds, and special cleaners. Soap, which is a product of fat and a base such as NaOH, used to be the chief laundry detergent. However, a large number and variety of synthetic detergents (syndets or nonsoaps) have long since displaced soap as the leader in this field. (For a discussion of the cleansing action of soaps and syndets, see Chapter 14.) The chemical composition of a typical solid detergent is given in Table 12.5. Note that special chemical ingredients are added to most detergent formulations to improve their cleansing efficiencies. Such detergent additives include a bleach of one sort or another. Both chlorine and boron compounds are commonly included in formulations to oxidize stubborn stains to colorless products. However, many householders prefer to add their own bleach. Commercially available liquid bleach is a dilute solution of sodium hypochlorite (NaOCl). To a large extent, stain-removal procedures are based on solubility patterns or chemical reactions. Chemicals used to remove some common stains are listed in Table 12.6.

TABLE 12.5
Composition of a typical solid detergent

Component	Weight %
sodium alkylbenzene sulfonate (detergent)	18
sodium tripolyphosphate (builder)	50
water and inorganic filler	19.7
anticorrosion agent	6
dedusting agent	3
foam booster	3
optical brightener	0.3

TABLE 12.6
Some common stains and stain removers

Stain	Stain remover
antiperspirants	ammonium hydroxide
asphalt	benzene, carbon disulfide
berry (fruit)	hydrogen peroxide
blood	cold water, hydrogen peroxide, sodium hypochlorite
chocolate	tetrachloroethylene
coffee	sodium hypochlorite
grass	amyl acetate and benzene (50:50) or sodium hypochlorite or alcohol
ink	oxalic acid, methanol, water
lipstick	isopropyl alcohol, isoamyl acetate, chloroform
mildew	hydrogen peroxide, sodium hypochlorite
mustard	alcohol, sodium hypochlorite
nail polish	acetone
perspiration	ammonium hydroxide, hydrogen peroxide
rust	water, methanol, oxalic acid
scorch	hydrogen peroxide
soft drinks	sodium hypochlorite
tobacco	sodium hypochlorite

Most dishwashing machine detergents are alkaline inorganics, such as the carbonates, silicates, and phosphates. Some products include a detergent that is stable enough to resist the very strong alkaline medium. However, hand dishwashing products are 10–15 percent water solutions of soap containing a disinfectant such as the iodophors (polymers that complex iodine molecules and dissolve in water) and the quats (a compound having four alkyl groups attached to a nitrogen or phosphorus atom). Besides laundry and dishwashing detergents, there are many other chemical cleaners available for home use. These are the special cleaners for rugs, windows, ovens, drains, toilet bowls, and many other items. Rug cleaners are very similar to hand dishwashing products, but with a higher active concentration; window cleaners are dilute solutions of a detergent in isopropyl alcohol and water; oven cleaners and drain cleaners are essentially pure sodium hydroxide (also called *lye*); and, toilet bowl cleaners are strong solutions of hydrochloric acid or acid salts such as sodium hydrogen sulfate. Some of these products can severely burn the skin and can cause blindness if improperly handled.

Beauty agents Chemical preparations that are applied to skin and hair to make ourselves beautiful and pleasing to others are called *cosmetics*. They include chemicals that color hair and skin, disinfect our body surfaces, remove hair or hold it in place, and reduce unwanted body odor. The compositions of some typical cosmetics are given in Table 12.7. Many of these products interact chemically with the protein called *keratin*, which is the essential constituent of hair and skin. In fact, toughness of both skin and hair is due to the bridges between different protein chains, such as hydrogen bonds and disulfide linkages (see Chapter 11).

Most *hair shampoos* contain a mild detergent, such as sodium lauryl sulfate $[CH_3(CH_2)_{11}OSO_3^-Na^+]$, and a builder such as a phosphate ion that can complex the calcium ions that otherwise precipitate and form a scum on the hair. Although these shampoos clean hair of dirt and oil, their very chemical nature makes the hair hard to manage in many cases. Because the problem is caused by the anionic component of the shampoo, it can be remedied by a *rinse* using a cationic detergent

$$\left(H-\overset{|}{\underset{|}{C}}-CH_2-CH_2-COO^-\ H^+ \right).$$ The so-called *hair permanents* involve the appli-

cation of a reducing agent, such as thioglycolic acid ($SH-CH_2COOH$), to disrupt the disulfide linkages in hair, followed by oxidizing agents, such as H_2O_2, $KBrO_3$, and perborates ($NaBO_2 \cdot H_2O_2 \cdot 3H_2O$), for the subsequent re-formation of some of the broken bonds (see Chapter 11). Of course, various additives are present in both the oxidizing and the reducing solutions to control pH, odor, and color, and for general ease of application. Modern *hair sprays* contain essentially the plastic resin, polyvinylpyrrolidone (PVP),

PVP

blended with a plasticizer (dimethyl phthalate), and a solvent-propellant mixture (ethanol and Freon 12). Such a mixture when sprayed on hair furnishes a film with sufficient strength to hold the hair in place after the solvent has evaporated.

In recent years, *hair bleaches* and *colors* have become very popular. Hair bleaches are solutions of hydrogen peroxide made basic with ammonia to enhance the oxidizing power of the peroxide. They destroy the hair pigments (brown-black *melanin*) by oxidation. Hair-coloring formulations vary from temporary coloring (removable by shampoo), which is usually achieved by means of a water-soluble dye acting on the surface of the hair, to semipermanent dyes that penetrate the hair fibers to a great extent. These often consist of cobalt or chromium complexes of dyes dissolved in an organic solvent. Permanent dyes, on the other hand, are generally "oxidation" dyes similar to fabric dyes. They penetrate the hair and then are oxidized to give a colored product that is permanently attached to the hair by chemical

TABLE 12.7
Compositions of some typical cosmetics

Cosmetic preparation	Ingredients (wt %)
hair-washing lotion	thioglycolic acid (5.7) ammonia (2) water (92.3)
hair-coloring lotion (blond formulation)	p-phenylenediamine (0.3) p-methylaminophenol (0.5) p-aminodiphenylamine (0.15) o-aminophenol (0.15) pyrocatechol (0.25) resorcinol (0.25) inert solvent (98.40)
hair spray	polyvinylpyrrolidone (resin) (4.7) dimethyl phthalate (plasticizer) (0.2) silicone oil (sheen) (0.10) ethanol (solvent) (25.00) Freon 11 (propellant) (45.00) Freon 12 (propellant) (25.00) perfume (trace)
cold cream	almond oil, (35) beeswax (12) lanolin (15) spermacetti (8) rose water (30)
lipstick	dye (coloring agent) (4–8) castor oil, paraffins, or fats (solvent) (50) lanolin (emollient) (25) carnauba wax (for rigidity) (18) beeswax (for rigidity) (18) perfume (pleasant taste) (1.5)
suntan lotion	monoglyceryl p-aminobenzoate (3) mineral oil (25) sorbitan monostearate (emulsifier) (5) polyoxyethylene sorbitan monostearate (emulsifier) (5) water (solvent) (62) perfume (pleasant odor) (trace)
antiperspirant deodorant (cream preparation)	aluminum hydroxychloride (astringent) (20) sorbitan monostearate (water absorbent) (5) polyoxyethylene sorbitan monostearate (water absorbent) (5) stearic acid (amine precipitant) (15) propylene glycol (fatty acid precipitant) (5) water (for consistency) (50)

bonds. These hair dyes generally are derivatives of phenylenediamine, which itself dyes hair black. Its derivatives, *p*-aminodiphenylamine-sulfonic acid and *p*-phenylenediamine-sulfonic acid, are used in blond dye formulations (Table 12.7):

phenylenediamine
(black hair dye)

p-aminodiphenylamine-sulfonic acid

(blond hair dyes)

p-phenylenediamine-
sulfonic acid

Other beauty preparations include creams, lipstick, and suntan lotions that are applied to the skin. Many commercially available skin preparations contain lanolin (a complex mixture of esters from hydrated wool fat) which is an excellent skin softener (emollient). Creams are generally emulsions, that is, colloidal suspensions of one liquid in another. For example, *cold cream* is a suspension of rose water in a mixture of almond oil and beeswax stabilized by the presence of lanolin; *vanishing cream* is a suspension of stearic acid in water stabilized by the presence of potassium stearate (soap). Today, creams of various sorts are used as the base for other cosmetic preparations, in which case further ingredients are added to give additional properties to the creams. As an example, hydrated aluminum chloride is added to prepare a cream deodorant.

Lipstick consists of a suspension of coloring agents in a mixture of high molecular weight hydrocarbons or their derivatives, or both, perfumed to give an odor and pleasant taste. The color usually comes from a dye or "lake," that is, a precipitate of a metal ion such as Fe^{3+}, Co^{2+}, Ni^{2+} with an organic dye.

Suntan products range from lotions, which selectively filter out the higher-energy ultraviolet rays of the sun, to preparations which essentially dye the skin a tan color. A common ingredient in such preparations is *p*-aminobenzoic acid which absorbs strongly in the ultraviolet region of the spectrum. The *p*-aminobenzoic acid is emulsified with a mixture of alcohols, an oil, and water by a high molecular weight, fatty acid ester. Preparations for the relief of the pain of sunburn are solutions of local anesthetics such as benzocaine and other ingredients that include lanolin for softening of the burned tissue.

Two other chemical preparations that are applied to the skin are the familiar *deodorants* and *disinfectants*. They are not beauty aids in the true sense. The most widely advertised deodorants are the chemicals used to "dry up" perspiration or act as astringents. Hydrated aluminum salts, such as $AlCl_3 \cdot 6H_2O$ and $Al_2(OH)_5Cl \cdot 2H_2O$, are astringents in that they can reduce or close up the openings of the sweat glands by affecting the hydrogen bonds that hold protein molecules together. A typical spray deodorant will have the aluminum salt and minor ingredients dissolved in an alcoholic solution. In addition to these antiperspirants, deodorants that have an odor to mask the odor of sweat and that remove odorous compounds by combining with them are available. Such deodorizing agents include the odor-masking oils and perfumes, and oxidizing agents such as zinc peroxide and a variety of mild antiseptics.

Disinfectants are chemicals used to kill pathogenic microorganisms before they can infect the skin. Most commonly used disinfectants are the alcohols (70%

ethanol or 50% isopropanol), which kill germs apparently by hydrogen bonding with water—a process that dehydrates the cellular structure of the germ. These are the only disinfectants used in many aftershave preparations. The oldest known disinfectant, however, is a water solution of phenol known as *carbolic acid*. Today, several phenol derivatives are included as milder disinfectants in soaps, deodorants, facial creams, and other cosmetics. However, one derivative, *hexachlorophene,*

hexachlorophene

which is extremely effective against staphylococci and streptococci bacteria, has been banned in all but prescription uses following reports of brain cell damage in test baby monkeys.

QUESTIONS

12.41 List the categories of chemical products in your home.

12.42 Which of the chemical products in your household are
   a. harmless?
   b. hazardous to human health?

12.43 What is the difference in the composition of machine and hand dishwashing products?

12.44 Give examples of special cleansing agents. Why are they needed?

12.45 Name some of the cosmetics that you use every day. Explain their functions.

12.46 What do you suppose is the mechanism by which chemical preparations remove unwanted hair?

12.47 Write the names of the important ingredients in
   a. hair permanents
   b. hair sprays
   c. hair colorings

12.48 What are the beauty preparations that are applied to the skin? Give examples.

12.49 Explain the modes of action of commercial deodorants and antiperspirants.

12.50 How is a sunburn treated?

SUGGESTED PROJECT 12.3

Examining chemical products at home (or why suds are suspect)

This project is intended to familiarize the student with several household chemical products. Read the labels on laundry and dishwashing products, cosmetic preparations, and special cleansing agents that you use every day. Record the type of chemical product and the active ingredients. Use the Merck Index to determine the use and toxicity of each of the active ingredients. Which products contain the same active ingredients?

MEDICINES AND DRUGS Most households regularly keep a number of drugs that are intended to relieve indigestion, constipation, and headaches, to cure ailments and infections, to prevent pregnancy, to calm anxiety, and to induce sleep. An examination of the contents of medicine cabinets will often reveal both over-the-counter nonprescription drugs such as antacids, aspirins, laxatives, decongestants, and cough and cold preparations as well as prescription drugs such as birth-control pills, sleeping pills, tranquilizers, energizers, and appetite suppressants. In some households with people that are drug addicts, illegal, under-the-counter, illicitly traded drugs such as LSD, marijuana, and heroin may be found. This discussion begins with over-the-counter drugs that can be obtained from your pharmacy. Illegal drugs are discussed separately (p. 332); for a discussion of birth-control pills, see Chapter 11.

Aspirin (acetylsalicylic acid),

salicylic acid   acetylsalicylic acid
(aspirin)

is the most widely used drug in the world. Experiments have shown that aspirin produces its effect on the central nervous system and that salicylic acid from the hydrolysis of aspirin is the active chemical. Besides being an analgesic (pain reliever), salicylic acid is an antipyretic (fever reducer). However, it is not effective for severe pain, such as that of migraine headaches, and prolonged use, as for arthritic pain, can lead to gastrointestinal disorders. Irrespective of the manufacturer, most aspirin tablets contain 5 grains of acetylsalicylic acid held together in an inert binder such as starch or clay. Buffered aspirin (*Bufferin*) contains antacids to prevent gastric hyperacidity caused by the acetylsalicylic acid.

Most commercially available headache remedies are *combination pain relievers* and include other drugs besides aspirin. The most usual combination is the *APC tablet* which contains aspirin, caffeine, and phenacetin. Caffeine is a mild stimulant and does not enhance the effect of aspirin in any significant way. Phenacetin (acetophenetidin) has about the same effectiveness as aspirin in reducing fever and relieving minor aches and pains, but has been implicated in kidney diseases and blood abnormalities. Perhaps because of these possible side effects, phenacetin has been dropped from some preparations such as *Anacin* and replaced by acetaminophen

caffeine   phenacetin   acetaminophen

in some other preparations such as *Excedrin*. Acetaminophen is comparable to aspirin in relief of pain and reduction of fever and is recommended to people who are allergic to aspirin. It is available by itself under several trade names some of which are *Apamide*, *Lyteca*, and *Tempera*.

Antacids are probably the most widely used drugs after aspirin. Although a number of varieties of antacid pills are on the market, they all have one common function — to relieve a condition known as *hyperacidity* (too much HCl in stomach). They are generally insoluble bases or salts with weakly basic anions capable of neutralizing a portion of the hydrogen ions (from acid) in the stomach. The chemistry of the commonly used antacids is given in Table 12.8. The preferred antacids are those that do not reduce the stomach acidity very much and thus avoid any "acid rebound" response from the stomach. A number of antacid preparations contain other drugs. For example, *Alka-Seltzer* contains aspirin which is an analgesic and antipyretic, *Bromo-Seltzer* contains potassium bromide which is a depressant and sleep-inducer, and *Pepto-Bismol* contains an antidiarrhea drug.

Among the prescription drugs are the depressants, stimulants, tranquilizers, and psychotomimetic (hallucinogenic) drugs. All these drugs affect the central nervous system. Some induce way-out hallucinations, psychoses, and colorful visions and their use is regulated by federal law. Some are used in the treatment of mental illness, although the causes of the various types of mental illness are not fully understood. Let us look at some of these chemicals with which many people pollute their bodies.

The oldest known *depressant* is ethyl alcohol, which generally slows down both physical and mental activity. (However, a small amount of it acts sometimes as a stimulant, perhaps by relaxing tensions and relieving inhibitions.) *Anesthetics* such as ethyl ether, chloroform, and cyclopropane (see Chapter 9) are depressants and act to produce a loss of feeling and sensation, usually accompanied by unconsciousness. *Sedatives* such as sodium bromide and chloral hydrate $[CCl_3CH(OH)_2]$ are drugs used to quiet and relax a patient without producing sleep. Drowsiness and sleep can result if the dosage is increased above that producing sedation. *Hypnotics* are depressant, sleep-inducing drugs, otherwise known as "downers." The most

TABLE 12.8
The stomach chemistry of some antacids

Antacid	Formula	Stomach reaction
aluminum hydroxide (Tums)	$Al(OH)_3$	$Al(OH)_3 + 3HCl \longrightarrow AlCl_3 + 3H_2O$
calcium carbonate (chalk)	$CaCO_3$	$CaCO_3 + 2HCl \longrightarrow CaCl_2 + H_2O + CO_2$
dihydroxyaluminum sodium carbonate (Rolaids)	$NaAl(OH)_2CO_3$	$NaAl(OH)_2CO_3 + 4HCl \longrightarrow$ $NaCl + AlCl_3 + 3H_2O + CO_2$
magnesium hydroxide (milk of magnesia)	$Mg(OH)_2$	$Mg(OH)_2 + 2HCl \longrightarrow MgCl_2 + 2H_2O$
magnesium oxide (magnesia)	MgO	$MgO + 2HCl \longrightarrow MgCl_2 + H_2O$
sodium bicarbonate (baking soda)	$NaHCO_3$	$NaHCO_3 + HCl \longrightarrow NaCl + H_2O + CO_2$
sodium citrate	$Na_3C_6H_5O_7$	$Na_3C_6H_5O_7 + 3HCl \longrightarrow 3NaCl + H_3C_6H_5O_7$

widely used hypnotics are the *barbiturates,* derivatives of barbituric acid:

barbituric acid   pentobarbital   phenobarbital

The barbiturates are used medically for sedation and sleep. They are legally available by prescription only but are also a part of the illegal drug market. When mixed
with alcohol the depressant effect of barbiturates is synergistic and can be fatal.
Many deaths reported as due to an overdose of sleeping pills were probably caused
by drug combinations such as the alcohol–barbiturate.

Drugs that produce both sedation (narcosis) and relief of pain (analgesia) are
called *narcotics.* Many drugs produce these effects, but in the United States only
those that are also *addictive* are referred to as narcotics and regulated by federal
law. Such drugs include the alkaloid morphine, which is obtained from opium
poppy plants, and its derivatives codeine and heroin, which were discussed earlier
in this chapter (p. 317). The physiological action of these narcotics is similar except
that codeine has less tendency to induce sleep and heroin has more tendency to
produce stronger feeling of euphoria for a longer period of time. All these drugs are
strongly addictive and many deaths have occurred due to illegal use of these drugs.
The mechanism of addiction is not clearly understood. It is believed that a liver
enzyme called cytochrome P-450 is essential to the detoxification of such drugs.
Just as an overdose of insulin requires the diabetic to take some sugar, the buildup
of an overabundance of cytochrome P-450 in the liver could require the drug addict
to consume more drugs.

Tranquilizers affect the emotional state of an individual by calming or quiet-
ing without decreasing the level of consciousness. Compounds, such as reserpine,
chlorpromazine (Thorazine), and meprobamate,

reserpine

chlorpromazine   meprobamate

are available by prescription for treatment of anxiety, tension, and various neuroses. These compounds are also used, in combination with other drugs, to relieve pain.

Earlier we have encountered caffeine, a mild *stimulant*, as one of the drugs present in many combination pain relievers. A stronger stimulant, amphetamine or benzedrine induces excitability, restlessness, tremors, dilated pupils, increased pulse rate and blood pressure, hallucinations, and psychoses. Although this compound is sometimes prescribed for weight reduction, mild depression, or narcolepsy, it is very much a part of the illegal drug scene under names such as pep pills, speed, bennies, and uppers.

Before we take up the chemistry of hallucinogenic drugs let us compare the known toxicities of some of the chemicals found in common drugs. The usual biochemical practice is to compare the dosage that would be lethal to 50 percent of a population of test animals, usually rats. This dosage, which is termed LD_{50}, is for oral administration and is expressed in terms of amount of drug per kilogram of animal body weight. The LD_{50} for humans is usually estimated from animal studies and therefore it can only be approximate at best. Further, the data given in Table 12.9 relate only to short-term (acute) toxicities and give no indication of long-term effects.

TABLE 12.9
Toxicities of some chemicals found in common drugs[a]

Chemical	LD_{50}[b]
amphetamine	0.085[c]
aspirin	1.75
caffeine	0.200
epinephrine (adrenaline)	0.050
ethyl alcohol	13.7
ethylene glycol	8.54
glycerol	31.5
isopropyl alcohol	5.8
methanol	—[d]
nicotine (sulfate)	0.055
phenacetin	1.700
phenobarbitol	0.600
phenol	0.53

[a] Data taken from *Merck Index*, 8th ed., 1968.
[b] LD_{50} values are for oral administration of the drug to rats (g/kg of body weight) and relate only to short-term (acute) toxicities.
[c] For oral administration of the drug to rabbits.
[d] No LD_{50} is given for methanol.

QUESTIONS

12.51 Write an equation for the hydrolysis reaction of aspirin.
12.52 What is meant by combination pain reliever? Give three examples.
12.53 Why is acetaminophen preferred over phenacetin as a drug?
12.54 How does an antacid function in the stomach?
12.55 Give examples of prescription drugs that are used as
   a. sedatives
   b. hypnotics

    c. stimulants
    d. tranquilizers
12.56 What is heroin? How is it prepared?
12.57 How is the mechanism of drug addiction explained at present?
12.58 What is the basis for comparing the toxicities of various chemicals?

SUGGESTED PROJECT 12.4

Active ingredients in your medicine chest (or skeletons in the closet)

This project is intended to familiarize the student with the active ingredients present in household medicines and drugs. Examine the labels on all the medicines in your home and record the type of medicine and the active ingredients. Use the Merck Index to determine their toxicities and side effects, if any. Tabulate your results. What drugs contain the same active ingredients?

ILLEGAL DRUGS   There are a number of drugs that are illegal in the United States and in many other countries (Fig. 12.8). Some of these drugs are called *hallucinogenic*, psychotomimetic, or psychedelic drugs because they induce way-out hallucinations, psychoses, and colorful visions. The best known of such drugs include mescaline, lysergic acid diethylamide (LSD), and tetrahydrocannabinol (the active component of marijuana):

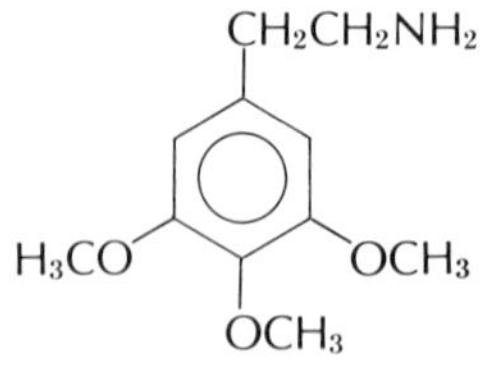

mescaline

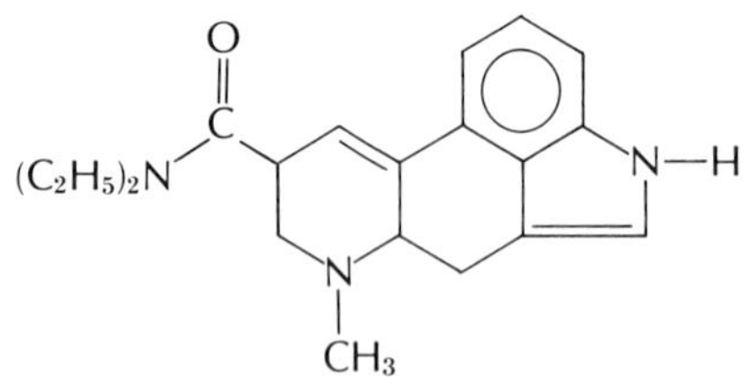

LSD

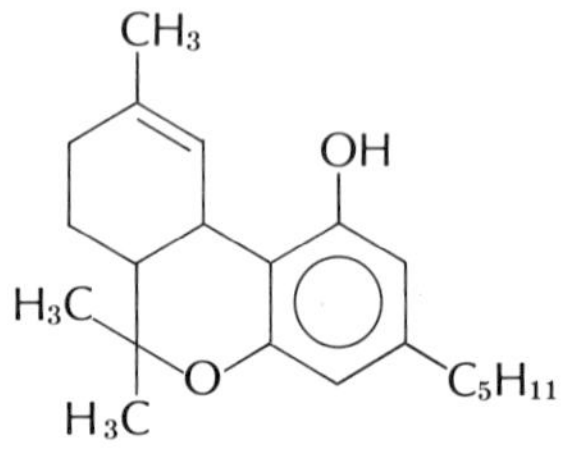

tetrahydrocannabinol

Mescaline is one of the oldest known hallucinogens and is isolated from the peyote plant. *Marijuana* or "pot" is the dried resinous leaves and seed pods of the weed *Cannabis stiva* (Indian hemp) which grows wild over most of the world. *LSD* is derived from lysergic acid ($C_{16}H_{16}N_2O_2$), a crystalline alkaloid obtained from the *ergot* fungus that infects various cereal plants. Chemically, information on these extremely powerful drugs is meager. What causes the observed changes in visual perception, depersonalization, and other brain disturbances is yet to be determined.

It has, however, been shown that LSD interacts with the compounds serotonin and norepinephrine that are considered to be responsible for transmitting impulses across the synapses* in the brain. But the biochemistry of these brain chemicals is still largely unknown. Although, as yet the theory of the hallucinogenic mechanism is being debated, the dangers of using LSD are becoming clear. In children of mothers who had been on LSD, birth defects, mental illness, and death have been documented. Spontaneous recurrence of hallucinations without ingesting the drug again often occur weeks or even months later.

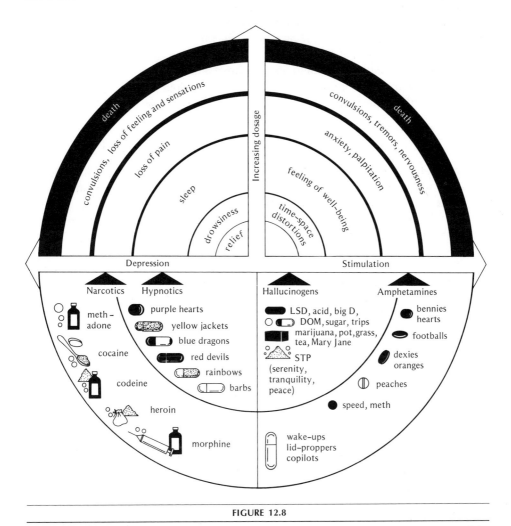

FIGURE 12.8

Commonly abused drugs and their physiological effects. Adapted from *Drugs of Abuse,* courtesy of the Drug Enforcement Administration, Washington, D.C.

Because marijuana is totally nonaddictive there is continuing debate whether or not it should be legalized. It is argued that marijuana is no more (or even less!) harmful than alcohol or tobacco. In fact, a U.S. Government report released in 1972 gave alcohol abuse as the nation's greatest drug problem, with an estimated 5 per-

---

* Synapses are the very tiny gaps formed when one nerve cell terminates in close proximity to the beginning of another nerve cell or the cells of an organ or gland.

cent of the adult population being alcoholics. Although little scientific evidence concerning the effects of marijuana is available, evidence concerning the effects of alcoholism and tobaccoism is well documented.

ALCOHOLISM   The principal component of alcoholic beverages is ethyl alcohol which has a toxic effect on the human body. However, the toxic effect is mild enough to render it safe for consumption as a stimulant. As soon as a person consumes small amounts of an alcoholic drink, the ethyl alcohol is absorbed by the stomach mucosa and passed through the stomach lining into the bloodstream. In an average adult, a pint of whiskey at a time can raise the alcohol concentration in blood to about 0.3 percent. Concentrations above this amount are extremely dangerous, because they lead to coma and possible heart failure.

Over 90 percent of the ethyl alcohol absorbed into the tissues is slowly oxidized to acetaldehyde, then to acetic acid, and finally to carbon dioxide and water:

$$C_2H_5OH \xrightarrow{O_2} CH_3\overset{\overset{\displaystyle O}{\|}}{C}-H \xrightarrow{O_2} CH_3\overset{\overset{\displaystyle O}{\|}}{C}-OH \xrightarrow{O_2} CO_2 + H_2O$$

Although almost every cell in the body can accomplish the oxidation, it occurs primarily in the liver. The intoxicated person's staggering gait, stupor, and nausea are caused by the presence of acetaldehyde, but the chemical reactions involved are not fully understood. It is believed that there is an overabundance of the detoxifying enzyme cytochrome P-450 in the liver of an alcoholic and that this requires the alcoholic to consume more alcohol.

Chronic alcoholism is often cured by the compound *disulfiram* that blocks the oxidative steps beyond acetaldehyde. The accumulated acetaldehyde causes nausea and induces vomiting which discourages further drinking of alcohol.

TOBACCOISM AND SMOKING   Tobacco does contain the alkaloid *nicotine* (2–8% by wt),

nicotine

which is a deadly poison when swallowed even in milligram amounts (Table 12.9). Unlike marijuana, tobacco is mildly addictive, but then not many people eat tobacco. However, those people with the habit of cigarette smoking find it very hard to stop smoking.

Cigarette smoke contains a carbon monoxide concentration greater than 20,000 ppm. Even though during inhalation this is diluted to a level of about 400–500 ppm, it is still quite a high level. The toxic effect of CO on the body is caused primarily by the formation of carboxyhemoglobin in the blood. Hemoglobin normally functions in the blood as a transport system to carry oxygen from the lungs to the body cells and CO_2 from the cells to the lungs. However, the affinity of CO for hemoglobin is several times (~200) greater than that of O_2. As a result, carboxyhemoglobin rather than oxyhemoglobin is formed and the ability of the blood to transport oxygen is reduced. The higher the amount of hemoglobin bound up in the form of carboxyhemoglobin, the more serious is the effect for different categories of smokers and nonsmokers, as indicated in Table 12.10. Heavy cigarette smoking therefore contributes to respiratory failure and heart disease.

In addition to the toxic carbon monoxide, tobacco smoke has been shown to contain potentially carcinogenic (cancer-causing) compounds such as 3,4-benzopyrene (see Chapter 9). Thus, the use of tobacco has been linked to lung cancer as well.

TABLE 12.10
Blood carboxyhemoglobin levels of smokers and observed health effects[a]

Category of smoker	Ambient air CO concentration (ppm)	Blood level of carboxyhemoglobin (%)	Observed health effects
never smoked	5	1.3	no apparent effect
ex-smoker	5.6	1.4	no apparent effect
pipe and/or cigar smoker only	7.5	1.7	effect on behavioral performance
light cigarette smoker, noninhaler	11.3	2.3	central nervous system effects
light cigarette smoker, inhaler	20.6	3.8	central nervous system effects
moderate cigarette smoker, inhaler	33.8	5.9	cardiac and pulmonary functional changes
heavy cigarette smoker, inhaler	40.0	6.9	cardiac and pulmonary functional changes
	(60–500)	(10–80)	(headache, fatigue, drowsiness, coma, respiratory failure, death)

[a] Data selected from "Air Quality Criteria for Carbon Monoxide," U.S. Department of Health, Education and Welfare, 1970, pp. 8–9; *Environmental Science and Technology,* **5,** 213 (1971).

CARCINOGENS AND CANCER Chemicals that are found to be the cause of cancer (any abnormal cell growth that does not respond to normal growth-control mechanisms) are said to be carcinogenic and are called *carcinogens.* Such chemicals now include a large number of diverse compounds, mostly of organic origin. Some of these are listed in Table 12.11. Much of our early knowledge of the carcinogenic action of chemicals such as 3,4-benzopyrene and 2-naphthylamine came from studies in which the cancer was linked to a person's occupation which, in turn, was linked to some chemical he came in contact with daily. Subsequently, the lower animals, particularly mice and rats, were used for studies of carcinogenicity and the observed carcinogenic effects extrapolated to humans.

In general, there are two groups of carcinogens: those compounds, such as 3,4-benzopyrene, that cause cancer at the point of contact and those compounds, such as 2-naphthylamine and benzidine, that cause cancer in an area remote from the point of contact. In the latter group, some other compounds made from the original chemical are considered to be the carcinogens. Some carcinogens are relatively nontoxic in a single large dose, but quite toxic, often increasingly so, when administered continuously. For example, the development of cancer in the connec-

TABLE 12.11
Some carcinogens, their sources, and structures

Compound	Structure	Source
carbon tetrachloride	CCl_4	dry-cleaning agent, solvent
dioxane	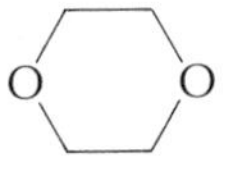 (1,4-dioxane ring with two O)	industrial solvent, cosmetics, glues, deodorants
cyclamate	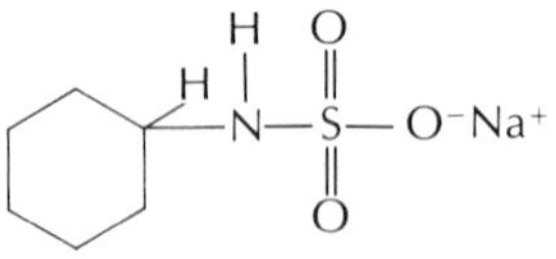 cyclohexyl—N(H)—S(=O)(=O)—O$^-$Na$^+$	artificial sweeteners (now banned)
2-naphthylamine	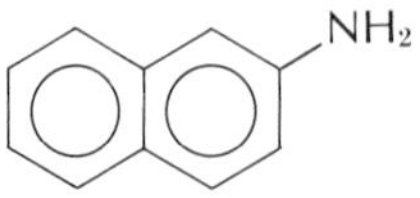 naphthalene ring with —NH_2	optical bleaching agent
benzidine	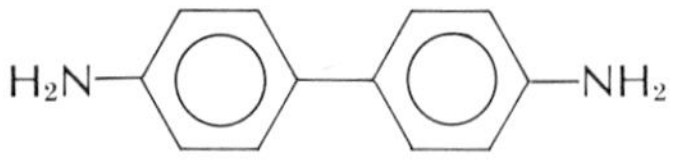 H_2N—(ring)—(ring)—NH_2	polymer and dye industries
N, N-dimethyl-p-phenylazoaniline	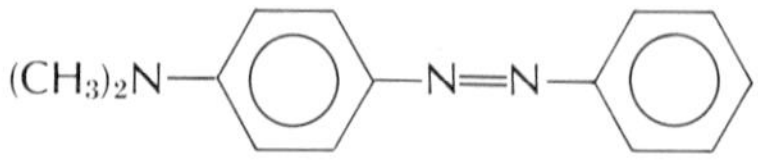 $(CH_3)_2N$—(ring)—N=N—(ring)	dye industries, food color (now banned)
N-ethyl-N-nitroso-n-butylamine	C_2H_5—N—CH_2—$(CH_2)_2$—CH_3 with N=O on the nitrogen	insecticide, gasoline and lubricant additive
N-2-fluorenyl-acetamide	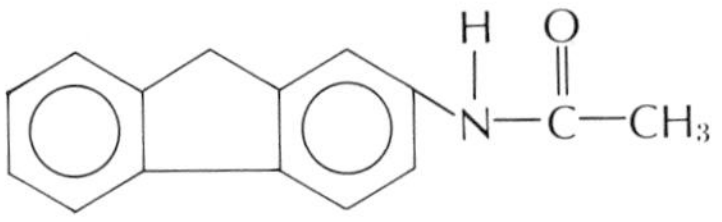 fluorenyl—N(H)—C(=O)—CH_3	herbicide
aflatoxin	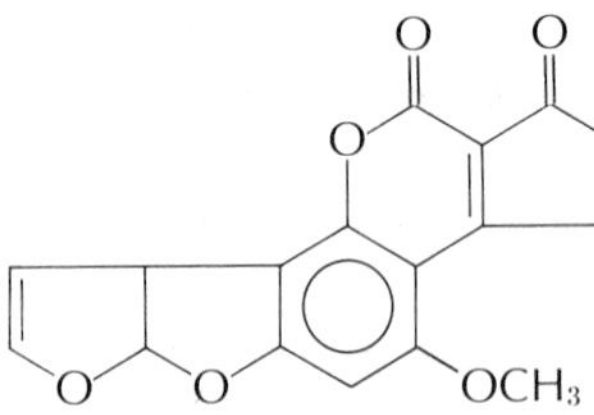 (polycyclic structure with two C=O, O atoms, and OCH_3)	molds on peanuts and other plants
3, 4-benzopyrene	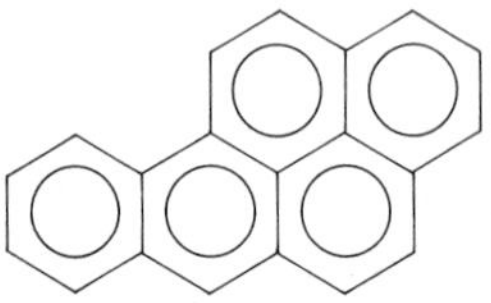 (polycyclic aromatic hydrocarbon)	cigarette and coal smoke

tive tissues in humans, from the activation of the first cell to the clinical manifestation of the cancer, takes from 20 to 30 years. The importance of geographic location in determining the intake of a potentially carcinogenic compound is signified by the fact that cancer does not occur with the same frequency in all parts of the world. Published reports indicate that the industrialized world has a higher incidence of cancer, particularly lung cancer.

There are several theories to account for the structural causes of cancer. One theory postulates that the carcinogen combines with growth-control proteins and renders them inactive. However, the specific growth proteins involved are not yet known for any of the carcinogens. Another theory suggests that carcinogens react with and alter nucleic acids. In consequence, the proteins formed on the messenger RNA are sufficiently different to alter the cell's function and growth rate. Despite considerable efforts, a cure for cancer is nowhere in sight and the number of deaths due to cancer is increasing.

ALLERGENS AND ALLERGIES   Substances such as foods, cosmetics, drugs, pollens, and molds are known to cause inflamed eyes, congested sinuses, sneezing, and the running noses that accompany hay fever in people. Such substances are called allergens and the condition is known as allergy. Generally speaking, an allergy is an adverse response to a foreign substance or to a physical condition that produces no obvious ill effects in most other organisms, including man. Most allergens are highly complex substances and have molecular weights of 10,000 or more. For example, the principal allergen (called *ragweed antigen E*) of ragweed pollen is a protein with a molecular weight of about 38,000. Picograms (1×10^{-12} g) of this allergen injected into an allergic person are enough to induce a response.

To understand the chemical cause of an allergy such as, say, pollen allergy, which is very prevalent in the midwest regions of the United States, one has to look at the *antibodies* in our cells and bloodstream that react with and protect us from harmful parasites and bacteria. These antibodies are globulin proteins and are known as immunoglobulins (Ig). Today, four antibodies, designated IgA (of molecular weight 180,000), IgE (of molecular weight 196,000), IgG (of molecular weight 150,000), and IgM (of molecular weight 950,000), are known. Of these, the antibody IgE is formed in the nose, bronchial tubes, and gastrointestinal tract, and binds firmly to specific cells, called mast cells, in these regions. In a nonallergic person, IgE comprises only 0.001–0.01 percent of the globulin proteins in blood serum, but, in an allergic person, 5–15 times as much IgE may be found in his blood serum and 100 or more times in his nasal secretions. When the allergens (antigen E) come in contact with the IgE antibody attached to the mast cells, antigen–antibody complexes are formed and they, in turn, release the so-called allergy mediators from special granules in the mast cells. What causes the release of these mediators is not known with certainty, but it is known that one of these mediators, and the most potent one found so far, is *histamine*, which is formed by the breakdown of the amino acid, histidine:

histidine   histamine

The chemical mediators such as histamine must be released from the cell to cause the symptoms of allergy. They cause dilation of blood capillaries and make the capillaries more permeable to blood fluids so that these fluids can readily leak out of the capillaries and cause swelling of the tissues. In addition, they can cause contraction and spasm of smooth muscles as in bronchial asthma, produce skin swellings as in hives, and stimulate the glands that secrete watery nasal fluids, mucus, tears, saliva, and so on as in hay fever.

Today, more than 50 *antihistamines* and several other drugs are available for treating allergies. An important antihistamine is pyribenzamine.

$$\bigcirc\text{--CH}_2\text{N}\underset{\underset{\text{CH}_2\text{CH}_2\text{N(CH}_3)_2}{|}}{\bigcirc}$$

pyribenzamine

It acts competitively by occupying the receptor sites normally occupied by histamine on cells. Another drug, *disodium cromoglycate,* acts, not by blocking the action of histamine (as do the antihistamines), but by blocking the release of histamine and other mediators from the granules. Although these drugs are effective in treating allergies, they cause a number of troublesome side effects such as drowsiness, blurred vision, nausea, nervous disorder, and depression. Because no drug that will relieve the allergy and completely avoid the side effects is available, many allergics resort to desensitization therapy. This requires many small subcutaneous injections, spaced in time, of a blocking antibody that preferentially reacts with the allergen so that it cannot react with the IgE allergy-sensitizing antibody. The chain of events leading to the release of histamine or other allergy-producing mediators is thus broken. However, desensitization therapy is costly and inconvenient, since 20 or more injections are required to get what usually is a partial cure. Efforts are therefore being made to find drugs with less troublesome side effects than those often associated with antihistamines.

MUTAGENS AND MUTATIONS Chemicals that cause mutations, that is, changes in the genes in the chromosomes of the offspring, are called mutagens. Indeed, many of us have heard of the horrible thalidomide tragedy, which occurred in the 1960s. Thalidomide, a tranquilizer and sleeping pill, which appeared to be remarkably safe on the basis of animal studies, was found to cause birth defects in children whose mothers used it during their pregnancy. Although thalidomide was subsequently found not to be mutagenic, the disaster focused worldwide attention on chemically induced mutations. Extensive investigations were undertaken on the action of chemicals in causing mutations in bacteria, viruses, molds, fruit flies, mice, rats, human white blood cells, and so on. Some of the results of these investigations are summarized in Table 12.12. Note that some of the mutagens such as 3,4-benzopyrene and aflatoxin are proven carcinogens as well (see Table 12.11). In fact, one intriguing theory considers that some compounds cause cancer because they are first and foremost mutagenic.

At present, there is evidence for chemical mutation in plants and lower animals only. It should be emphasized that there is no conclusive evidence that any chemical induces mutations in the germinal cells of humans. Efforts in determining the effects of mutagenic chemicals in man have been impaired due to the extreme rarity of mutations, the long time intervals between generations, and the multitude of chemicals with which man comes in contact. If and when it is demonstrated that chemicals can produce transmissible alteration of chromosomes in human germinal cells, the door will open to the manipulation and alteration of man's genetic heritage.

TABLE 12.12
Some mutagens and their known effects

Compound	Source	Known effects
aflatoxin	molds	mutations in bacteria, viruses, fungi, wasps, mice, human cell cultures
3,4-benzopyrene	cigarette and coal smoke	mutations in mice
caffeine	coffee, tea, cola	chromosome changes in bacteria, fungi, onion root tips, fruit flies, human tissue cultures
captan	fungicide	mutagenic in bacteria and molds; chromosome breaks in rats and human tissue cultures
dimethyl sulfate	chemical industry	potent mutagen in bacteria, viruses, fungi, higher plants, fruit flies
LSD	illegal drug scene	chromosome breaks in somatic cells of rats, mice, hamsters, white blood cells of humans and monkeys
maleic hydrazide	plant growth inhibitor	chromosome breaks in many plants and cultured mouse cells
mustard gas	chemical warfare	mutations in fruit flies
nitrous acid		mutations in bacteria, viruses, fungi
ozone		chromosome breaks in root cells of broadleaf plants
acetone, cyclohexane, ethyl acetate, toluene, hexane	glue solvents (glue sniffing)	4% more human white blood cells showed breaks and abnormalities (6% versus 2% normal)
triethylenemelamine	anticancer drug, insect chemosterilants	mutagenic in fruit flies and mice

QUESTIONS

12.59 **What are illegal drugs? Give examples.**

12.60 **Discuss the possible effects of the use of LSD and marijuana.**

12.61 **How does alcohol react in the human body? What drug is used to cure chronic alcoholism?**

12.62 **Name the most toxic ingredient in tobacco.**

12.63 **Why is cigarette smoking dangerous?**

12.64 List some of the common carcinogens. Suggest a possible mechanism for their action.

12.65 What is an allergic reaction? What causes it?

12.66 What are antibodies? How many antibodies have been identified? How do they react with allergens?

12.67 Discuss the action of histamine as an allergy mediator. How do the antihistamines function?

12.68 Describe the available evidence for chemical mutations in plants and lower animals.

12.69 Why is it difficult to determine the effects of mutagenic chemicals in man?

12.70 What evidence is there to suggest that mutagens may be carcinogens and vice versa?

MAN VERSUS MAN

War is one lethal form of competition among populations of a single species, predominantly man. At the current rate of human population growth (70 million/year) it has not been an effective population control mechanism. Nearly all weapons of war, such as guns, bombs, and rockets, use chemicals in some form. Examples of such chemicals used in earlier wars are gunpowder (a mixture of sodium nitrate, charcoal, and sulfur), cordite (a mixture of nitroglycerin, guncotton, and petroleum jelly), and dynamite (nitroglycerin on kieselguhr, a diatomaceous earth). However, the words "chemical warfare" are used in a fairly restricted way today to exclude conventional weapons such as guns and bombs. Modern chemical warfare had its beginning when poisonous gases such as chlorine and phosgene ($COCl_2$) were used for the first time in World War I. Following World War I the development of new chemical and biological warfare agents continued apace. Despite huge stockpiles, the use of the more lethal of these agents has been largely avoided since that time.

Although the individual components of an arsenal of chemical weapons change over the years there are two major categories of chemical warfare agents, namely, *casualty agents* and *harassing* or *incapacitating agents*. Casualty agents are designed to inflict injury and death to enemy personnel and include choking gases, nerve gases, blood agents, and blister agents. Harassing agents are designed to incapacitate enemy personnel for a limited period of time (minutes or hours). They include the tear gases and vomiting agents which are also used to control unruly prisoners and rioters. Table 12.13 lists a variety of members of the arsenal of chemical weapons and their effects on humans. In spite of the deadly potential of the chemical arsenals in various countries, it is gratifying to note that the more deadly of these chemicals have never been used.

Fire, though not a chemical, has been used as a weapon in war to kill the enemy personnel through heat. During some of the Greek wars substances such as pitch (asphalt) and sulfur were burnt to produce not only heat but also suffocating gases and to consume life-giving oxygen. A much more dangerous incendiary agent used in World War II is *napalm*. Originally, it consisted of aluminum soaps of *naphthenic* and *palmitic* acids dissolved in gasoline. The modern Napalm-B consists of polystyrene dissolved in a mixture of gasoline and benzene. It is designed so that it is only partly burned when it reaches the target where the thick *jelly* splatters, spreading fire and producing great volumes of deadly carbon monoxide which blocks the transport of oxygen in the blood. It is estimated that more than one-half of the city of Tokyo was wiped out by napalm "fire storms."

The symbol of war in our age is the atomic bomb (see Chapter 6) that ended World War II. During that war two atomic bombs were dropped over the Japanese cities of Hiroshima and Nagasaki. Both cities (and their population) were destroyed

TABLE 12.13
The arsenal of chemical weapons

Military symbol	Name	Formula	Function
—	chlorine	Cl_2	choking agent
CG	phosgene	$COCl_2$	choking agent
CX	phosgene oxime	$Cl_2C{=}NOH$	blister agent
H	mustard gas	$S{\big\langle}^{CH_2CH_2Cl}_{CH_2CH_2Cl}$	blister agent
L	lewisite	$^{Cl}_{H}{\rangle}C{=}C{\langle}^{H}_{AsCl_2}$	blister agent
HN	nitrogen mustards	$CH_3N(C_2H_4Cl)_2$ $C_2H_5N(C_2H_4Cl)_2$ $C_2H_4ClN(C_2H_4Cl)_2$	blister agent
AC	hydrogen cyanide	HCN	blood agent
CK	cyanogen chloride	ClCN	blood agent
—	carbon monoxide	CO	blood agent
GA	tabun	$(CH_3)_2N{-}P{\big\langle}^{CN}_{OC_2H_5}$ with ${=}O$	nerve agent
GB	sarin	$CH_3{-}\underset{\underset{O}{\|}}{\overset{F}{\underset{\|}{P}}}{-}OCH(CH_3)_2$	nerve agent
GD	soman	$CH_3{-}\underset{\overset{\|}{O}}{\overset{F}{\underset{\|}{P}}}{-}O\overset{CH_3}{\underset{\|}{CH}}{-}C(CH_3)_3$	nerve agent
CN	α-chloroacetophenone	⟨benzene ring⟩$-COCH_2Cl$	tear agent
CS	o-chlorobenzal-malononitrile	⟨chlorobenzene ring⟩$-\overset{H}{\underset{\|}{C}}{=}C{\big\langle}^{C{\equiv}N}_{C{\equiv}N}$	tear agent
DM	diphenylaminechlorarsine	⟨phenazarsine ring with N–H top, As–Cl bottom⟩	vomiting agent

by the fantastic amount of heat, devastating shock waves, and deadly gamma radiation generated by the explosions. The radioactive fallout was spread around the biosphere by rain and wind, producing far-reaching effects, and conveyed to man in a variety of ways, as shown in Fig. 12.9. The effect of radiation on living matter has since been investigated thoroughly. Intense radiation, like that from an atomic explosion, can kill instantaneously. Less intense radiation kills slowly by disrupting the molecular constituents in cells and preventing their proper functions. Even low levels of radiation can be harmful by breaking the nucleic acid molecules that control vital body functions. This may be how some forms of cancer are induced. If the disruption of nucleic acids happens to be in the germ cells, the harmful effects may show up as birth defects in the offspring.

Because the atomic bomb is limited in size and power, the hydrogen (H) bomb was designed. The H bomb is a thermonuclear weapon built around an atomic bomb and is believed to be a thousand times as powerful as the atomic bomb (see

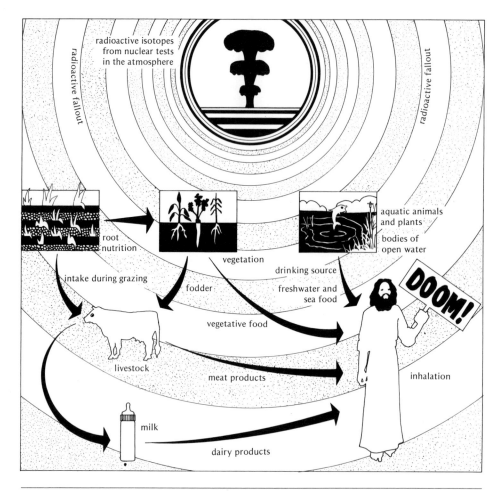

FIGURE 12.9

Transmission of radioactive fallout to human beings. Redrawn, by permission, from *Wastes in Relation to Agriculture and Forestry*, 1968, p. 18. United States Department of Agriculture Miscellaneous Publication No. 1065.

Chapter 6). The use of H bombs in a full-scale nuclear war would result in utter catastrophe and could certainly wipe out all civilization, if not all the biosphere and all life.

QUESTIONS

12.71 What is meant by the term chemical warfare?
12.72 List the major categories of weapons in a chemical arsenal. Give examples.
12.73 What is napalm? How does it work as a weapon?
12.74 Discuss the variety of ways by which the transmission of radioactive fallout to man may occur.
12.75 What are the effects of radiation on living matter?

13

The earth and its soil

. . . no vestige of a beginning, no prospect of an end.

JAMES HUTTON (1726–1797)
Scottish geologist

Our physical environment is the planet earth, which, considered in its entirety, contains all three states of matter. It is conveniently divided into areas involving solid earth (lithosphere), liquid water (hydrosphere), and gaseous air (atmosphere). The chemistry of these areas influence the living matter of the world (biosphere) tremendously. On the other hand, living organisms exert a powerful influence themselves upon the chemistry of most areas, particularly the lithosphere and the hydrosphere. In this and the succeeding chapters (Chapters 14 and 15), we shall study the chemistry of these environments and examine how man-made changes in the physical environment are affecting life on our planet.

Most living organisms are related only indirectly to the solid earth, primarily through plants growing on the earth. Man, in addition to his reliance on the earth as a source of minerals for his civilization, depends on the very thin but very significant layer of soil on the solid earth for food resources. For this reason, we shall begin our study of the physical environment with a discussion of the lithosphere — the outer parts of the solid earth. In general, the term *lithosphere* refers to minerals encountered in the earth's crust. In the environmental sense, the soil is probably the most significant part of the lithosphere. In this chapter, we shall be concerned with the earth and its minerals, and the soil, its chemistry and pollution.

THE EARTH AND ITS MINERALS

Most of what we know about the structure and composition of the earth comes from observations of earthquakes and analysis of exposed surfaces, mines, and drill holes of a depth to about 5 mi. Earthquake waves travel at varying speeds through the earth, depending on the characteristics of the material through which they pass. By comparing the time at which the waves from an earthquake arrive at different points of observation, it has been ascertained that there are three major discontinuities in the earth's interior. The presently accepted model of the earth consists of four major parts: (1) the *crust* (lithosphere), which extends down for about 4 to 20 mi to the surface of the first discontinuity; (2) the *mantle*, which extends for about 1800 mi below the crust; (3) the *outer core*, which extends from the mantle to a depth of about 3100 mi where it meets the inner core; and (4) the *inner core*, which extends from the outer core to the center, about 4000 mi from the surface (Fig. 13.1).

The chemical character of the crust includes igneous rocks, shale, sandstone, and limestone. The mantle consists mostly of heavy metal silicates (93%) and some sulfides and oxides. The core is essentially iron–nickel alloy with the outer part probably molten. This predicted chemical character is in accord with the composition of meteorites and with the composition of the material brought up from the interior of the earth by the volcanic action. It also accounts for the calculated average density of the earth (5.5 g/ml compared to 2.8 g/ml for the earth material in the accessible crust) and the different velocities of the earthquake waves. However, man has not yet even succeeded in penetrating all the way through the crust — the deepest penetration achieved is only to depths of 4 to 5 mi.

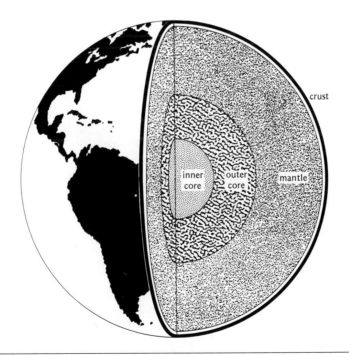

FIGURE 13.1

A model of the earth showing the crust, the mantle, and the outer and inner core regions. (Not to scale.)

THE EARTH'S CRUST Having considered the structure of the earth it is possible now to examine the crust in greater detail. The entire crust is composed of low-density (2.8 g/ml) materials, but it can be subdivided into (1) continental crust and (2) oceanic crust. The portion of the crust that stands above sea level as "continents" is composed of three layers (Fig. 13.2). The uppermost layer consists of sediments and sedimentary rocks; the second layer is composed of granite or granitelike rocks; and the third and lowest layer consists of basalt. The granite layer is associated only with continental land masses and is missing from the crust underlying the oceans (Fig. 13.2). The oceanic crust is composed almost entirely of the rock basalt.

Sediments are loose agglomerates such as soil, mud, and clay arising from erosion or weathering of rocks by water, wind, successive freezing and thawing, and other agents. Sediments that are carried away by rivers and streams as suspensions precipitate ultimately to the bottom of the river bed or are deposited in a delta at the mouth of the river. These deposits gradually become transformed by a cementing process, aided by underground heat and pressure, into sedimentary rocks. *Sedimentary rocks* are usually layered or stratified (Fig. 13.3), each layer representing a certain time period during which sediment was deposited by the action of water. The layers are of great help to geologists in determining the ages of fossils of extinct animals and plants, almost all of which are found in sedimentary rocks.

Granite and *basalt* are *igneous rocks* produced directly from solidification of molten rock, which is called *magma* or *lava*. Most volcanic lavas, when they solidify, form basalt, which is darker and more fine-grained than granite and has a somewhat different composition. When the chemical compositions of basalt and granite are compared (Table 13.1), it turns out that basalt contains more iron, magnesium, and calcium and less silicon, oxygen, potassium, and sodium. Normally, basalt forms the

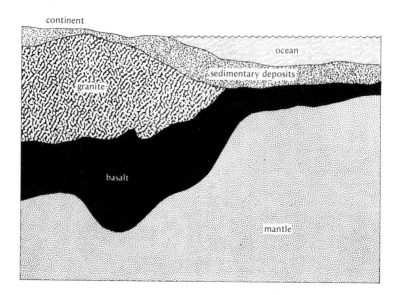

FIGURE 13.2

A cross-sectional representation of the earth's crust. (Not to scale.)

bottom-most layer of the earth's crust; however, both basalt and granite are some-times found exposed on the surface, perhaps as the result of weathering away of overlying sedimentary rocks.

Sedimentary and igneous rocks may be transformed by intense heat and pressure, or by chemical action within the earth's crust, into new and different types of rocks that are known as *metamorphic rocks*. Familiar examples of such meta-morphosis are marble and slate. Marble is produced by subjecting limestone to heat and pressure; slate arises from the transformation of shale by heat into a mixture of mica (complex aluminum silicates) and chlorite (complex magnesium silicates) that comprises slate.

In general, rocks consist of mixtures of two or more minerals. *Minerals* are solids that possess definite chemical compositions and orderly internal arrangements of atoms (called crystal forms). A crustal rock generally contains two or three domi-nant minerals and several minor ones. The major rock categories and their important mineral constituents are given in Table 13.2. Each of the three major rock categories includes many rock types (e.g., limestone, sandstone, granite, and marble) due to small variations in available materials and the environmental conditions that pre-vailed during their formation.

The most abundant minerals in the earth's crust contain primarily oxygen and silicon atoms. Oxygen accounts for about half the total mass contained in rock-forming minerals and silicon, and the next most abundant atom, for about one-quarter. They are usually combined in the form of silicate ions (SiO_4^{4-}):

FIGURE 13.3

Stratification of rock. Photograph courtesy of American Airlines.

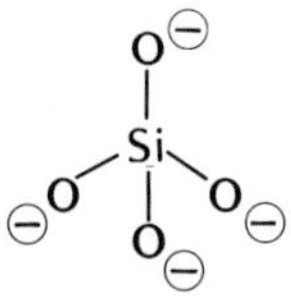

$$O^{\ominus}$$
$$|$$
$$Si$$
$$^{\ominus}O \quad | \quad O^{\ominus}$$
$$O_{\ominus}$$

that are, in turn, combined with cations such as aluminum, iron, and calcium to form the common minerals of the earth's crust. Only six metals—aluminum (8.13%), iron (5.00%), calcium (3.63%), sodium (2.83%), potassium (2.59%), and magnesium (2.09%)—exceed 1 percent of the total mass of crustal rock material. However, industrially important crustal resources include not only the metallic ores but many nonmetallic minerals and rocks as well. For example, crushed stone, sand, and

TABLE 13.1
Relative abundances of the elements in granite and basalt

Element	Relative abundance (%)	
	Basalt	Granite
oxygen	48.5	44.9
silicon	24.6	33.9
aluminum	7.8	7.4
iron	7.8	1.4
calcium	7.8	0.9
magnesium	3.9	0.24
sodium	1.5	2.5
potassium	0.5	4.5

TABLE 13.2
Some common rock types and their major mineral constituents

Rock Classification	Type	Important minerals
sedimentary	dolomite	dolomite ($CaCO_3 \cdot MgCO_3$)
	limestone	calcite ($CaCO_3$)
	sandstone	quartz (SiO_2)
igneous	basalt	feldspars such as orthoclase ($KAlSi_3O_8$), albite ($NaAlSi_3O_8$), anorthite ($CaAl_2Si_2O_8$), pyroxenes such as enstatite ($MgSiO_3$), diopside [$CaMg(SiO_3)_2$]
	dunite	olivine ([Mg, Fe]$_2SiO_4$)
	granite	quartz (SiO_2) muscovite [$KAl_2Si_3AlO_{10}(OH)_2$] feldspars (see above)
metamorphic	quartzite	quartz (SiO_2)
	marble	calcite ($CaCO_3$)

gravel are used in building and road construction and diamonds are used for cutting and drilling hard materials besides being displayed as decorative gems. In fact, the total value of sand and gravel mined each year in the United States exceeds that of all metallic ores. Coal, oil, and natural gas — found within layers of ancient rock — are our major sources of energy (see Chapter 16).

The environmental impact of the methods of extracting these materials is far-reaching: Lands are stripped bare of soil and vegetation; air and water are often polluted. The severity of the damage varies with the methods of extraction, the physical and chemical properties of the material, and the location of the deposit. The natural distribution of crustal resources is selective so that the environmental impact from the removal of crustal resources is not widespread.

The *oceanic crust* is uniformly thin and rests directly on the upper part of the mantle. It is composed almost entirely of the rock basalt, on top of which is found a thin layer of sedimentary deposits. The bulk of the material on the ocean floor is composed of a mixture of *clay* and *ooze*. The clay is made up of finely divided iron, aluminum and magnesium silicates, together with tiny particles of mica (potassium aluminum silicates), and quartz (silicon dioxide). Most of the clay is derived from sediments produced on land areas by weathering and transported to the ocean by rivers and streams. Ooze consists of the skeletons of tiny marine organisms known as *plankton;* chemically, they are made up chiefly of $CaCO_3$ and SiO_2, and settle from the surface waters where living plankton are found. Large areas of the ocean floor also contain potato-shaped nodules that consist of manganese dioxide formed from the oxidation of manganese ion by hydroxyl ions and dissolved oxygen in the seawater:

$$2Mn^{2+} + 4OH^- + O_2 \longrightarrow 2MnO_2\downarrow + 2H_2O$$

THE EARTH VERSUS THE MOON The moon is the earth's natural satellite. Before the recent intensive study of the moon, the only planet that could be studied in detail was the earth. In comparison with the earth, the moon is a relatively inactive

TABLE 13.3
Distribution of elements in the crust of the earth and of the moon

Element	Average weight (%)	
	Earth's crust	Moon's crust
oxygen	46.6	39.2
silicon	27.72	20.0
aluminum	8.13	5.6
iron	5.00	14.6
calcium	3.63	10.6
sodium	2.83	0.33
potassium	2.59	0.13
magnesium	2.09	3.6
titanium	0.44	5.4
phosphorus	0.105	0.06
manganese	0.095	0.18
zirconium	0.0165	0.044
vanadium	0.0135	0.005
chromium	0.010	0.18
nickel	0.0075	0.009

planetary body. Following the Apollo expeditions conducted by the United States during 1969–1973, samples of the lunar surface returned to earth were subjected to careful analysis. In Table 13.3 the average percentages of the major elements found in the moon samples, which varied in size from rocks weighing several pounds to fine soil-like material, are compared with the average percentages of the same elements in the earth's crust. Clearly, the composition of the rocks from the moon is considerably different from similar rocks (basalts) on earth. The difference in chemical composition between earth and moon rocks indicates that the initial composition of their magma was different. Hence, it is unlikely that the two bodies were ever one.

QUESTIONS

13.1 What is meant by the physical environment?
13.2 What is the lithosphere? What is its most significant part?
13.3 Name the major parts in the presently accepted model of the earth and explain their predicted chemical character.
13.4 Compare the chemical compositions of granite and basalt.
13.5 Distinguish between a rock and a mineral.
13.6 Discuss the differences in the compositions of continental and oceanic crusts.
13.7 How are sedimentary rocks produced?
13.8 What is the difference between igneous rock and metamorphic rock?
13.9 How are marble and slate produced in the earth's crust?
13.10 What resources are extracted from the earth in your community?
13.11 Discuss the environmental impact of extracting crustal resources.
13.12 What are the fundamental differences in the composition of lunar and earth surfaces?

SUGGESTED PROJECT 13.1

The minerals near you (or mapping the mines)

Prepare a list of the minerals that are (1) available and (2) mined in your state. Show the location, the name, and the composition of each of the minerals on a map of your state. Obtain information on:

a. what is being done with the extracted minerals
b. the operation of the mines and their effect on the surrounding areas

SOIL—ITS CHEMISTRY AND POLLUTION

Life on the land is directly or indirectly dependent on the thin layer of small rock particles, sand, clay minerals, and decaying organic matter that we call the *soil*. Soil is formed from rocks by the mechanical action of winds, water, and other agents, and the chemical action of substances such as oxygen, water, and carbon dioxide. Without the soil and the food which it produces, most living matter could not exist. In agricultural areas, there is a direct correlation between the standard of living and the quality of soil. Indeed, the soil is such an important part of the biosphere that it is sometimes called the *pedosphere*—the portion of the solid earth or lithosphere in which life exists. In this section, some aspects of the chemistry involved in the formation, fertility, and the structure of soils will be examined. Further, we shall examine the chemistry of soil pollutants such as agricultural chemicals and emphasize the importance of soil conservation.

NATURE AND COMPOSITION OF SOIL Generally speaking, good soil has a loose texture and consists of solid mineral and organic matter, water, and air spaces. The proportions of the four major soil components vary from place to place and from time to time, but the breakdown given in Table 13.4 is fairly representative. The solid fraction of a typical productive soil is approximately 10 percent organic matter and 90 percent inorganic matter. Some soils, for example, peat soils, may contain as much as 95 percent organic material. Other soils contain as little as 1 percent organic matter. The organic portion of soil consists of the remains of plants in various stages of decay. Typical soils support high populations of bacteria, fungi, and worms.

Mineral matter in soil The mineral portion of soil is formed from parent rocks by the weathering action of physical, chemical, and biological processes. Physical weathering first disintegrates the massive rocks into smaller fragments by mechanical stresses. After physical weathering has fragmented the rocks, chemical processes become important. The processes most frequently involved are hydrolysis and oxidation, but there are others as well. *Hydrolysis*, the reaction of rock minerals with water, or more accurately, with the ions derived from water, is the most important of all chemical weathering reactions. A reaction of great interest is that of the abundant aluminosilicate minerals (feldspars, micas):

$$4KAlSi_3O_8 + 11H_2O \longrightarrow Al_4Si_4O_{10}(OH)_8 + K_4SiO_4 + 7H_2SiO_3$$

orthoclase   kaolinite   potassium *metasilicic*
(feldspar)   (a clay mineral) *ortho*silicate   acid

The rate of weathering by hydrolysis is increased in an acid environment and at higher temperature. Chemical weathering by reaction of minerals with carbonic acid is important in rock containing carbonates of calcium (limestone) or of calcium and magnesium (dolomite). *Oxidation* is secondary in importance to hydrolysis, but is distinctive for rock minerals containing iron(II):

$$2Fe^{2+} + \tfrac{1}{2}O_2 + 2H^+ \longrightarrow 2Fe^{3+} + H_2O$$

It is responsible for the common red colors of iron-containing rocks. These chemical processes are also responsible for the occurrence of acid mine drainage, which pollutes water and soils.

Some rock fragments and minerals resulting from physical and chemical weathering require the presence of living organisms for their decomposition to soil. Bacterial decomposition of the remains of plants and of other organisms releases mineral acids such as H_2CO_3, HNO_3, and H_2SO_4, and various organic acids to the soil. Bacteria aid in this decomposition and also act as catalysts for reactions

TABLE 13.4
Major soil components

Component	Volume (%)
mineral matter	35–45
organic matter	15–5
water (soil solution)	15–35
air (O_2, N_2, CO_2)	35–15

such as the oxidation of iron. Thus, substances derived from biological activity contribute to the continued weathering of rocks and formation of soil.

As the agents of weathering do their work on the upper portion of a soil, a layered structure develops. The layers, called *horizons*, are characteristic of soils and, taken together, constitute the *soil profile* (Fig. 13.4). The top layer, typically several inches in thickness, is known as the *topsoil*. It contains most of the organic matter and is the layer of maximum biological activity in the soil. It is also the layer in which metal ions and clay particles are subject to considerable leaching. The next layer, typically 2–3 ft in depth in a temperate climate, is known as the *subsoil* and is the recipient of material such as organic matter, salts, or clay particles leached from the topsoil. Below the subsoil is the horizon composed of weathered parent rocks from which the soil originated. It is the layer of least weathering and little biological activity, and tops bedrock (primary rock) which may or may not be the material from which the soil was formed.

As expected, minerals composed of the major elements in the earth's crust constitute most of the mineral fraction of the soil. Because three-quarters of the earth's crust consists of silicon and oxygen, minerals containing these two elements are especially prevalent in soil. Silica (SiO_2) is a common soil mineral component. Among the silicates, orthoclase ($KAlSi_3O_8$), albite ($NaAlSi_3O_8$), and epidote [$4CaO \cdot 3(AlFe)_2O_3 \cdot 6SiO_2 \cdot H_2O$] are very common components of soil minerals. Iron oxides, particularly goethite ($FeO \cdot OH$) and magnetite (Fe_3O_4), make up much of the mineral fraction of many soils; manganese oxides and titanium oxides are found in abundance in some soils. The clay minerals, consisting largely of hydrated aluminum and iron silicates, are also widely occurring and chemically significant components of soils. They serve to bind cations and hold them from leaching by water so that cations such as Ca^{2+}, Mg^{2+}, K^+, and NH_4^+ are available as plant nutrients.

Organic matter in soil Plants manufacture carbon-containing organic compounds from the CO_2 in the air, the water percolating through the soil, and chemicals released by weathering. When plants decay, organic matter accumulates

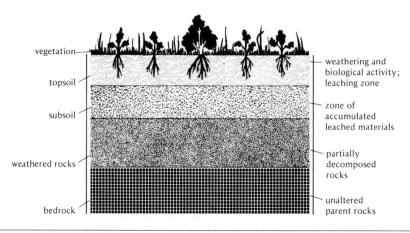

FIGURE 13.4

Soil profile showing soil horizons. (Not to scale.) Redrawn, by permission, from Stanley E. Manahan, *Environmental Chemistry*, p. 266. Copyright © 1972 by Willard Grant Press, Inc.

in the soil. Most of the organic matter that accumulates consists of a water-insoluble relatively nonbiodegradable material known as *humus*. The accumulation of humus is favored by decreasing temperatures and the nonavailability of oxygen. Usually soils containing over 90 percent organic matter are formed in areas where plants grow and decay in soil saturated with cold water. Besides serving as a source of food for microorganisms in soil, the organic matter plays a very important role in determining soil productivity. Along with clay minerals, it plays the major role in determining the structure and properties of soils.

Water and air in the soil The spaces between the mineral and organic soil particles are occupied by variable proportions of water and air. The water, together with its dissolved substances, is called the *soil solution* which is the direct source of most of the nutrients required by plants. Roughly one-quarter of the volume of a typical soil is composed of air-filled pores the oxygen of which is necessary for the respiration of aerobic organisms, including plants, growing in the soil. Soil air generally contains less oxygen than atmospheric air. On the other hand, soil air may be several hundred times more concentrated in carbon dioxide than the atmosphere as a result of decomposition of organic matter and other metabolic processes of soil organisms. Although soils high in organic matter may hold a relatively large amount of water compared to other soils, the decay of organic matter in soil may increase the level of dissolved carbon dioxide in groundwater and displace the equilibria involved in processes that maintain the carbon dioxide cycle in nature (see Chapter 15).

The practice of *hydroponics* or solution culture indicates that only the soil solution and air are essential to the growth of plants. What, then, is the purpose of solid mineral matter or of any kind of organic matter that is supposedly essential to a true soil? As we shall see next, the solid mineral and organic matter in soils determines their fertility and structure.

STRUCTURE AND FERTILITY OF SOILS The ability of a soil to serve as a reservoir for the chemical nutrients required for plant growth and to make these nutrients easily available to plants in an assimilable form is known as the fertility of a soil. Soil structure, on the other hand, refers to the manner individual soil particles are attached to each other to form larger aggregates, the nature of which determines, among other things, the water retention and aeration properties of a soil. In a natural soil, soil fertility cannot exist without the proper structure. Loss of soil productivity is associated with a deterioration in soil fertility or structure or both and is brought about by either the addition or the removal of some substance from a soil. Both the solid mineral and organic matter in a soil play a dominant role in the physical and chemical foundations of fertility and structure and hence in the productivity of a soil.

Ion exchange reactions in soils The mineral and organic portions of soils have the ability to absorb cations (positive ions) on their surfaces and to exchange them with other cations in the soil solution. Clay minerals attract, adsorb, and exchange cations because of the presence of negatively charged sites on their surfaces. Organic materials exchange cations because of the presence of carboxyl and other basic functional groups. Cation exchange with soil is the mechanism by which potassium, calcium, magnesium, and essential trace-level metals are made available to plants. In the process plants exchange hydrogen ions for the metal ions in soil:

$$Ca^{2+}(soil) + 2H^+(plant) \rightleftharpoons Ca^{2+}(plant) + 2H^+(soil)$$

Thus cation exchange equilibria are deeply involved in almost every phase of soil fertility.

Acid–base reactions in soils   Soil can be acidic, neutral, or alkaline with a pH (see Chapter 7) somewhere within the range of 4 to 8.5, but the majority of soils are acidic and the majority of plants grow best in a soil with a pH between 6 and 7. The hydrogen ion exchange reaction together with the leaching of metal ions from the soil by water containing dissolved carbon dioxide tends to make the soil acidic:

$$Ca^{2+}(soil) + 2CO_2 + 2H_2O \rightleftharpoons 2H^+(soil) + Ca^{2+}(plant) + 2HCO_3^-$$

If the soil becomes too acidic for optimum plant growth, it is normally restored to productivity by liming, that is, by the addition of powdered limestone ($CaCO_3$):

$$2H^+(soil) + CaCO_3 \longrightarrow Ca^{2+}(soil) + CO_2 + H_2O$$

Soil alkalinity may occur due to the presence of basic salts such as Na_2CO_3, particularly in regions of limited rainfall or poor drainage, but may be corrected by the addition of aluminum or iron sulfate:

$$2Al^{3+} + 3SO_4^{2-} + 6H_2O \longrightarrow 2Al(OH)_3\downarrow + 6H^+ + 3SO_4^{2-}$$

Sometimes sulfur is used to acidify alkaline soils because it is readily oxidized by bacterial action to sulfuric acid. The large quantities of sulfur now being removed from petroleum to prevent air pollution by sulfur dioxide may make the sulfur treatment much more common in the future.

Soil fertility depends not on the total amount of a nutrient element present in a soil, but on its amount in available forms. The acidity of a soil determines to a large extent the availability of nutrient elements to plants, because the solubility of available forms of the elements varies with change in soil pH value. Recall that pH was defined in Chapter 7 as a measure of the hydronium ion concentration in a solution. The effect of pH will differ for various nutrient elements, depending on the form of the element most easily assimilated by plants, and with the presence of other cations. For example, the availability of phosphorus in the most readily assimilated form of $H_2PO_4^-$ is affected at both low and high pH by the presence of high concentrations of iron and calcium, respectively. Iron and calcium contribute to low phosphorus availability through precipitation of highly insoluble iron and calcium phosphates so that maximum phosphorus availability occurs at soil pH values between 6 and 7. Because of the solubility of the oxides in an acidic environment, the dissolved concentration of some elements such as Fe, Al, and other trace elements (except B and Mo) may become so high at low pH that they become toxic to plants. Further, a low pH is indicative of both extensive replacement of adsorbed cations by hydrogen and of the absence of reserves of exchangeable cations in soil minerals. That is to say, it is a sign of an infertile soil, for all processes leading to infertility are accompanied by the replacement of nutrient cations with hydrogen and the gradual acidification of the soil.

PLANT NUTRIENTS AND CHEMICAL FERTILIZERS   Soil fertility and productivity depend on a number of factors, the most important of which is the essential plant nutrients the soil can supply. All plants require at least 17 nutrient elements for their growth — 9 of these elements, the *macronutrients*, are needed in relatively large quantities, whereas the rest, the *micronutrients*, are needed only in trace amounts. These elements are listed in Table 13.5 together with their major sources in the soil and the form in which they can be assimilated by plants. Note that the form of an element in a source is not necessarily the form that is directly usable by a plant. A soil may be deficient in one or more essential elements for a variety of reasons, the most significant of which are leaching and crop removal. Deficiencies in nutrient elements are made up by adding available forms directly to the soil in fertilizers. However, this breaks the soil–plant nutrient cycle and may result in the *eutrophication* of natural waters such as streams and lakes; that is, ecological aging may be stimulated by increased levels of mineral nutrients (see Chapter 14).

Micronutrients in soil The nutritional needs of growing plants were originally considered to be limited to the macronutrients used in large quantities by plants. However, as methods of detection became more sensitive, the need for minute quantities of certain elements, the trace or micronutrient elements, in plant growth was established. Today boron, chlorine, copper, iron, manganese, molybdenum, cobalt, vanadium, zinc, strontium, and iodine are considered essential plant micronutrients. These elements are needed by plants only at very low levels, and most of them are considered to function as components of essential enzymes. For example, iron is a part of a plant enzyme that participates in the production of chlorophyll, the green pigment essential to photosynthetic activity. It is entirely possible that other micronutrients will be found as techniques for growing plants in environments free of specific elements improve.

Macronutrients in soil The elements generally recognized as essential for plants are carbon, hydrogen, oxygen, nitrogen, phosphorus, potassium, calcium, magnesium, and sulfur. Of these, carbon, hydrogen, and oxygen are obtained from air and water. Nitrogen may be obtained by some plants directly from the atmosphere through the action of nitrogen-fixing bacteria. The other essential macronutrients must be obtained from soil. Of the macronutrients obtained from soil, nitrogen, phosphorus, and potassium are the most likely to be lacking. They are called the *fertilizer elements* and are often applied together in commercial fertilizer

TABLE 13.5
Essential plant nutrients[a]

Element	Assimilable form	Source
MACRONUTRIENTS		
carbon	CO_2, CO_3^{2-}, HCO_3^-	air, organic matter, mineral solids
hydrogen	H^+, OH^-	water–soil solution
oxygen	O_2	air
nitrogen	NH_4^+, NO_3^-, NO_2^-	air, organic matter
phosphorus	HPO_4^{2-}, $H_2PO_4^-$	organic matter, mineral solids
potassium	K^+	mineral solids
calcium	Ca^{2+}	mineral solids
magnesium	Mg^{2+}	mineral solids
sulfur	SO_3^{2-}, SO_4^{2-}	organic matter, mineral solids
MICRONUTRIENTS		
iron	Fe^{2+}, Fe^{3+}	mineral solids
manganese	Mn^{2+}, Mn^{4+}	mineral solids
boron	BO_3^{3-}, $B_4O_7^{2-}$	mineral solids
molybdenum	MoO_4^{2-}	mineral solids
copper	Cu^+, Cu^{2+}	mineral solids
zinc	Zn^{2+}	mineral solids
chlorine	Cl^-	mineral solids
cobalt	Co^{2+}	mineral solids

[a] Data from R. G. Gymer, *Chemistry: An Ecological Approach*, Harper & Row, New York, 1973, p. 502.

mixtures. Because of their importance, some of the ecological aspects of commercial fertilizers furnishing these nutrients are considered here.

Nitrogen fertilizers make up nitrogen losses from the soil resulting from crop removal, leaching of soluble forms, erosion, and bacterial denitrification processes. Prior to the production of synthetic fertilizers, farm manure was extensively used. Manure is essentially organic matter and it must undergo biodegradation with conversion of the nutrient elements to simple inorganic forms such as NO_3^- (for nitrogen) before the plants can assimilate nitrogen:

$$R—NH_2 + H_2O \longrightarrow R—OH + NH_3$$
$$NH_3 + 2O_2 \longrightarrow H^+ + NO_3^- + H_2O$$

Synthetic fertilizers are much more efficient means of adding nutrients to soil because they are in a more readily available form than are organic fertilizers. Most commercial nitrogen fertilizers are synthetic products prepared from ammonia that is manufactured from atmospheric nitrogen by the Haber process (see Chapter 8). Some common inorganic nitrogen fertilizers are anhydrous liquid ammonia, ammonium nitrate, ammonium sulfate, ammonium phosphate, calcium nitrate, potassium nitrate, and urea. They are all water soluble, and the nitrogen is totally available to plants. When nitrogen is applied to soils in the ammonium form, nitrifying bacteria perform an essential function in converting it to available nitrate ion.

The total amount of nitrogen in a representative soil is less than 0.5 percent and mostly in the organic form. It is converted slowly to the available forms, and only a very small percentage of the total soil nitrogen content is available to plants at any given time. Overapplication of nitrogen fertilizers can, therefore, result in ecological problems. One such problem is the high runoff of nitrates to natural waters which causes eutrophication of lakes and streams. Certain bacteria are capable of reducing nitrate ion to toxic nitrite ion:

$$NO_3^- + 2H^+ + 2e = NO_2^- + H_2O$$

Nitrite inactivates hemoglobin, particularly in infants, causing a condition known as methemoglobinemia (blue babies). A number of cases of this disease, including several deaths, have been attributed to nitrate in farm well water. Although feedlot wastes are a major source of nitrate pollution, fertilizers have been implicated in such pollution, particularly in agricultural areas.

Phosphorus fertilizers make up phosphorus losses from soil resulting from crop removal and fixation processes that make the phosphorus unavailable to plants. As with nitrogen, the total amount of phosphorus in a soil is quite low (<0.1%); it is made up of varying proportions of inorganic phosphate minerals and organic phosphorus compounds. Of the total amount of phosphorus, only about 1 percent is available phosphorus (mostly as the soluble $H_2PO_4^-$ ion) and the rest is in chemical forms that are water insoluble and, therefore, unavailable to plants. Complicating the problem of phosphorus fertilization is the tenacious fixation of added soluble phosphorus in the soil due to reaction with soil components. For example, iron, aluminum, or calcium ions present in the soil react with the dihydrogen phosphate ($H_2PO_4^-$), commonly found in the more soluble fertilizers, forming highly insoluble metal phosphates. Because of these soil-fixation reactions, leaching or runoff of phosphate fertilizers from the land cannot be a major source of phosphate eutrophication in natural waters unless the phosphate is carried away on eroded soil particles themselves.

Phosphorus, because of its low total amount in soil and its low availability to plants, is involved in more cases of soil infertility than any other element. By far the largest reservoir of phosphorus in nature is the phosphate rocks of the lithosphere. A common phosphate fertilizer, known as *superphosphate,* is made from rock phosphate by treatment with dilute H_2SO_4:

$$Ca_3(PO_4)_2 + 2H_2SO_4 \longrightarrow Ca(H_2PO_4)_2 + 2CaSO_4$$

The superphosphate products are much more soluble than the parent phosphate minerals. Ammonium phosphates, particularly liquid ammonium polyphosphates, are becoming very popular as phosphate fertilizers because of their ability to chelate iron and other micronutrient metal ions and to make them more available to plants.

Potassium fertilizers make up potassium losses from the soil resulting from crop removal and leaching of soluble forms. Unlike nitrogen and phosphorus, potassium is present in great total amounts in soil (~2%), but a large proportion of it is in unavailable forms in some silicate minerals. Exchangeable potassium held by clay minerals is relatively more available to plants. Potassium fertilizer components consist of naturally occurring potassium salts, generally KCl. One problem encountered with potassium fertilizers is that some crops absorb more potassium than is really needed for their maximum growth. This is particularly the case when nitrogen fertilizers are added to soils to increase productivity. The higher the productivity of the crop, the more potassium is removed from soil. Because potassium is essential for some carbohydrate transformations, crop yields are generally reduced in potassium-deficient soils.

Soils deficient in other macronutrients such as calcium and magnesium are relatively uncommon. Soil sulfur deficiencies have been reported in a number of regions of the world. Most fertilizers contain sulfur (as SO_4^{2-}), but the recent trend to fertilizers with higher concentrations of P and N but less S is resulting in more instances of sulfur deficiency.

The development of the chemical fertilizer industry is no doubt partly responsible for the so-called Green Revolution, that is, increased crop yields from land already in production. However, chemical pesticides and other crop protectors that combat insects and plant diseases (see Chapter 12) are equally responsible for the comfortable crop surpluses many nations of the world enjoy today. It is estimated that up to 90 percent of our food crops would be lost without effective chemical controls. Unfortunately, chemical pesticides find their way into soil and act as soil pollutants.

TOXIC SUBSTANCES AND SOIL POLLUTION There are many substances that can produce changes in the chemical, physical, or biological properties of a soil and adversely affect the quality and quantity of crops. Among them are both natural and synthetic substances that can be toxic to soil organisms, plants, or herbivores. Such substances include chemical pesticides, which have been shown to affect soil organisms, and inorganic salts, which besides being toxic, have been shown to affect soil structure.

In a natural soil, the preservation of both fertility and structure is necessarily linked with the preservation of the population of living organisms in a soil. Soil organisms feed on the organic matter in the soil and recycle nutrients and build soil structure. Pesticides that are applied to agricultural crops find their way into soil both during and after application and contaminate the soil. Because most pesticides are organic by nature, they have a considerable affinity for organic matter, including the living organisms in the soil. Pesticides act as soil pollutants because, by reducing both the population and the number of species of living organisms in the soil, they affect soil structure and fertility. In the case of nonpersistent pesticides, there is little permanent damage done and the affected soil organisms can reestablish themselves in the soil. However, the persistent pesticides, such as the chlorinated hydrocarbons, create more serious problems. These problems include their accumulation in soil animals (earthworms) and transfer in potentially toxic amounts to birds and other predators farther up the food chain.

As pointed out elsewhere, high concentrations of salts are directly toxic to plants and soil organisms. In addition, the presence of large amounts of Na^+ results in breakdown of the soil aggregates necessary for proper soil drainage and aeration. Exchangeable sodium can, however, be removed by replacement with Ca^{2+} from calcium salts or H^+ from sulfuric acid. Soil polluted by salts is often reclaimed by construction of an extensive drainage system followed by leaching of the salts with large quantities of nonsaline irrigation water.

Besides the pollution from toxic pesticides and fertilizers, soil suffers serious contamination from industrial wastes, saline waters, radionuclides, and from more permanent and less readily recognized pollutants such as concrete, asphalt, tin, iron, lead, aluminum, and plastics. Salinity of soils is a major problem where irrigation is a major component of agriculture. Strip mining operations and acid mine drainage (chiefly sulfuric acid due to rain seeping through coal mine wastes rich in sulfur compounds) are said to have altered the topsoil characteristics of millions of acres of land.

The importance of *soil conservation* was realized long before ecology became a popular movement. In addition to the control and abatement of soil pollution, soil conservation programs are concerned with accelerated soil erosion and crop removal. When erosion accelerates to the point where the topsoil is lost faster than new topsoil can be formed from the subsoil, then nutrients are removed and both soil fertility and soil structure suffer. Accelerated soil erosion is prevented by maintaining the necessary amount of organic matter in the soil so that stable soil aggregates with the proper soil structure and fertility are formed. Often the soil is depleted of organic matter by continued crop removal. Because all plants mine the soil beneath them for nutrients and return these same nutrients to the soil when they die, continued harvesting of any crop will eventually deprive the soil of one or more of the essential nutrients unless fertilizers are used to replenish the supply. The worldwide campaign to promote soil conservation is proof that we have not forgotten the importance of "good" soil in our preoccupation with water and air pollution (see Chapters 14 and 15).

QUESTIONS

13.13 List the major components of soil.

13.14 What are the significant chemical processes involved in the formation of mineral matter in soil?

13.15 What is meant by soil profile?

13.16 Name some of the more important soil minerals.

13.17 What is meant by soil solution? In what way does it contribute to the growth of plants?

13.18 How do you account for the fact that soil air contains less oxygen and more carbon dioxide than atmospheric air?

13.19 How is the fertility of a soil related to soil structure?

13.20 What is an ion exchange reaction? Give an example of such a reaction in the soil.

13.21 List the processes that tend to make the soil acidic. How is the excess acidity in soil removed?

13.22 What is meant by the available form of a plant nutrient?

13.23 Distinguish between macronutrients and micronutrients. Give examples of both.

13.24 What causes a deficiency in soil nutrients and how is it remedied?

13.25 Name some of the common synthetic inorganic fertilizers. In what respects are these fertilizers different in their action from natural fertilizers such as farm manure?

13.26 Discuss the effects of overapplication of synthetic fertilizers.

13.27 List some of the toxic substances found in soil and explain how they pollute the soil.

13.28 What is meant by soil conservation and how is it achieved?

SUGGESTED PROJECT 13.2

Your soil—acidic or basic (or How does your garden grow?)

Add a tablespoon of topsoil from your backyard into a glass of water. Stir the mixture well and then allow it to stand for about 5–10 min. Test the supernatant soil solution with pH or litmus paper. If necessary, consult your instructor or a garden supplier for ways and means of improving the quality of your soil.

SUGGESTED PROJECT 13.3

Effect of fertilizers on drinking water (or contaminating capers)

Examine the labels on at least 10 different bags of fertilizers in a garden supply store. Record the names and compositions of the fertilizers as given on the labels. Compare your list with those of other students.

Because fertilizers are water soluble, they may penetrate the groundwater supply. If the regulations of the U.S. Public Health Service are available in your library, find out the maximum available limits for these compounds in drinking water. If commercial test kits are available, you may check for phosphates and nitrogen compounds in your water supply.

14

The hydrosphere

About all the chemical processes which occur in nature, whether in animal or vegetable organisms, or in the nonliving surface of the earth, take place between substances in solution.

WILHELM OSTWALD (1853–1932)
German physical chemist

Covering three-quarters of the earth's surface is that part of our environment which we call the *hydrosphere* and which consists of solutions of molecular and ionic species in water. It includes the oceans, rivers, lakes, and springs, as well as subsurface waters, ice caps, and glaciers. It also includes atmospheric moisture which, in its visible forms, constitutes the clouds, fogs, and mists. Besides being bound to the atmosphere by the waters exchanged between them, the hydrosphere plays a major role in maintaining the heat balance between the earth and the sun.

Human exploration of the hydrosphere is in its infancy, and the precise origin of the hydrosphere is still a matter of speculation. It is presumed that all the water was originally confined in the interior of the earth, possibly as hydrates of various minerals, and was then expelled in large quantities as steam by volcanic activity. This steam, on cooling in the atmosphere, condensed as rain which gradually accumulated to form the oceans, rivers, lakes, and aquifers (water-bearing beds). Each of these water reservoirs is the subject of specialized study. The ocean is the realm of *oceanography;* the lakes, *limnology;* and the ice caps, *glaciology.* The relatively rapid movement of water in the atmosphere, in rivers, and in shallow depths of the soil is the particular field of study known as *hydrology.*

Before we start our discussion of water supply, pollutants, and the treatment of polluted water, we shall examine the water molecule itself and learn something about its structure, properties, and environmental role.

WATER AND ITS ENVIRONMENTAL ROLE

Under normal conditions on earth, water is a liquid. Just like all liquids, evaporation occurs from the surface of all open bodies of water at a rate closely related to the temperature. The vapor accumulates in the atmosphere, drifts and travels long distances with wind currents, condenses into clouds, and finally rains or snows on the earth. Much of the moisture falling on the land makes its way back to sea via flowing streams or as groundwater moving through the soil. In the process some of it may be intercepted and used by humans and plants only to be released again as vapor. This slow but endless movement of water from the sea back to the sea is called the natural water cycle or *hydrologic cycle* (Fig. 14.1).

However, water is a liquid with some unusual properties. A very striking and important property of water is that the solid form of water (ice) is less dense than the liquid. Thus, ice forms on the surface of lakes and insulates the lower layers of water, enabling fish and other aquatic organisms to survive the winters of the temperate zones. On the other hand, the same property has dangerous consequences for living cells. As ice crystals are formed, the expansion ruptures and kills the cells. Nevertheless, the intricate chemistry of life is molded around water as the medium. Other unusual properties of water include the unexpectedly high (in terms of its molecular weight) boiling point (100°C), heat of vaporization (540 cal/g), freezing point (0°C), and heat of fusion (80 cal/g). The abundance of water accounts in part for its widespread use as a coolant. In addition, water is an effective coolant for many natural and man-made processes because its changes of state absorb or give off large amounts of energy.

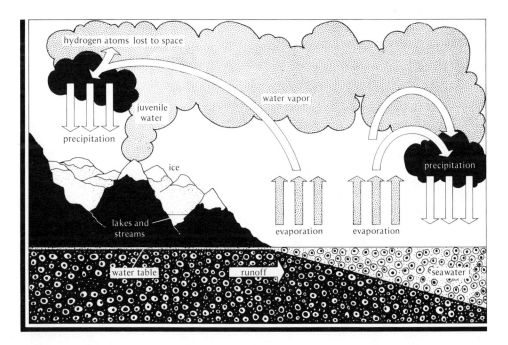

FIGURE 14.1

The hydrologic (water) cycle. Adapted from H. L. Penman, "The Water Cycle," *Scientific American*, p. 102 (September 1970). Copyright © 1970 by Scientific American, Inc. All rights reserved.

All of these unusual properties of water depend on the unique structure of the water molecule, H_2O. Recall that water molecules are polar and that they are strongly associated by the attraction of the positive end (H) of one molecule for the negative end (O) of another (see Chapter 4). The resulting bond is called a *hydrogen bond,* because it appears that hydrogen atoms are, to some extent, bonding oxygen atoms together in the resulting network of molecules (Fig. 14.2). In the liquid state, hydrogen bonding is random and the molecules are fairly close together. However, when water freezes the molecules take on a much more ordered arrangement with large hexagonal holes (Fig. 14.3) that account for the fact that ice is less dense than liquid water. The unexpectedly high boiling and freezing points may also be explained by the forces of hydrogen bonding that are able to keep the molecules in the more condensed state in spite of high kinetic energies.

Although the total water on earth is enormous in quantity ($\approx 1.5 \times 10^{18}$ tons), most water in oceans, ice caps and underground aquifers is inaccessible. Only a small fraction (0.02%) of the total water on the earth's surface is directly accessible to man. Even this slight fraction of the earth's water is impure and nonuniformly distributed. In fact, some areas of the earth are seriously deficient as a result of natural inequalities and man's use of water.

Because water dominates the processes of life, it is an essential of our environment. In man, water forms the medium in which digestion, food assimilation, and waste elimination proceed. It sustains the integrity of cell walls, transports blood, and holds and helps transport the ionic species that generate nerve signals which allow the human brain to function. It is estimated that the average human body consists of about 65 percent water and that the blood alone is about 78 percent water. Clearly, the biological role of water by itself is sufficient to make water one of our primary concerns at all times.

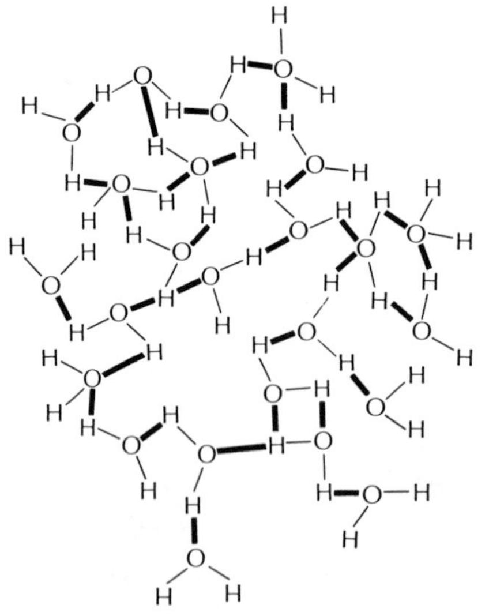

FIGURE 14.2

Hydrogen bonding in water.

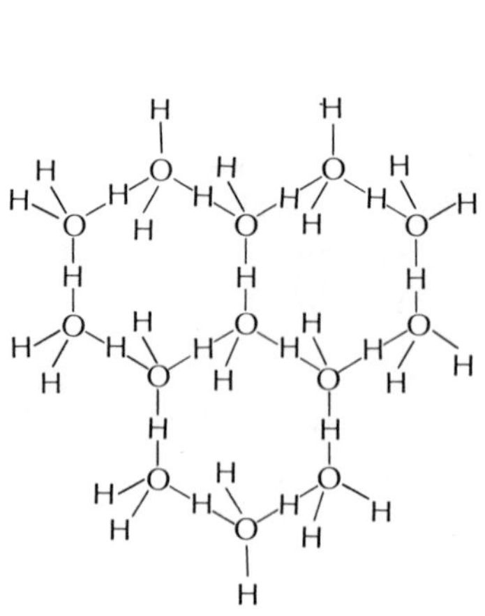

FIGURE 14.3

Structure of ice.

Equally important is the role of water in our physical environment. As a vapor, it brings moisture to the continents and influences the heat balance and temperature of our environment; as a liquid, it erodes and shapes our land, transports and concentrates minerals, carries away human wastes and waste heat, and moderates the earth's climate; as a solid, it pulverizes rocks by expanding when it freezes and thereby creates soil as well as storing winter moisture. In all three states, water infiltrates, covers, and carves the rocks of the lithosphere.

Perhaps the most essential use of water is simply for drinking, to sustain the body fluids necessary for life. However, the quantity of water used for drinking is relatively low (less than 0.1%). Most of the water goes to agriculture, industry, and power-generating facilities. The average per capita daily consumption of water in the United States is estimated at 2000 gal. About 45 percent of this is used for irrigation, 28 percent for power generation, and 20 percent for industrial processes. The remaining 150 gal or so per person is used as municipal water for sanitation, cleaning, cooking, and drinking. However, most of the water used is not necessarily lost and can be recycled and used again.

QUESTIONS

14.1   **What is meant by the term hydrosphere?**
14.2   **How would you account for the natural water cycle?**
14.3   **List some of the unusual properties of water and their consequences.**
14.4   **What is hydrogen bonding? How does it explain the unusual properties of water?**
14.5   **Discuss the environmental role of water.**

NATURAL WATERS — THEIR IMPURITIES AND PURIFICATION

All water, as it occurs in nature, is impure. The sources and natures of the impurities are listed in Table 14.1. The impurities include suspended matter, microbiological organisms, and dissolved minerals and gases. The dissolved mineral content is the basis for classifying natural waters into freshwater, inland brackish water, and seawater (Table 14.2). Water containing less than 0.1 percent total dissolved solids is generally considered to be fresh, but the U.S. Public Health Service has prescribed 0.05 percent as the recommended upper limit for drinking water.

Because natural water always contains dissolved or suspended impurities, it must be purified before use. The extent of purification depends on the ultimate use of water. City water purification plants are designed primarily to remove disease-producing bacteria, suspended matter, and objectionable odor. Figure 14.4 is a schematic of a municipal water-treatment plant. Finely divided suspended solids are removed by treating the water with substances that form a flocculating agent, composed of a gelatinous solid which is highly adsorbent to dirt particles and some bacteria. This flocculating agent settles to the bottom of sedimentation tanks and carries most suspended matter with it. A commonly used flocculating agent is $Al(OH)_3$, which is formed *in situ* by the reaction

$$Al_2(SO_4)_3 + 3Ca(OH)_2 \longrightarrow 2Al(OH)_3 \downarrow + 3CaSO_4 \downarrow$$

After the flocculation treatment, the water is passed through filtering beds of gravel and sand where any remaining flocculating agent and suspended particles are re-

TABLE 14.1
Impurities in natural waters[a]

Source of impurities	Nature of impurities	
	Dissolved impurities	Suspended and/or colloidal impurities
atmosphere	H^+, HCO_3^-, SO_4^{2-}, O_2, N_2, CO_2, SO_2	dust
mineral soil and rock	Na^+, K^+, Ca^{2+}, Mg^{2+}, Fe^{2+}, Mn^{2+}, F^-, Cl^-, NO_3^-, HCO_3^-, CO_3^{2-}, SO_4^{2-}, various phosphates, CO_2	sand, clay, mineral soil particles
living organisms and their decomposition products	H^+, Na^+, NH_4^+, Cl^-, HCO_3^-, NO_3^-, N_2, O_2, CO_2, H_2S, CH_4, various organic wastes some of which produce odor and color	algae, diatoms, bacteria, fish and other organisms, topsoil, viruses, organic coloring matter.

[a] Data from A. Turk, J. Turk, and J. T. Wittes, *Ecology, Pollution, Environment,* Saunders, Philadelphia, 1972, p. 115.

TABLE 14.2
Classification of natural waters

Type of natural water	Percent of the earth's total water supply	Percent of dissolved solids
freshwater	0.5	<0.1
brackish water	2.5	0.1–3.5
seawater	97.0	>3.5

TABLE 14.3
Criteria for public drinkable water supplies[a]

Criterion	Permissible level	Desirable level
total residue	500 mg/liter	<200 mg/liter
turbidity, odor		virtually absent
dissolved oxygen—applicable to free-flowing streams (not lakes or reservoirs)	≥ 4 ppm (monthly mean) ≥ 3 ppm (individual sample)	near saturation, 8–10 ppm
pH (range)	6.0–8.5	7.4–7.8
hardness (as $CaCO_3$)	60–150 mg/liter	<60 mg/liter
chlorides	250 mg/liter	<25 mg/liter

[a] Data from I. Sunshine, Ed., *Handbook of Analytical Toxicology,* The Chemical Rubber Co., Cleveland, Ohio, 1969.

moved. In many installations the water is aerated by spraying into the air or passed over finely divided carbon to remove odor-causing substances before liquid chlorine is added to kill any remaining bacteria. Table 14.3 lists some standard permissible levels for drinkable water supplies. Less strict permissible levels are imposed on industrial water supplies which are not for public use.

Many communities add some form of fluoride salt (such as NaF, Na_2SiF_6, H_2SiF_6, and SnF_2) to the municipal water supply. Fluoride in concentrations of about 1 ppm can reduce the incidence of dental caries in children. Recent evidence seems to indicate that tooth decay in adults may also be retarded by the use of fluorides in water. Although there appears to be no hazard to human health, fluoridation of water has remained a controversial issue in many communities.

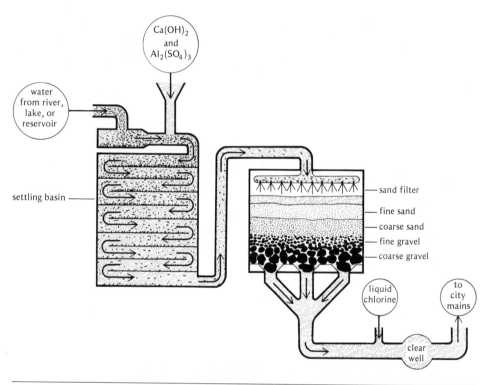

FIGURE 14.4

A schematic diagram of a municipal water-purification plant. Redrawn, by permission, from Otto W. Nitz, *Chemistry: An Introduction*, p. 189. Copyright © 1971 by Willard Grant Press, Inc.

QUESTIONS

14.6   What are natural waters? How are they classified?
14.7   Explain how aluminum sulfate and calcium hydroxide are used to help purify water.
14.8   How is the water in your locality purified for human consumption?
14.9   Why is the fluoridation of water a controversial matter?

SUGGESTED PROJECT 14.1

Evaluating municipal water treatment (or don't go near the water)

Visit the nearest water-treatment plant and examine the facilities now being used at that plant. Sketch a "flow chart" showing the treatment pathways. At what stages is the water sampled for bacteriological analysis? Is the treatment and analysis provided really adequate? If not, suggest ways and means to improve the facilities.

SEAWATER AND ITS DESALINATION

All but about 3 percent of the water that covers three-quarters of the earth's surface is found in the ocean. It is estimated that seawater occupies a volume of about 3.3×10^8 cubic miles and has a weight of about 1.5×10^{18} metric tons. It contains, on the average, about 3.5 percent dissolved materials by weight, almost all of which are ionic salts (electrolytes). (A listing of the major ions and their relative abundance in seawater is given in Table 14.4.) In consequence, the most immediate and well-known property of seawater is its saltiness or *salinity*. Evaporation of a sample of seawater to dryness causes the ionic salts to crystallize into a residue that is mostly NaCl but that also contains other salts such as KBr, $MgCl_2$, Na_2SO_4, $MgSO_4$, and $CaCl_2$. However, only magnesium, bromine, and common salt are now being extracted in significant quantities (see Chapter 5). Complete recovery of certain of the elements from seawater would yield the following tonnages: Cl_2, 26.7×10^{15}; Na, 14.2×10^{15}; Mg, 1.8×10^{15}; Br_2, 9.2×10^{13}; I_2, 6.6×10^{10}; U, 6.6×10^8; Au, 8.5×10^6; Ra, 1.65×10^2. Research is badly needed to establish what is profitable, marginal, or unlikely in the realm of mineral extraction from seawater.

The increasing demand for freshwater, due to increasing population and arid climatic conditions, has placed the emphasis on *desalination* of seawater to produce potable water rather than on recovery of minerals. Although desalted seawater is presently three times more expensive than municipal water from conventional freshwater sources, we can expect the cost of freshwater purification to increase if

TABLE 14.4
Major ionic components of seawater and their relative abundance

Ion	Relative abundance [ppm (by wt) of seawater]
Cl⁻	19,000
Na⁺	10,600
SO₄²⁻	2,600
Mg²⁺	1,300
Ca²⁺	400
K⁺	400
Sr²⁺	100
HCO₃⁻	100
Br⁻	65
F⁻	1
I⁻	0.05

pollution continues to increase, and the cost of purifying salt water to decrease as technology is improved. Because the obvious answer to many of the world's water supply problems is to tap the almost inexhaustible quantities of water in the oceans, we shall discuss some of the desalination techniques under development.

DESALINATION PROCESSES   Water may be separated from salt water by a change in state, that is, either by evaporation to the gas or by freezing to the solid. The use of *distillation* to obtain freshwater from seawater has been developed to a considerable state of refinement. As a matter of fact, about 98 percent of the total world's desalted water is produced by distillation processes. However, there are some difficult technological problems that must be solved before economical distillation plants become a reality. These problems include the corrosion effect of hot seawater on exposed metal surfaces, boiler-scale formation and the resulting inefficient heat-transfer operation, and, worst of all, the "waste heat" that heats up our environment.

Freezing processes take advantage of the fact that when ocean water freezes, as it, indeed, does in polar regions, crystals of pure water are formed. Actually, the freezing operation is superior to distillation because of the negligible corrosion and scaling usually encountered at lower temperatures, and the relatively small amount of energy transfer involved (540 cal of heat energy must be supplied for vaporizing 1 g of water, whereas 80 cal must be removed for freezing 1 g of water). However, a major problem with freezing processes involves the lack of efficient methods for separating the ice crystals from the brine.

In an entirely different approach, under the influence of electric charge, chemical attraction, or selective membranes, the ions in the salt water are compelled to move out of the stream of water flow. Because it is relatively slow when applied to large flows of water and involves new techniques that are still under development, this process is expensive. *Electrodialysis* is the most advanced of these

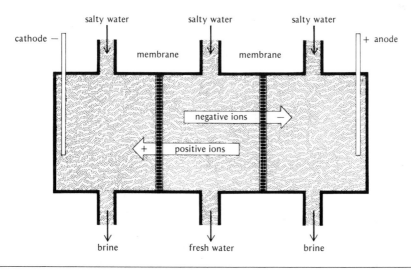

FIGURE 14.5

An electrodialysis cell for the desalinization of water. Redrawn, by permission, from Richard H. Wagner, *Environment and Man*, p. 102. Copyright © 1971 by W. W. Norton & Company, Inc.

methods and is based upon a combination of the principles of *osmosis* and *electrolysis*. Osmosis is the process whereby water will move through a semipermeable membrane* from a region of relatively pure water into a region containing a concentrated solution. The dialysis cell (Fig. 14.5) contains two different types of ion-selective membranes — one allows the passage of negative ions and the other the passage of positive ions. An electric current is used to drive the ions through the membranes, and the compartment between the membranes is thus depleted of salt.

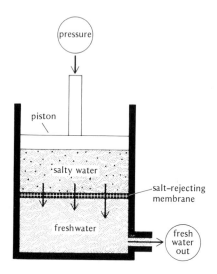

FIGURE 14.6

Reverse osmosis.

Another desalting technique of some potential value is called *reverse osmosis*. The technique involves applying sufficient pressure to force the water from the salt solution back through a membrane (Fig. 14.6). Because the salt is in effect filtered out of the seawater, the process is sometimes referred to as *hyperfiltration*. Membranes used thus far are mostly cellulose acetate, and the method works better with brackish (mildly salty) water than with straight seawater.

QUESTIONS

14.10 List five major ions that are present in seawater. Which of these are being extracted in significant quantities?

14.11 What is meant by salinity and how is it caused?

14.12 What methods are currently employed in desalination of seawater?

14.13 During the distillation of seawater, why does the salt remain behind while the water vaporizes?

14.14 What is meant by the terms osmosis, electrodialysis, and hyperfiltration.

---

* A membrane that permits the passage of only certain types of molecules.

HARD WATER, SOFT WATER, AND DETERGENTS

The ease or difficulty with which soap forms a lather in water was the original basis for classifying water into two categories: soft and hard water. The hardness of water is due to the presence of appreciable quantities (usually more than 0.005%) of dissolved minerals. Clearly, the hardness of water in a particular locale depends on the rock strata with which the water comes in contact. Although naturally hard water is fit to drink, its high mineral content creates problems in steam boilers and in washing machines. The deposition of hard scales in steam boilers and hot water pipes, and the formation of insoluble scums (sticky curdy precipitates) with soap are related to the insolubility of certain magnesium and calcium salts. Calcium and magnesium ions are two of the positive ions present in largest amounts in hard water; the other prevalent ion, Na^+, does not present any problems, because all common sodium salts are soluble.

Boiler scale consists primarily of calcium carbonate and magnesium hydroxide. The calcium carbonate is formed by the thermal decomposition of soluble calcium bicarbonate:

$$Ca(HCO_3)_2 \xrightarrow{\text{heat}} CaCO_3 \downarrow + H_2O + CO_2 \uparrow$$

The removal of CO_2 increases the pH, that is, produces a higher concentration of OH^- ions, and, in turn, causes the precipitation of $Mg(OH)_2$. Sometimes, the hard scale deposit contains calcium sulfate because some hydrated forms of $CaSO_4$ are less soluble in hot water than in cold water. Clearly, all the reactions leading to boiler scale result from an increase in temperature. Not only can scale lead to the almost complete blockage of pipes, it may cause such uneven and inefficient heating in boilers as to result in their breakdown.

To understand the formation of insoluble scums with soap, one has to look at the cleansing action of soap. Ordinary soap is sodium stearate ($NaC_{18}H_{35}O_2$). It dissolves partially in water forming the stearate ion, $C_{18}H_{35}O_2^-$, which is the active cleaning agent in a soap:

hydrocarbon end   ionic end
(dissolves in oils)   (dissolves in water)

On one end of this long chainlike ion is an oxygen atom on which the negative ionic charge is localized; it is thus polar and dissolves in water. The other end of the ion is the uncharged hydrocarbon end, which is nonpolar and similar in character to soiling agents such as oils or greases. A link is thus formed between the two that allows the soiling agents to be lifted from the soiled item and suspended in the wash water. If we represent the ionic end of the molecule as a circle and the hydrocarbon end in the usual zig-zag fashion, we can illustrate the cleansing action of soap as shown in Fig. 14.7. However, when a soap is dissolved in hard water, the negative stearate ion reacts with the hard-water ions Ca^{2+} and Mg^{2+}, forming the insoluble salts, $Ca(C_{18}H_{35}O_2)_2$ and $Mg(C_{18}H_{35}O_2)_2$, respectively, that precipitate out of solution. These precipitates make up the familiar "bathtub ring" and are responsible for the "tattletale gray" of the family wash.

One answer to the problems caused by hard water was the discovery of the *synthetic detergents*. Another answer was the development of a variety of water-softening agents and devices for removing objectionable minerals from water, that is, for *softening the water*. In view of their importance in water pollution control, we shall discuss them in some detail.

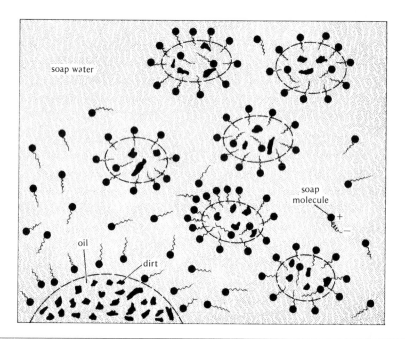

FIGURE 14.7

The cleansing action of soap. Redrawn, by permission, from John W. Hill, *Chemistry for Changing Times*, p. 188. Copyright © 1972 by Burgess Publishing Company.

SYNTHETIC DETERGENTS   Unlike soaps that are derived from animal fats, synthetic detergents (abbreviated syndets) are derived from petroleum chemicals. Originally they were made cheaply from benzene, a tetramer (meaning "four parts") of propylene, sulfuric acid, and sodium carbonate. The product, called *alkylbenzenesulfonate* (ABS), has the structure

$$CH_3 - \left(CH - CH_2 \right)_3 - CH - \bigcirc - SO_3^- \; Na^+$$
$$\qquad\quad |\qquad\qquad\quad |$$
$$\qquad\quad CH_3\qquad\qquad CH_3$$

and produces the desirable cleansing action in all kinds of water. However, it is nonbiodegradable and accumulates in waterways and water supplies, producing foam accumulations on lakes and rivers. Therefore, biodegradable detergents with structures such as

$$CH_3(CH_2)_9CH - \bigcirc - SO_3^- \; Na^+$$
$$\qquad\qquad\quad |$$
$$\qquad\qquad\quad CH_3$$

were synthesized. Unlike the branched (nonlinear)-chain molecules, molecules with linear chains of carbon atoms are readily broken down by microorganisms in the sewage-treatment plants.

Besides the surfactant or cleansing molecules, the modern synthetic detergents may contain a number of other chemicals. These include (1) builders that are water softeners and detergent aids, (2) enzymes that remove protein-based soil and stains, (3) bleaches that increase whiteness, and (4) optical brighteners that absorb ultraviolet light and convert it to visible light to make clothes appear brighter. Some even contain perfumes that "sweeten up" the washed clothes. Of these components, builders are at the center of the environmental controversy that now surrounds detergents. The most commonly used builders are polyphosphates, such as sodium tripolyphosphate ($Na_5P_3O_{10}$). They ionize in water to give the tripolyphosphate (TPP) ion:

$$\left[\begin{array}{ccc} :\ddot{O}: & :\ddot{O}: & :\ddot{O}: \\ | & | & | \\ :\ddot{O}-P-\ddot{O}-P-\ddot{O}-P-\ddot{O}: \\ | & | & | \\ :\ddot{O}: & :\ddot{O}: & :\ddot{O}: \end{array} \right]^{5-}$$

The TPP ions form soluble complexes with Ca^{2+} and Mg^{2+} ions that interfere with detergent action and thus soften the water. In addition, the unshared electron pairs of TPP extract hydrogen ions (H^+) from water, leaving OH^- and a basic solution necessary for good detergent action.

Although phosphorus-containing builders are highly effective, they also happen to be plant nutrients that contribute to the eutrophication problem in our lakes and rivers (see p. 380). Therefore, in 1970 the Federal Water Quality Administration called for the immediate reduction of phosphate levels in detergent formulations to minimum practical levels. Subsequently, phosphates were partly replaced with the sodium salt of nitrilotriacetic acid (NTA):

$$Na^+\ {}^-OOCCH_2-N\begin{array}{c} {}^{CH_2COO^-\ Na^+} \\ {}_{CH_2COO^-\ Na^+} \end{array}$$

Although the ion formed from NTA was nearly as effective as the TPP ion, it was found that NTA in combination with heavy metals such as Cd and Hg caused congenital abnormalities in rats and mice. Therefore, NTA builders were abandoned after temporary use in some detergent formulations. Other builders, such as sodium metasilicate (Na_2SiO_3) and sodium perborate ($NaBO_2 \cdot H_2O_2 \cdot 3H_2O$) have found limited application chiefly due to the strongly alkaline solutions they form on reaction with hot water:

$$Na_2SiO_3 + 2H_2O \longrightarrow 2NaOH + H_2SiO_3$$
$$NaBO_2 + H_2O \longrightarrow NaOH + HBO_2$$

Thus, although phosphates have helped alleviate the problem of hardness in water, they have opened up environmental problems. Another approach to solve the hard-water problem is to use a water softener so that no builder is needed at all.

WATER-SOFTENING AGENTS AND DEVICES It is recognized that there are two types of hardness—temporary and permanent. *Temporary hardness* is due to the presence of the dissolved bicarbonates, $Ca(HCO_3)_2$ and $Mg(HCO_3)_2$, and can be eliminated by boiling the water. In the process the bicarbonates are decomposed into insoluble carbonates, $CaCO_3$ and $MgCO_3$, that precipitate out of solution. On the other hand, *permanent hardness* is caused by the presence of dissolved chlorides and sulfates of calcium and magnesium. These salts are unaffected by boiling so that other methods have to be used for softening permanently hard water. Such methods include distillation, precipitation, sequestration, and ion exchange.

In the *distillation* process water is boiled and the steam, which contains no ions, is condensed into ion-free water. At present, distillation is too expensive for large-scale use. In the *precipitation* method both Ca^{2+} and Mg^{2+} are removed as the insoluble carbonates and phosphates by the addition of Na_2CO_3 and Na_3PO_4:

$$Ca^{2+} + CO_3^{2-} \longrightarrow CaCO_3\downarrow$$
$$3Mg^{2+} + 2PO_4^{3-} \longrightarrow Mg_3(PO_4)_2\downarrow$$

A disadvantage of this method is the formation of an insoluble product that might show up as a scum. No such scum is produced if the offending ions are made to form soluble complexes instead of being precipitated as insoluble salts. Although the ions remain in solution as complexes they can no longer form precipitates with soaps. The technique is known as *sequestration*. A commonly used sequestering reaction is

$$Ca^{2+} + (NaPO_3)_n \longrightarrow [Ca(NaPO_3)_n]^{2+}$$

The sequestering agent is sodium polyphosphate which is marketed under the trade name Calgon. Today, many commercial water-softening products contain large amounts of sequestering agents. In fact one function of the phosphates found in some detergents is that of sequestering.

The most effective method for softening water is the *ion exchange* method. Here the ions causing hardness are removed and substituted by nonoffending ions such as Na^+ and K^+ (Fig. 14.8). The process is carried out by percolating the hard water through a container filled with a finely divided substance, usually a resin,

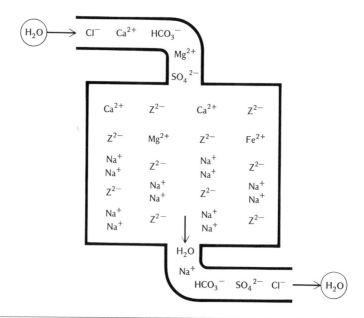

FIGURE 14.8

Principle of a zeolite ion exchanger. Redrawn, by permission, from Otto W. Nitz, *Chemistry: An Introduction*, p. 194. Copyright © 1971 by Willard Grant Press, Inc.

capable of performing the ion exchange. Naturally occurring minerals with this ability are complex sodium aluminum silicates with formulas such as $Na_2Al_2Si_4O_{12}$ (abbreviated Na_2Z), called *zeolites*. The reaction can be represented as

$$Na_2Z + Ca^{2+} \rightleftharpoons CaZ + 2Na^+$$

When the water softener becomes exhausted by depletion of its sodium ions, it can be regenerated by contact with saturated NaCl solution. This treatment shifts the equilibrium to the left and converts most of the resin back into Na_2Z. In addition to naturally occurring zeolites, equally effective, synthetic, ion exchange resins are available. They are used in most commercial water softeners. However, it should be noted that a zeolite water softener does not remove all minerals.

For many industrial purposes, it is necessary to remove all mineral ions. This is done by using synthetic ion exchange resins that work in pairs. The water first flows through a so-called cation exchange resin that captures all metal ions and releases H^+ in their place; the water next flows over an anion exchange resin that captures all Cl^-, SO_4^{2-}, HCO_3^-, and so on and releases OH^- in their place. The released H^+ and OH^- ions in large part combine to form water:

$$Ca^{2+} + 2HR \rightleftharpoons CaR_2 + 2H^+$$
$$SO_4^{2-} + 2R'OH \rightleftharpoons R_2'SO_4 + 2OH^-$$
$$2H^+ + 2OH^- \longrightarrow 2H_2O$$

The water thus formed is called *deionized water*. If the two resins are kept separate while in use (Fig. 14.9), both can be regenerated, the first by the use of an acid and the second by use of a basic solution. When the regeneration is not desirable or practical, the two resins may be contained in the same tank (mixed-bed demineralizer).

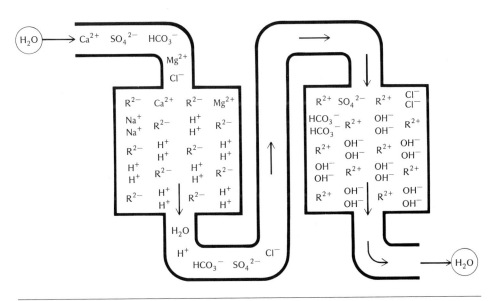

FIGURE 14.9

Principle of a demineralizer. Redrawn, by permission, from Otto W. Nitz, *Chemistry: An Introduction*, p. 194. Copyright © 1971 by Willard Grant Press, Inc.

QUESTIONS

14.15 Distinguish between soft water and hard water.

14.16 Describe the common problems that arise from hard-water use.

14.17 Discuss the cleansing action of soap and detergents.

14.18 What are synthetic detergents? How do they contribute to water pollution?

14.19 What are optical brighteners? Could you explain their action?

14.20 Distinguish between temporary and permanent hardness of water.

14.21 What is the difference between water softening and water demineralizing?

14.22 Compare the advantages and disadvantages of the various methods used for softening hard water.

SUGGESTED PROJECT 14.2

Detergents and water pollution.

Examine the labels on at least 10 different boxes of detergents in a local grocery store. Write down their names and active ingredients. Compare one product with another for the presence of the same active ingredients. Determine to what extent these ingredients have contributed to water pollution in your area.

SUGGESTED PROJECT 14.3

Your water — hard or soft?

Prepare a soap solution by dissolving a teaspoon of ordinary soap shavings in a pint of deionized or distilled water from your chemistry department. Using a dropper, add 10 drops of this solution to 1 pint of deionized or distilled water contained in a clean quart-size bottle. Stopper and shake the bottle vigorously to obtain a foam layer. If no stable foam is formed, add more drops of the soap solution and shake. In this fashion, continue to add soap solution (5–10 drops at a time) until there is a stable foam that persists for 1 min after shaking. Record the number of drops of soap solution needed to produce a satisfactory foam. Use this amount of foam as a reference to find out the number of drops of soap solution needed with (1) the cold water and (2) the hot water available in your home. In each case, compare the number of drops of soap solution needed with the number of drops used to obtain the reference condition. Is the water in your home hard or soft?

Make a similar comparision using detergent solution instead of soap. Compare the behavior of soap and detergent. You may repeat these tests with special detergents or soaps available in your household.

POLLUTION OF THE WATER ENVIRONMENT

Historically, man's earliest concern with *water quality* (fitness for the beneficial uses of mankind) resulted from the establishment of the microbiological causation of disease and the role of water-borne pathogens in spreading epidemics of typhoid, cholera, hepatitis, and bacillary dysentery. The water-treatment practices were, therefore, aimed at maintaining water purity in terms of bacterial contamination, usually as determined by the quantitative count of indicator organisms such as the coliform bacteria.* Very recently, however, there has been increasing concern over the extent to which our supplies of freshwater have been polluted by many different types of substances, especially in highly urbanized regions with densely concentrated industry. The U.S. Environmental Protection Agency estimates that almost one-third of the stream miles in the nation are polluted in that the federal water-quality criteria are violated. An indication of the wide diversity of water pollutants is given by the varied outward signs of their effects, such as bad-tasting water, emission of disgusting odors, floating scums, and masses of aquatic weeds growing unchecked. Presently there is developing a total ecosystem approach to water quality in order to avoid turning our streams, rivers, lakes, and estuaries into open sewers and to preserve the normal flora and fauna and the esthetic qualities of natural waters.

The most extensive pollutants of surface waters are the *sediments* produced by the naturally occurring process of erosion. However, most of the dangerous water pollutants are the wastes arising out of human activity. Wastes such as sewage, industrial effluents, litter, and so on are directly dumped into waterways, whereas wastes such as animal manure, insecticides, and mine tailings are eventually washed into waterways. These wastes contain thousands of undesirable chemical compounds that are of either inorganic or organic origin. In this section we shall study these two categories of water pollutants and their effects on the water quality. We shall also examine the effects of adding excess heat to a body of water (thermal pollution).

INORGANIC WATER POLLUTANTS   This category of water pollutants includes mineral acids, inorganic salts, and finely divided metals or metal compounds that are occasionally radioactive as well. Their presence in water may increase the water's acidity, salinity, toxicity, or radioactivity.

Acid mine drainage is the primary source of pollutants that increase water acidity. The actual pollutants present in mine drainage are sulfuric acid and soluble iron compounds, formed in reactions that involve air, water, and the pyrite (FeS_2) present in coal seams. It is estimated that nearly 5 million tons of H_2SO_4 per year are drained from coal mines in the United States, most of them once worked but now abandoned. Further, coal mining is not the only source of acidic water pollutants. Other mining operations and industries make their contributions as well. The known effects of water acidity include excessive corrosion of plumbing systems, boats, and related structures (pH < 6.0), damage to agricultural crops (pH < 4.5), and destruction of aquatic life (pH < 4.0).

Industrial effluents that contain inorganic salts are the major source of pollutants that increase water salinity. Other sources of salinity include: (1) water used in irrigation that dissolves large amounts of minerals from the soil; (2) salt brines released from mines or oil wells; (3) tidal flow from the sea; and (4) salt used on highways to melt winter ice and snow. It is common knowledge that salty water is not

---

* These are benign organisms that live and multiply in large numbers in the large intestine, but die rapidly in natural waters. Billions of these bacteria are present in feces and their presence in water is an indication of fecal discharge into the water. If fecal contamination is recent, it can be assumed that pathogenic organisms may be present along with the harmless coliform bacteria. The absence of coliform bacteria implies that recent intestinal discharges are not present in the water and presumably the water is free of pathogens. Note that the bacteria responsible for the decomposition of the organic matter of sewage have no sanitary or public health significance, because they are not found in the intestinal tracts of man or animals, nor are they pathogenic.

suitable for human consumption. Besides those related to human consumption, adverse effects on aquatic animal and plant life have resulted from large amounts of salinity in water. One of the most serious long-term effects of increased salinity of waters involves the use and reuse of water in irrigation thus allowing residual salts to return to the general water supply in a more concentrated form and/or to accumulate in the soil. In addition to total salt content, the nature of the individual salt components affects the soil, leading to problems in the agricultural industry.

Inorganic salts such as phosphates and nitrates are plant nutrients and their enrichment in water is known as *eutrophication* (from the Greek *eutrophia,* meaning "well nourished"). This enrichment leads to other slow processes collectively referred to as the *natural aging of lakes*. In a strict sense, eutrophication is a naturally occurring biological process and is not a matter of water pollution. However, it does become a pollution problem when human activities accelerate the process and the resultant aging of lakes. High concentrations of phosphates produce rapid plant growth, which first becomes apparent as algae blooms* and ultimately simply adds more organic waste to the water. As a result of the decay of these plants and associated organisms, the oxygen content of the water is greatly decreased and fish and other aquatic animals may find it impossible to exist. The serious consequences of this complex ecological process are illustrated by the eutrophication of Lake Erie.

Numerous studies of the phosphorus content of various types of water have shown that large amounts (of the order of 1–10 ppm) of phosphorus are present in domestic wastewater (sewage) as a result of the use of synthetic detergents (see previous discussion, p. 374). This fact has led to public demands that phosphates be removed from household detergents. Although there is still no absolute evidence that the elimination of phosphates from laundry detergents will solve the eutrophication problem, detergent manufacturers are taking steps to reduce or eliminate phosphates from their products. However, the development of nonphosphate detergents has proved to be difficult.

The toxic properties of numerous inorganic compounds, particularly those of some of the *heavier metallic elements,* are well known. The most toxic, persistent, and abundant of these compounds in the environment at this time appear to be those of lead and mercury. These metals are known to accumulate in the bodies of organisms, remain for long periods, and behave as cumulative poisons. A number of years ago some serious outbreaks of lead poisoning took place in Europe as a result of the use of lead pipes for interior and exterior plumbing. Very little lead pollution of this type is known in the United States because iron pipe is used in exterior water distribution systems and galvanized iron or copper pipes are used in households. Public water supplies in the United States rarely contain clinically significant quantities (~ 0.05 mg/liter) of lead.

The occurrence of mercury poisonings is well documented. Death and serious injury have resulted from the eating of mercury-contaminated fish or mercury treated seed grains, particularly a number of times in Japan. In the 1960s it was discovered that inorganic industrial and metallic mercury discharged into the Great Lakes, chiefly by the chlor–alkali industry, appeared as dimethyl mercury [$(CH_3)_2Hg$] and methyl mercury ion [CH_3Hg^+] in fishes. Subsequent laboratory studies indicated that inorganic mercury reacted with microorganisms (presumably present in the

---

* The term *bloom* is used when the concentration of an individual species exceeds 500 individuals per milliliter of water.

sludge on the lake bottom) to produce alkyl (organic) mercury compounds. Inorganic mercury salts by themselves are toxic, but organic mercury salts found in living materials are extremely toxic because the organic component in the compound enables the mercury to be transported across cell membranes. Once inside living tissues, mercury diffuses quickly to nerve cells, causing severe damage such as paralysis in humans. It is, therefore, surprising that organic mercury compounds still find applications as mildewcide in the paint, paper, and pulp industries and as fungicide dressings for seeds in the agricultural industries.

Apart from the toxicity of the heavy metals and their compounds is the possibility that some of these metals may possess highly unstable nuclei whose *radioactive emissions* are highly injurious or even lethal to living organisms. Perhaps the greatest problem is caused by the unwanted uranium tailings which are piled up in uranium-mining areas, only to be eroded by rainfall into the general water supply. It is estimated that 10 million tons of uranium tailings are piled up in the Colorado River Basin, and some waters now contain concentrations of radium-226 that are twice the maximum level prescribed for human consumption. Radium-226 and thorium-230 are the radioactive decay products of uranium. Chemically, they resemble calcium and so tend to be absorbed by the bones when taken into the body. However, their presence in the bone tissue seriously curtails the red blood cell formation leading to anemia or more serious disorders.

ORGANIC WATER POLLUTANTS Unlike inorganic wastes, organic wastes that pollute our water resources are mostly *oxygen-demanding compounds*. Typically they come from such sources as sewage, both domestic and animal; industrial wastes from food-processing plants; and effluent from slaughterhouses and meat-packing plants. Most compounds involved in this type of pollution contain carbon as their most abundant element and are easily broken down or decayed by bacteria in the presence of oxygen:

$$C + O_2 \longrightarrow CO_2$$

Thus the available dissolved oxygen in water is consumed by bacterial activity. The process results in a quick depletion of dissolved oxygen. When the dissolved oxygen concentration drops below the level necessary to sustain a normal biota (the animal and plant life of a particular region considered as a total ecological entity) that body of water is classified as polluted. Normally, a dissolved oxygen level lower than 6 ppm (6 mg/liter) indicates polluted water.

Because oxygen-demanding wastes rapidly deplete the dissolved oxygen in water, it has become necessary to estimate the amount of these pollutants in a given body of water. A quantity that is determined for this purpose is called the *biochemical oxygen demand* (abbreviated as BOD). In a water sample, the BOD indicates the amount of dissolved oxygen consumed during the oxidation of oxygen-demanding wastes. Usually, the amount of oxygen consumed is determined by a chemical analysis of the dissolved oxygen concentration of the water before and after incubation, say for 5 days at 20°C. A BOD of 1 ppm is characteristic of nearly pure water. When the BOD value reaches 5 ppm the water is considered to be of doubtful purity. However, the range of BOD values characteristic of organic wastes (such as untreated municipal sewage, food-processing wastes, and runoff from barnyards and feed lots) is 100–10,000 ppm. Obviously, these pollutants must be highly diluted upon entering water if the dissolved oxygen is not to be rapidly and completely depleted. In general, public health authorities object to runoffs entering streams if the BOD of the runoff exceeds 20 ppm.

Besides the disappearance of plant and animal life, the oxygen depletion of water may shift water conditions from those favoring aerobic activity (oxygen required) to those that support anaerobic activity (oxygen not required). The products of decomposition under these conditions are quite different (Table 14.5). Clearly, the products under anaerobic conditions are bad-smelling and toxic. Together with the unpleasant odor of decaying fish or algae, they contribute to undesirable air pollution as well.

TABLE 14.5

End products of the decomposition of organic wastes under aerobic and anaerobic conditions

Aerobic conditions	Anaerobic conditions
carbon dioxide (CO_2)	methane (CH_4)
ammonia (NH_3) and nitric acid (HNO_3)	ammonia (NH_3) and amines ($R-NH_2$)
sulfuric acid (H_2SO_4)	hydrogen sulfide (H_2S)
phosphoric acid (H_3PO_4)	phosphine (PH_3)

Synthetic organic compounds constitute another serious group of organic water pollutants (see Chapter 9). The enormous quantities (of the order of trillions of pounds) of these substances which are produced and used every year have given rise to some bizarre effects. The areas of particular concern are their toxicities and resistance to natural degradation. The compound known as DDT (see Chapter 9) illustrates the problems that can result from the use of synthetic organic compounds.

Because of the tremendous success of DDT in controlling pests, it was used on a large-scale for nearly two decades following World War II. However, in the early 1960s, it was noticed that DDT was not easily broken down into harmless compounds. Thus, natural processes were able to spread DDT from target areas to nontarget areas and ultimately throughout the environment. A 1971 report of the National Research Council of the National Academy of Sciences suggested that the oceans are the ultimate accumulation site for as much as 25 percent of all DDT used to date. In the past DDT has been applied directly to water in the form of aerial sprays to control water insects in ponds, lakes, and rivers. DDT may also reach water in surface runoff from soil. When streams or rivers become contaminated with relatively high levels of DDT, it is usually through wind or water erosion of treated soils. Even though large amounts of DDT have reached the waterways, and ultimately the oceans, the actual concentration in the oceans is quite low because of the dilution effect exerted by the huge volume of ocean water involved. Concentrations of DDT in the parts-per-billion range for freshwater and parts-per-trillion for ocean water are typical.

A source of great concern regarding DDT is the ability of many plants and animals to concentrate it (and other similar compounds) in their body tissues. When the process of biological amplification* is repeated through several links in a food chain, very high concentrations of DDT residues can be deposited in species at the top of the chain. An outstanding example of this is the bald eagle, the American national bird, which seems to be in danger of extinction in spite of federal laws pro-

---

*A process by which living organisms extract certain substances from the environment, concentrate them, and pass them, in increasing amounts, up the food chain (Fig. 14.10).

tecting it from hunting or capture. Although the use of DDT is now banned, the DDT that was used on a large scale for agricultural purposes has washed into ponds and streams where it has become concentrated in the fish that constitute much of the eagles' diet (Fig. 14.10). The DDT absorbed within the eagles' bodies upsets their calcium metabolism and causes them to lay thin-shelled eggs that cannot withstand the rigors of incubation.

The organic pollutants increasingly found in the sea are the complex mixtures of aromatic and aliphatic hydrocarbons which constitute what is familiarly known as *oil*. It is estimated that the total oil influx into the ocean is between 5 and 10 million tons annually. Less than 10 percent of this comes from accidental spillage; the other 90 percent comes from routine activities involving drilling rigs, oil tankers, refineries, and filling stations. Some of the spilled oil evaporates into the atmosphere, some oil emulsifies with water and becomes dispersed in the turbulent seas, and some oil degrades by spontaneous photooxidation into black tarry lumps of a dense asphaltic substance that frequently wash up on beaches. Both water birds and water plants have been killed by smothering coats of oil which washed ashore in some areas. Oil films retard the rate of oxygen uptake by water, thus producing lower dissolved oxygen levels that affect marine life. They also hamper photosynthesis by marine plant life because of considerable reduction in sunlight transmission through oil slicks. Besides these effects which are caused by coating and asphyxiation, there are serious toxic effects of oil pollution. The aromatic hydrocarbons abundant in oil are acute poisons for all living beings; the aliphatic hydrocarbons produce anesthesia, narcosis, cell damage, and even death in animals exposed to high concentrations. Moreover, these compounds can pass through many members of a marine food chain without becoming altered; that is, they are subject to food-chain amplification and eventually reach man who is found at the top of many food chains.

There is reason to believe that many biological processes that are of importance to the survival of marine life and that occupy key positions in their life processes are mediated by extremely low concentrations of chemical carriers or messengers in seawater. It has now been demonstrated that some marine predators are attracted to their prey by organic compounds present in the seawater at concentrations in the parts-per-billion range. Similar chemical attractions and repulsions play important roles in behavior connected with finding food, escaping from predators, locating habitats, and sexual attraction. High-boiling, saturated aromatic compounds are likely to interfere with these activities because they may block the taste receptors of organisms or mimic the natural stimuli, inducing false responses by the organisms. Such interference could have disastrous effects on the survival of many marine species.

THERMAL WATER POLLUTION Thermal pollution of water originates primarily with the practice of using water as a coolant in many industrial processes and returning it thermally enriched to the original sources (see Chapters 6 and 8). Nearly all the waste heat entering our waterways comes from industrial sources. Eighty percent of the water required to quench all the waste heat on earth is used by electric power plants. Used coolant water frequently has a temperature 10°C higher than the river or stream to which it is returned. The added heat raises the temperature of the natural waters with the result that the amount of dissolved oxygen in the water is decreased (Fig. 14.11). What is worse is that the heated water discharged into lakes and rivers encourages the formation of stratified layers and reduces the periodic mixing that renews oxygen in deep waters. Thermally polluted water ponds set in motion certain processes that make an increasing demand on the already low, dissolved

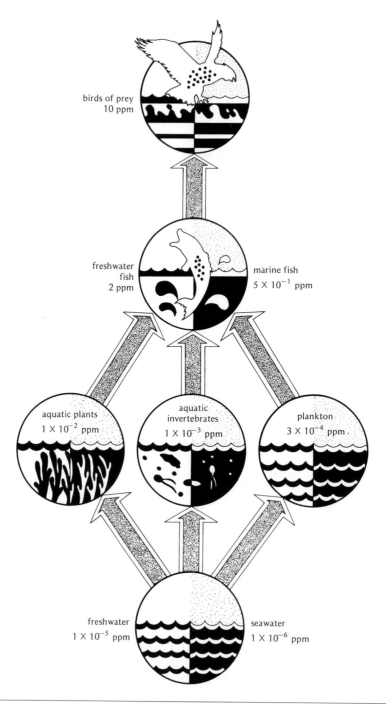

birds of prey
10 ppm

freshwater
fish
2 ppm

marine fish
5×10^{-1} ppm

aquatic plants
1×10^{-2} ppm

aquatic
invertebrates
1×10^{-3} ppm

plankton
3×10^{-4} ppm

freshwater
1×10^{-5} ppm

seawater
1×10^{-6} ppm

FIGURE 14.10

Typical food-chain concentration of DDT. The increasing concentration of DDT is shown by dots.

oxygen levels in the water. For example, increased temperature speeds up chemical reactions, including respiration in living systems. Thus, fish in thermally enriched water require more oxygen because of an increased respiration rate. Microorganisms of decay become more active and they accelerate their use of oxygen to oxidize organic matter. Because the available oxygen in such water is low, fish and microorganisms die, and the resulting anaerobic decay produces foul-smelling, polluted pond water. Thus, water that is thermally enriched is polluted in the sense that it is unfit for the beneficial uses of mankind.

The normal life patterns of aquatic organisms can be completely disrupted by changes in water temperatures. Aquatic species alter their behavior from one temperature range to another, just as they respond to fluctuating levels of sediment, minerals, and dissolved oxygen. Extremes of temperature, like extremes of chemical concentrations, will kill organisms outright. A problem of thermal pollution not related to aquatic life is the reduced cooling capability of warmed waters. Various approaches are under study to limit the pouring of heat into our waterways. One approach is to use the heat to extend the growing season or otherwise enhance agricultural and aquatic production. Catfish and oyster farming, for instance, can be done with warm water. Another approach is to divert the heat to the atmosphere through cooling towers. Wet towers dissipate heat by evaporation; dry towers save water by transferring heat directly to air. However, they are unbearably expensive. The present approach appears to be the use of oceans which, like the atmosphere, provide a much greater sink for waste heat than do surface waters. This is precisely the reason why nuclear power plants, which waste more heat than do fossil-fuel plants, are being built along the coastlines in many countries. For the present the dissipation of heat to the atmosphere and oceans will lessen the impact of heat on lakes and rivers. But in the long run, there is no part of our environment, no matter how large, that can cope with endless amounts of waste heat.

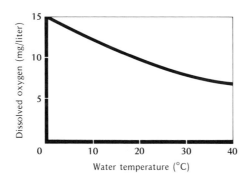

FIGURE 14.11

Dissolved oxygen versus water temperature. Redrawn, by permission, from Richard H. Wagner, *Environment and Man*, p. 137. Copyright © 1971 by W. W. Norton & Company, Inc.

QUESTIONS

14.23 What is meant by water quality?

14.24 List the general effects on water attributed to inorganic chemicals and mineral substances. Give examples.

14.25 What is meant by eutrophication? What causes it and what has been done to prevent it?

14.26 Name the significant heavy metal pollutants in water. Describe their sources and discuss the effects attributed to them.

14.27 What are the main sources of radioactive pollution.

14.28 List the common types of water pollutants that are classified as oxygen-demanding wastes.

14.29 What is meant by biochemical oxygen demand? How is it determined?

14.30 What is meant by having a good BOD? Discuss.

14.31 Give an example of a synthetic organic compound that has become a water pollution problem.

14.32 In what way does DDT reach natural water supplies?

14.33 What are the effects of oil pollution?

14.34 Why is heat considered to be a water pollutant?

14.35 Discuss the effects of thermal pollution of our waterways.

14.36 Estimate the extent of water pollution in your region of the country. How does your region compare to the country as a whole?

SUGGESTED PROJECT 14.4
Determining acidity of mine drainage (or minding the mine next door)

Visit a neighboring coal mine and collect samples of mine drainage from various locations. Test the acidity of the samples using pH papers. Use the pH scale shown in Fig. 7.5 to compare the acidity of the mine drainage samples with the acidity of common substances.

WATER POLLUTION CONTROL AND WASTEWATER TREATMENT

The water pollution problem is to some degree due to indifference and excessive cost-consciousness on the part of industry. A lack of suitable sewage treatment in many cities also contributes to this problem. Although the Water Quality Act of 1965 has set federal standards on interstate water purity, cleaning up dirty water that runs from state to state has been extremely difficult in the past because each state had its

TABLE 14.6
Average removal efficiency of water pollutants by sewage-treatment plants[a]

Pollutant	Removal efficiency (%)	
	Primary treatment	Primary plus secondary treatments
degradable organics (BOD)	35	90
total organic matter	30	80
sediment	60	90
nitrogen	20	50
phosphorus	10	30

[a] Data from *Cleaning Our Environment: The Chemical Basis for Action,* The American Chemical Society, Washington, D. C., 1969.

own set of standards. Enforcement of these federal standards will, of course, be the test of their value. To this end, every industrial plant must be required to clean up its own effluent, either in its own special treatment plant or in conjunction with a municipal sewage treatment plant. Further, every community — cities, towns, villages, and even rural areas — must construct *adequate* wastewater treatment plants.

At present most communities have some sort of treatment plant — usually a holding tank or pond where the sewage is allowed to settle for a while to remove some of the solids as "sludge." In this so-called *primary treatment,* all the dissolved oxygen is used up and in course of time anaerobic odorous decomposition takes over. Therefore, effluent from a primary treatment plant still contains a lot of dissolved and suspended organic matter, including pathogenic bacteria. Strangely enough, this is the sole treatment for wastes from about one-quarter of the population in the United States! However, *secondary treatment* is added to the primary treatment of sewage from over 100 million people. In the process the effluent from the primary treatment passes through mechanical devices known as *activated sludge tanks* and/or *trickling filters.* Aeration is achieved simply by bubbling air through tanks of wastewater. The so-called trickling filters are giant beds of rocks, open to the atmosphere, through which the wastewater is trickled. Microorganisms present on the rocks and in the water oxidize the organic matter with the aid of oxygen gas that circulates freely with air between the rocks. Following oxidation, the wastewater undergoes further sedimentation, and it is then usually chlorinated (to kill microorganisms) and released.

Wastewater treatment plants are arranged to have sewage flowing from one stage to another. A typical arrangement of stages for primary and secondary wastewater treatment is shown in Fig. 14.12. Table 14.6 shows the relative efficiency of the treatments in removing contaminants from water. Clearly, the primary process alone is entirely inadequate. Even the secondary treatment fails to stop the steady

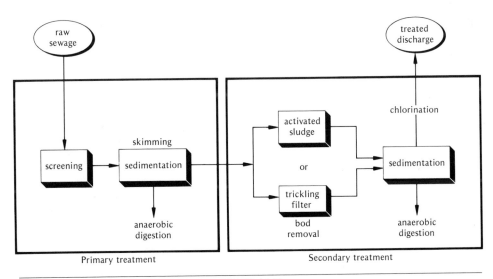

FIGURE 14.12

Typical stages in the primary and secondary wastewater treatment. From J. Calvin Giddings, *Chemistry, Man and Environmental Change,* p. 309. Copyright © 1973 by J. Calvin Giddings. Redrawn by permission of Harper & Row, Inc.

nitrogen and phosphorus nutrient flow to the waterways. Further treatment is clearly needed and a number of new experimental processes are being investigated. Such *tertiary treatment* must include the adsorption of contaminants by charcoal beds, chemical oxidation of organic matter by ozone, and chemical precipitation of phosphate ions by added aluminum, calcium, or iron ions. All these ions form insoluble, collectible precipitates of metal phosphates. However, financing tertiary treatment will be one of the major problems the nation will have to face in the near future.

Even if there is an end to water pollution by industries and municipalities, it will be necessary to stop wasting good water. For example, every time a toilet is flushed, about 5 gal of water sweep a relatively small amount of human waste into the sewage system. This waste of good water, especially in regions where it is scarce is probably the reason for the development of the "waterless john" — a closed-loop system that uses mineral oil as a reusable flushing substance. Here again, the system is too expensive as a replacement for a conventional toilet. Nevertheless, it has the added advantage that the wastes are not deposited in the sewer but burned in an almost pollution-free, 650°C incinerator that leaves only a residue of sterile ash.

QUESTIONS

14.37 What are the primary, secondary, and tertiary treatments of wastewater?
14.38 Discuss the ways and means by which waste of good water can be avoided.

SUGGESTED PROJECT 14.5

Inspecting your neighborhood sewage plant (or inherit the wind)

Visit the nearest sewage-treatment plant and examine the facilities now being used at that plant. Find out what treatment is provided and whether it is adequate. If you are convinced that the treatment is inadequate, suggest ways and means for improving the facilities.

15

The atmosphere

The invisible stuff around us is almost never still. Clinging to the earth by its own weight, the sun may boil it up into dense mountains but it immediately seeks its own level exactly in the manner of water; then it cascades downhill as wind, and it settles into quiet pools of calm air. . . . In my mind's eye I see the atmospheric sea as a gaseous symphony that gyrates and squirms over the surface of the earth. . . .

ERIC SLOANE
Author of Look at the Sky

Enveloping the earth is that part of our environment which we call the *atmosphere* and which consists of a mixture of gases called *air.* Within the air itself, we find in addition to nitrogen all of the substances commonly associated with life itself: oxygen, water, and carbon dioxide.

To understand the atmosphere, we have to understand gases. The behavior of gases has been studied extensively and was for the first time summarized during the last half of the nineteenth century in a form that is now known as the *kinetic theory of gases.* According to this theory, gases consist of molecules that are in constant motion through space, frequently colliding with each other and with the walls of the container, if any. The wall collisions account for the *pressure* exerted by gases; the constant molecular motions through space account for the *diffusion* phenomenon exibited by gases. The average speed of a molecule depends on its molecular weight as well as on the temperature. The number of collisions depends on the temperature and on the gas density. In a hot gas, the molecules move rapidly and take only a short time to go from collision to collision. In a dense gas, there are more molecules to collide with.

From the size of molecules and their spacing, it has been estimated that the molecules in air travel 10^{-7} or 10^{-8} m between collisions. An average air molecule is travelling 450 m/sec, so it must perform about 10^{10} collisions each second. Without collisions, a molecule in the earth's atmosphere would fall toward the ground. Instead, it is on the average held up by collisions from below; collisions from above tend to push the molecule toward the ground. The atmosphere is stable only if the collisions from below slightly outnumber the collisions from above—enough to balance the effect of gravity. Thus, we can visualize the air molecules forming a layer of atmosphere without falling to the ground. Neglecting variations in temperature, the density of the atmosphere must decrease with altitude because collisions are fewer in a less dense gas.

Near the top of the atmosphere, where molecules due to low gas density can travel long distances without colliding, light molecules such as hydrogen and helium can move so rapidly that they have a good chance of leaving the atmosphere entirely. That is why the earth's atmosphere has so little hydrogen and helium even though they are the most abundant elements in the universe (Table 15.1). The force of gravity on the moon is weaker than on the earth, which accounts for the moon's lack of an atmosphere: Over the ages all the gas molecules have escaped.

Air molecules collide not only with one another but with the ground, with walls, and with all of us. Through molecular collisions the atmosphere interacts chemically and physically with the solid earth and the water to produce the conditions we find fit for life. In this chapter we shall examine the nature and possible origin of the atmosphere, the more important atmospheric interaction phenomena, and the extent to which our activities have affected the atmosphere.

TABLE 15.1

Composition of air near the earth's surface

Nonvariable component	Content by volume[a]	Variable component	Normal content, by volume[a]
nitrogen	78.084%	water vapor	0.1–1%
oxygen	20.946%	carbon dioxide	0.033%
argon	0.934%	ozone	0.02–0.07 ppm
neon	18.18 ppm	ammonia	0.01 ppm
helium	5.24 ppm	sulfur dioxide	0.0002 ppm
methane	2.00 ppm	carbon monoxide	0.1 ppm
krypton	1.14 ppm	nitrogen dioxide	0.001 ppm
hydrogen	0.50 ppm	radon	—
nitrous oxide	0.50 ppm	dust	—
xenon	0.087 ppm		

[a] Contents expressed as parts per million (ppm) are merely 10,000 times the volume percentage.

ORIGIN AND STRUCTURE OF THE ATMOSPHERE

The atmosphere as we know it today was very likely formed during an extensive period after the earth had cooled sufficiently to permit life to develop. The earliest atmosphere is considered to have been a mixture of H_2, He, CH_4, NH_3, and Ar. With the passage of time the atmosphere evolved into a mixture of O_2, CO_2, N_2, and the rare gases; the CO_2 comprised a much higher percentage of the atmosphere than it does today (Fig. 15.1). The origin of CO_2 is attributed to widespread volcanic activity; the nitrogen resulted perhaps from oxidation of the ammonia present at the earlier stage. With the appearance of green plants, the CO_2 was used up in the photosynthesis process, a reaction whereby chlorophyll-bearing plants are able to convert CO_2 and H_2O into carbohydrates and oxygen in the presence of sunlight:

$$6CO_2 + 6H_2O \longrightarrow C_6H_{12}O_6 + 6O_2$$

It was then that the atmosphere began to approach its present composition (Table 15.1).

The atmosphere can be structured into several regions or layers, each region corresponding to an altitude range. The major layers of the atmosphere are shown in Fig. 15.2. The density and pressure of the atmosphere steadily decrease with increasing altitude until they become practically zero in the vastness of outer space. The outer boundary of the atmosphere is not known; however, it is known that more than 99 percent of the total mass (5.184 quadrillions of metric tons) of the atmosphere is below an altitude of 20 miles. Indeed, as Fig. 15.2 shows, the atmospheric pressure approaches zero at an altitude of about 30 miles.

The region we live in is the lowest layer and is called the *troposphere*. It is in this region, which extends to a height ranging from about 5 miles over the poles to about 11 miles over the equator, that common atmospheric phenomena, such as cloud formation, air movements, and wide temperature gradations, occur. Vertical air currents are common in the troposphere with the result that there is considerable air turbulence and mixing in this region. Weather as we know it is almost exclusively confined to the troposphere. With increasing altitude within the troposphere, the air temperature falls steadily until attainment of a temperature of about −55°C at the upper boundary at about 8 miles, known as the *tropopause,* in middle latitudes. The

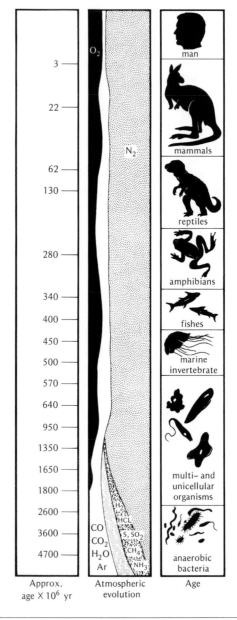

Approx. age X 10^6 yr	Atmospheric evolution	Age

FIGURE 15.1

Atmospheric evolution.

tropopause temperature is higher than this in the polar regions and lower (as low as −90°C) near the equator.

The next layer is called the *stratosphere*. This region is characterized by the presence of several layers or strata (hence the name) and differs from the troposphere in its lower air pressure, its lack of moisture, and the near absence of vertical air currents and turbulent mixing. The lower part of the stratosphere has a fairly constant temperature of about −55°C; above about 15 miles, however, there is a gradual rise in temperature until the attainment of a maximum of 0–10°C at the upper boundary at 25–30 miles, known as the *stratopause*. The increase in temperature is due to the heat given off by a photochemical reaction in which atmospheric oxygen is transformed into ozone by ultraviolet radiation of wavelength around 1800 Å from the sun:

$$3O_2 \xrightarrow{1800 \text{ Å}} 2O_3$$

This gives rise to a region of high ozone concentration, sometimes called the *ozone layer*. However, ultraviolet radiation of 2550 Å, also emitted from the sun, decomposes the ozone back into O_2:

$$2O_3 \xrightarrow{2550 \text{ Å}} 3O_2$$

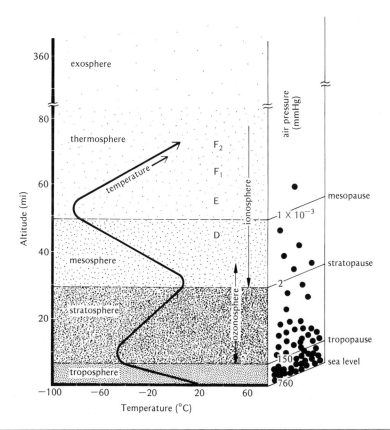

FIGURE 15.2

Major layers of the atmosphere.

Thus, there exists in this region a chemical equilibrium between the two allotropes. Fortuitously, the existence of the ozone layer, which screens out deadly ultraviolet radiation, permits life to exist on the earth. Apart from the presence of relatively high concentrations of ozone in the upper reaches and a lower water vapor content, the chemical composition of the stratosphere is the same as that of the troposphere.

Above the stratopause lies the *mesosphere,* which is marked by a steady decrease in temperature with altitude, to about −95°C at the upper boundary at 50–60 miles altitude, known as the *mesopause.* Most of the meteors entering the earth's atmosphere vaporize and "burn up" in this region to produce a luminous trail of ionized gases. It is also in this region that the so-called *D-layer* phenomenon is observed. This layer, present only during the daylight hours, is the lowest ionized layer in the atmosphere and consists of ionized monatomic oxygen, O^+, produced by absorption of ultraviolet radiation of 1216 Å by O_2. Otherwise, the composition of the mesosphere, even though the air in it is highly rarified by comparison, is substantially the same as that of the lower atmosphere.

The region at altitudes between 60 and 350 miles is called the *thermosphere.* It is characterized by a rapid increase in temperature with altitude; temperatures as high as 2000°C are known to exist in this region. However, these elevated temperatures have little significance because of the extreme "thinness" of the atmosphere at these altitudes. The molecules of air are so few and far between that, even though they may be moving at very high velocities, from a practical viewpoint little heat is imparted to objects such as the temperature-sensing devices on space vehicles. Within the thermosphere lie the so-called E-, F_1-, and F_2-layers; like the D-layer, they consist of charged gas particles (ions). The E-layer is lower than the F-layers; its properties arise from the absorption of solar X-rays that generate positively charged hydrogen and nitrogen ions. The F-layers are produced by the shorter-wavelength ultraviolet and X-rays and contain such positively charged species as O^+, O_2^+, and NO^+. Because they contain ions, the D-, E-, and F-layers together are referred to as the *ionosphere.* It is this ionosphere that causes radio waves to be reflected back to earth and makes radiobroadcasting possible.

The *exosphere* is the outermost region of the atmosphere, beginning at an altitude of about 350 miles and gradually merging into outer space. Because of the great thinness of "air" in this region, the very few particles that are present travel great distances between collisions. Very likely, some of the particles might simply continue traveling on into outer space, never again to return to earth.

QUESTIONS

15.1 What is meant by the term atmosphere?

15.2 How does the behavior of gases differ from the behavior of solids and liquids?

15.3 Give two examples to illustrate the diffusion phenomenon exhibited by gases.

15.4 What is the origin of the upward force on a balloon?

15.5 Account for the fact that atmospheric density decreases with altitude.

15.6 Calculate the mole percentages of nitrogen and oxygen in the atmosphere. How do your answers compare with the percentage by volume of each of these gases?

15.7 How has the atmosphere probably obtained most of its oxygen?

15.8 Describe the various atmospheric regions and their characteristic features.

15.9 In what respect is the ozone layer of the atmosphere important to us?

15.10 What is meant by ionosphere? In what way is the ionosphere technically useful to us?

SOME ATMOSPHERIC INTERACTION PHENOMENA

We have already noted that in the biosphere chemical elements move back and forth between living organisms and the physical environment. There is also a transfer of elements from one organism to another through food chains as well as transport of elements among the regions of the physical environment. The overall cyclic movement of an element is referred to as a *biogeochemical cycle*. Cycles that involve the atmosphere are the hydrologic cycle, the carbon dioxide cycle, the oxygen cycle, and the nitrogen cycle. Of these, the hydrologic and the carbon dioxide cycles are responsible for the components showing the largest concentration variability in air. Though minute compared to oxygen and nitrogen, the variable components in air are significant in connection with the transfer of solar energy to the earth's surface.

ATMOSPHERIC ROLE IN ENERGY TRANSFER As solar energy travels through the atmosphere, it interacts with atmospheric components. Part of it is absorbed by molecules such as O_2, O_3, H_2O, and CO_2; part of it is reflected, mostly by clouds, back to space; and part of it is scattered in all directions. Solar energy that is not reflected, scattered back to space, nor absorbed is transmitted to the earth's surface and constitutes more than 99 percent of the energy available for all activity on earth. Depending on the reflectivity and, hence, the nature of the earth's surface cover, a certain amount of energy that gets through the atmosphere is reflected back into the atmosphere while the remainder is absorbed. It is estimated that 29 percent of the incoming solar energy is reflected by the earth–atmosphere system and lost to space. Of the remaining energy, 20 percent is absorbed directly by the atmosphere and 51 percent by the earth's surface. The earth's surface is, therefore,

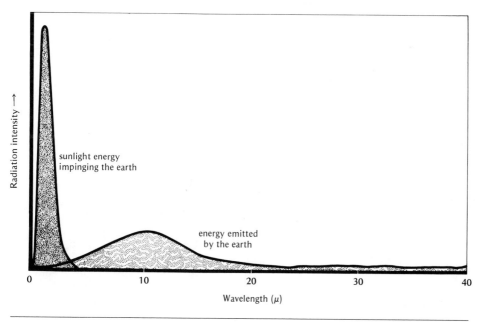

FIGURE 15.3

The energy (wavelength) distribution of the earth's incoming and outgoing radiation. From J. Calvin Giddings, *Chemistry, Man and Environmental Change*, p. 201. Copyright © 1973 by J. Calvin Giddings. Redrawn by permission of Harper & Row, Inc.

the prime recipient of solar energy. However, the earth's surface continuously re-radiates this absorbed energy to the atmosphere and then to space.

Unlike solar radiation that is light of the relatively shorter wavelengths (the ultraviolet and visible light), the terrestrial radiation consists of the longer infrared wavelengths (Fig. 15.3). Because solar and terrestrial radiation represent different portions of the electromagnetic spectrum, their respective interactions with atmospheric components are also different. Atmospheric components, such as carbon dioxide and water vapor and, to a much lesser extent, ozone, trap the terrestrial

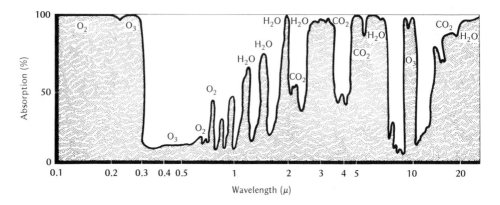

FIGURE 15.4

Absorption of solar and terrestrial radiation by atmospheric gases as a function of energy (wavelength). Redrawn, by permission, from A. Miller and J. C. Thompson, *Elements of Meteorology*, p. 69. Copyright© 1970 by Charles E. Merrill Publishing Co.

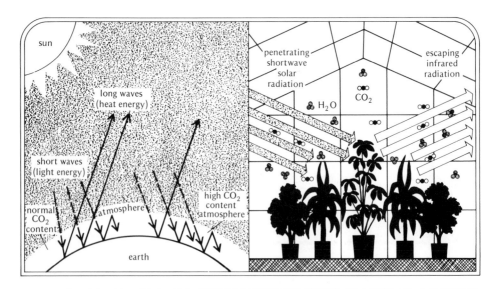

FIGURE 15.5

The greenhouse effect. Shortwave radiation (light) strikes the earth and is transferred into longwave radiation (heat). Because CO_2 absorbs longwave radiation, the more carbon dioxide contained in the atmosphere, the more heat is retained, and the warmer the atmosphere.

radiation (Fig. 15.4) and increase the atmospheric temperature, just as greenhouses retain terrestrial energy and increase the temperature within them. This phenomenon is known as the atmospheric *greenhouse effect* (Fig. 15.5). Clearly, higher concentrations of carbon dioxide and water vapor in the atmosphere can influence the solar energy transfer and eventually alter the climate and, consequently, the ecology, perhaps irreparably.

ATMOSPHERIC ROLE IN MATERIAL TRANSFER The major material transfers between the earth's surface and the atmosphere involve nitrogen, oxygen, water vapor, and carbon dioxide. Of these, nitrogen has its major natural reservoir in the atmosphere and, therefore, we shall consider the nitrogen cycle first.

The nitrogen cycle All of the nitrogen in plant and animal tissues originates in the atmosphere as part of the so-called nitrogen cycle, which is illustrated in Fig. 15.6. At first glance, this is rather hard to visualize because elemental nitrogen is fairly unreactive (see Chapter 5). Fortunately, nature has provided the following two paths whereby nitrogen from the air can be "fixed":

1. *Nitrogen-fixing bacteria* that lie on the roots of leguminous plants (clover, alfalfa) and that transform N_2 directly into organic nitrogen compounds. The precise mechanism of how this takes place is still largely speculative.

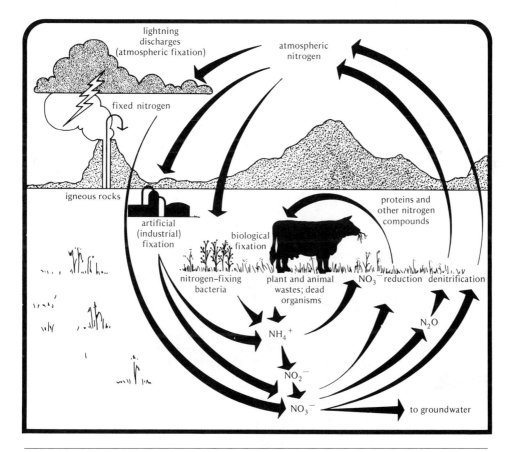

FIGURE 15.6

The nitrogen cycle in nature. Adapted from C. C. Delwiche, "The Nitrogen Cycle," *Scientific American*, pp. 138–139 (September 1970). Copyright © 1970 by Scientific American, Inc. All rights reserved.

2. *Lightning discharges* that cause chemical combination of atmospheric O_2 with atmospheric N_2 to form nitric oxide:

$$N_2 + O_2 \longrightarrow 2NO$$

The nitric oxide then reacts with more oxygen of the air to form the brown gas NO_2 which, in turn, is dissolved by rain-water and is carried down into the soil as nitrites (NO_2^-) and nitrates (NO_3^-):

$$2NO + O_2 \longrightarrow 2NO_2$$
$$2NO_2 + H_2O \longrightarrow NO_2^- + NO_3^- + 2H^+$$

The fixed nitrogen in the soil is next metabolized by plants. The plants, together with the nitrogen in them, are eaten by herbivorous animals which, in turn, are eaten by carnivorous animals. The animals give off wastes that are decomposed by bacteria to atmospheric nitrogen or ammonia or to other nitrogen compounds that enrich the soil. Finally, animals and plants die and decay, and their nitrogen returns either to the atmosphere or to the soil. The natural cycle is supplemented by the artificial fixation of nitrogen from the atmosphere using methods such as the Haber and Ostwald processes discussed in Chapter 8.

The oxygen cycle   The vital role of elemental oxygen in life processes is well known. The utilization of oxygen by living organisms forms part of what we naturally refer to as the oxygen cycle, which is illustrated in Fig. 15.7. Through

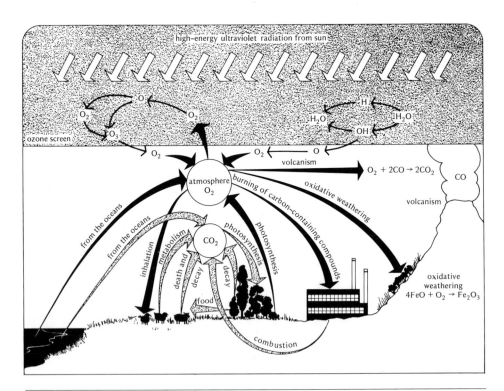

FIGURE 15.7

The oxygen cycle in nature. Adapted from Preston Cloud and Aharon Gabor, "The Oxygen Cycle," *Scientific American*, p. 115 (September 1970). Copyright © 1970 by Scientific American, Inc. All rights reserved.

inhalation (or by other methods in the lower animals) mammals, birds, and reptiles take in oxygen through the lungs. The oxygen is then transported by the hemoglobin of the blood to the cells of the different tissues, where it plays an essential role in *metabolism*. Carbon dioxide produced by metabolism is exhaled into the atmosphere:

$$\underset{\text{hemoglobin}}{Hb} + O_2 \longrightarrow \underset{\text{oxyhemoglobin}}{HbO_2} \text{ (in the lungs)}$$

$$HbO_2 \longrightarrow Hb + O_2 \text{ (released in the body cells)}$$

$$C_6H_{12}O_6 + 6O_2 \longrightarrow 6CO_2 + 6H_2O + \text{energy (in the body cells)}$$

$$Hb + CO_2 \longrightarrow Hb{-}CO_2 \text{ (in the body cells)}$$

$$Hb{-}CO_2 \longrightarrow Hb + CO_2 \text{ (released in the lungs)}$$

Carbon dioxide is also released by the decomposition of animals and plants following death and by the combustion of carbon-containing substances such as wood, coal, and gasoline. Plants consume atmospheric CO_2 in carrying out the process of photosynthesis, in which CO_2 and H_2O react in a complex series of steps, eventually yielding carbohydrates (sugars, starch, and cellulose) and releasing elemental oxygen back into the atmosphere:

$$6CO_2 + 6H_2O + \text{energy} \xrightarrow{\text{sunlight}} C_6H_{12}O_6 + 6O_2$$

$$nC_6H_{12}O_6 \longrightarrow (C_6H_{10}O_5)_n + nH_2O$$

The water cycle   Depending on the weather and atmospheric temperature, the amount of water vapor in air varies between values of a few tenths of 1 percent and 5 or 6 percent. At certain times the moisture in the atmosphere in desert regions is nearly zero; on the other hand, a condition of complete saturation exists inside clouds and fogbanks. The existence of water vapor in the atmosphere forms part of a larger process, called the *hydrologic (water) cycle*. The following are the essential features of this cycle: (1) *evaporation* from the oceans and lakes and from plants and moist soil; (2) *transportation* of water vapor in the atmosphere to distant locales by the winds and by simple diffusion; (3) *condensation* of the water vapor into minute droplets that make up clouds; (4) *precipitation,* whereby the moisture falls out of the clouds as rain onto the earth's surface; and (5) *runoff* and/or *percolation,* whereby the water is returned to the ocean by rivers or by underground seepage (see Fig. 14.1 on p. 365). Although moisture evaporation from water bodies is governed by the general rules applying to the vapor pressures of liquids and solutions, there is a limit on how much moisture air can hold at a given temperature.

The concept of *relative humidity* expresses the percentage of water vapor present in the air in comparison with saturated conditions. Keep in mind that the air in a closed vessel containing a liquid rapidly becomes saturated with vapor as a consequence of the vapor pressure exerted by the liquid. However, air in the normal outdoors, unlike that in the closed system just mentioned, usually carries less than a saturation quantity of moisture due to its physical distance from water. Even in the close proximity of water, saturation is prevented by a continuous mixing of air near the surface of the water with drier air from higher altitudes or from land areas. The condensation of moisture into tiny cloud droplets (or ice crystals) at high altitudes where low temperatures prevail also depends on the presence of tiny solid particles, such as smoke, meteor dust, or salt particles (originating from the spray of ocean waves), which serve as nuclei for their formation. If air is completely free of nuclei, it may become supersaturated with moisture.

The carbon dioxide cycle   Carbon dioxide is continually being cycled into and out of the atmosphere in a cycle involving the activities of plants and animals. In this cycle, plants use light energy to react CO_2 from the air with water and produce carbohydrates and oxygen. The carbohydrates are stored in the plant, and the oxygen is released into the atmosphere. When plants are oxidized by natural decomposition, burning, or consumption by animals, oxygen is absorbed from the air and CO_2 released back into the atmosphere:

$$6CO_2 + 6H_2O + energy \xrightarrow{\text{sunlight}} C_6H_{12}O_6 + 6O_2 \uparrow \text{ (in plants)}$$
$$C_6H_{12}O_6 + 6O_2 \longrightarrow 6CO_2 \uparrow + 6H_2O + energy \text{ (in animals)}$$

In a simplified fashion, this is the major carbon dioxide cycle of nature providing essentially a constant atmospheric CO_2 level if it is not upset by artificial means. Because CO_2 is soluble in water, there is also an exchange of CO_2 between atmosphere and oceans, the oceans constituting a huge reservoir of CO_2 (Fig. 15.8).

However, the amount of carbon dioxide in air has been increasing because of our intense industrial and agricultural activity involving the burning of fuels and destruction of forests, respectively. The burning of fossil fuels and the conversion of naturally occurring limestone into cement increase the amount of CO_2 in the atmosphere. On the other hand, clearing land, which decreases the available plants, decreases nature's ability to remove atmospheric CO_2. The net effect has been a gradual increase in atmospheric CO_2 level, of about 40 ppm during the last 70 years or so (from 285 ppm in 1900 to 325 ppm in 1970). One possible result of this change in the atmosphere is a change in the temperature of the earth. According to the

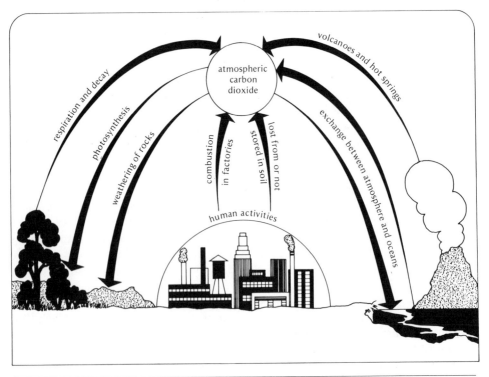

FIGURE 15.8

The carbon dioxide cycle in nature.

greenhouse theory (see p. 397), carbon dioxide in the air acts to trap heat. Although the worldwide average temperature did increase by 0.5°C between 1885 and 1940, a decline of 0.1°C has been noted since 1940. The one-way filter action of CO_2 leads to predictions that the temperature of the atmosphere and the earth will increase. On the other hand, a cyclic variation in airborne particulates that reflect sunlight (see p. 413) may decrease the temperature of the atmosphere and earth. In fact, the observed drop in the worldwide average temperature since 1940 is probably the net effect of particulate pollutants in the atmosphere and the greenhouse phenomenon.

Composition of dry air The composition of natural atmospheric dry air, from which absolutely all water vapor and carbon dioxide have been removed, is virtually constant over the entire world. Inspection of Table 15.1 (see p. 391) reveals that N_2, O_2, and Ar make up 99.96 percent of dry air. This constancy of the composition of clean dry air can be accounted for in two ways: (1) Because of the large amount of chemically inert elemental nitrogen (and therefore of air) and its constant stirring by winds, localized chemical and physical processes that tend to use up one or more components have little effect on the overall composition. (2) Natural oxidation, a major process in the removal of reactive oxygen, is balanced by photosynthesis, a process that restores oxygen. These, together with the chemical inertness of argon, suggest that the three major atmospheric gases are immune to any significant disturbance by human activities.

Components showing the largest concentration variability are water vapor, carbon dioxide, and ozone. Although these gases are present in trace amounts relative to nitrogen, oxygen, and argon, their actual amounts in the atmosphere are estimated to be 14×10^{12}, 2.5×10^{12}, and 3×10^9 metric tons, respectively (Table 15.2). In our previous discussion, we have already seen how changes in the atmospheric level of water vapor and carbon dioxide can affect the earth's climate

TABLE 15.2
Estimated amounts of atmospheric trace gases[a]

Trace gas	Amount (metric tons $\times 10^9$)
water vapor	14,000
carbon dioxide	2,500
neon	65
krypton	17
methane	4
helium	4
ozone	3
xenon	2
nitrous oxide	2
carbon monoxide	0.6
hydrogen	0.2
ammonia	0.02
nitrogen dioxide	0.013
nitric oxide	0.005
sulfur dioxide	0.002
hydrogen sulfide	0.001

[a] Compared to the so-called trace components, the actual estimated amount of the atmosphere's N_2, O_2, and Ar are 3.9×10^{15}, 1.2×10^{15}, and 0.67×10^{14} metric tons, respectively.

through their participation in the so-called atmospheric greenhouse effect. Most of the atmospheric ozone exists in the stratosphere, a layer 10–25 mi above sea level, and absorbs the deadly ultraviolet radiation from the sun, thus protecting living organisms on earth against the damaging molecular disruptions caused by high-energy ultraviolet radiation (Fig. 15.7). There has beeen widespread concern that the vital protective ozone shield will be damaged by the flight of supersonic transports (the SST's) in the stratosphere. Calculations by the American chemist Harold Johnston show that significant amounts of ozone would be removed by its reaction with nitric oxide and that the NO_2 thus formed would regenerate to NO by the reaction of NO_2 with oxygen atoms that are present in the same atmospheric layer:

$$NO + O_3 \longrightarrow NO_2 + O_2$$
$$NO_2 + O \longrightarrow NO + O_2$$

Johnston estimates that a fleet of only 500 SST's would reduce the present ozone level by half and, thus, greatly increase the penetration of damaging ultraviolet radiation to the earth's surface. Although the nature of the ozone problem facing us is clear, its full environmental impact will be determined only when the SST's are pressed into service.

Such gases as sulfur dioxide, hydrogen sulfide, and carbon monoxide are continually released into the air as by-products of natural occurrences such as volcanic activity, vegetation decay, and forest fires. Tiny particles of solids or liquids are distributed throughout the air by winds, volcanic explosions, and other similar natural disturbances. In addition, human activities release considerable amounts of nitrogen and sulfur oxides, carbon monoxide, hydrocarbons, and particulates into the air and contribute to what is so familiarly known as air pollution. We shall now explore the significant aspects of the changing global content of some of these gases, including the environmental consequences that may ensue therefrom.

QUESTIONS

15.11 What is meant by a biogeochemical cycle? List cycles in nature that involve the atmosphere.

15.12 Explain the atmospheric greenhouse effect with particular reference to its influence on the earth's climate.

15.13 What is meant by nitrogen fixation? How does it occur naturally? How is it made to occur artificially?

15.14 Explain the role of plants in the oxygen cycle.

15.15 Devise an experiment that could determine the approximate moisture content of the air in your locality.

15.16 What are the essential features of the hydrologic cycle?

15.17 Describe in detail how human activities tend to increase the amount of carbon dioxide in the atmosphere.

15.18 Although there is evidence that the carbon dioxide content of the air is increasing, the increase is extremely slow. Why?

15.19 How can you account for the fact that at the present time the three major atmospheric gases are immune to any significant disturbance by human activities?

15.20 What role does ozone play in our atmosphere? How may it be affected by the use of supersonic transports?

POLLUTION OF THE AIR ENVIRONMENT

Historically, air pollution has been recognized as a major problem of the urban people for more than three centuries. For instance, in his book entitled *Fumifugium,* published in 1661, the English diarist John Evelyn wrote:

It is this horrid smoake which obscures our Churches and makes our Palaces look old, which fouls our Clothes and corrupts the Waters, so as the very Rain, and refreshing Dews which fall in the several Seasons, precipitate the impure vapour, which, with its black and tenacious quality, spots and contaminates whatever is exposed to it.

Clearly, the addition of unwanted airborne matter, such as smoke, changes the composition of the earth's atmosphere, possibly harming life and altering materials. Isolated incidences of devastating air pollution, such as those which took place in 1948 in Donora, Pennsylvania, and in 1952, 1956, 1957, and 1962 in London, England, have revealed that coal smoke reacting with air generates deadly substances that when inhaled, caused severe lung irritation, a starvation of oxygen, and finally heart failure. Since then the normal smoggy conditions and the spasmodic air pollution crises from time to time in many urban areas, such as New York City and Los Angeles, have led many people to believe that the entire atmosphere is gradually becoming polluted as a result of human activities. In consequence, recently there has been a widespread interest in air pollution studies, including its long-range effects.

Although natural air pollution may result from uncontrollable nonhuman events such as volcanic eruptions and forest fires, it is the alteration of the outdoor atmosphere by human activities that is our immediate concern. It is estimated that each year, up to 1×10^{12} tons of emissions enter the global atmosphere from both natural and man-made sources, but only about 5×10^8 tons, that is, 0.05 percent of the total, arises from the activities and inventions of modern man. However, more than half of the global emissions occur as a result of human activity in the United States alone (Table 15.3). These emissions range from asbestos to zinc and come from a variety of sources: from industrial processes such as smelters, iron and steel mills, paper and pulp operations, petroleum refineries and chemical plants; from community operations such as garbage dumping, trash incineration, and agricultural burning; from power-generating facilities and space heaters; and from automobiles, trucks, and jet aircraft. Five types of substances, known as *primary pollutants,* account for more than 90 percent of the nationwide air pollution. They are carbon monoxide, nitrogen oxides, hydrocarbons, sulfur oxides, and particulates. Their major sources and amounts, as reported by the Environmental Protection Agency, are given in Table 15.3. According to this information, transportation is the main source of air pollution and carbon monoxide is the major individual pollutant. We shall examine these data more closely after discussing some general considerations.

The nearly 300 million tons of man-made pollutants that enter the air annually in the United States are not distributed uniformly in the massive amount of air (3×10^{15} tons) comprising the atmosphere of the continental United States at any one time. There is very little atmospheric diffusion of surface air beyond an altitude of 10,000 ft above ground level. Many pollutants never rise beyond 2000 ft. In addition, geologic and man-made barriers often limit lateral air movement as well, and they greatly decrease the mixing and diluting effects. Consequently, pollutant concentrations in the 10–50 ppm range are not unusual in certain areas of the

TABLE 15.3

**Major sources and amounts (millions of tons/year)
of air pollutants emitted in the United States in 1969[a]**

Sources	Carbon monoxide	Nitrogen oxides	Hydro-carbons	Sulfur oxides	Partic-ulates	Total amount from each source
transportation	111.5	11.2	19.8	1.1	0.8	144.4
fuel combustion in stationary sources	1.8	10.0	0.9	24.4	7.2	44.3
industrial processes	12.0	0.2	5.5	7.5	14.4	39.6
solid-waste disposal	7.9	0.4	2.0	0.2	1.4	11.9
miscellaneous	18.2	2.0	9.2	0.2	11.4	41.0
total amount of each pollutant	151.4	23.8	37.4	33.4	35.2	281.2

[a] Data from the *Second Annual Report of the Council of Environmental Quality*, Washington, D. C., August 1971, p. 212.

TABLE 15.4

**Tolerance levels and relative toxicities of the primary pollutants
based on proposed air quality standards for California[a]**

Pollutant	Tolerance levels (ppm)	Tolerance levels (μg/m³)	Relative toxicity (weighting factor)
carbon monoxide	32.0	40,000	1.00
hydrocarbons	—	19,300	2.07
sulfur oxides	0.50	1,430	28.0
nitrogen oxides	0.25	514	77.8
particulates	—	375	106.7

[a] Data from *Journal of the Air Pollution Control Association*, **20**, 658 (1970).

TABLE 15.5

Total emissions by source — mass versus weighted mass basis[a]

Pollutant source	Percentage of total emissions — Mass basis	Percentage of total emissions — Weighted mass basis
transportation	51.3	16.8
fuel combustion (stationary sources)	15.8	33.0
industrial processes	14.1	26.4
solid-waste disposal	4.2	2.9
miscellaneous	14.6	20.9

[a] Calculated using data given in Tables 15.3 and 15.4.

country; on occasions, pollutant concentrations of 50–100 ppm have also been observed in highly industrialized urban areas.

One pollutant may be much more harmful or dangerous than another. Therefore, it is necessary to evaluate pollutants not only in terms of tonnage but also in terms of the relative toxicities of the pollutants. The tolerance levels and relative toxicities of the primary pollutants, based on proposed air quality standards for California, are given in Table 15.4. Multiplication of the weight of each pollutant by the appropriate weighting factors given in Table 15.4 yields conclusions somewhat different from those based only on total weights of pollutants (Table 15.5). For example, transportation, which is the main source of air pollution on the weighted basis, becomes only the third most serious source when relative toxicities are used. Because the weighting factors are somewhat arbitrary, the conclusions they support should be considered tentative at this time.

MAJOR AIR POLLUTANTS FROM TRANSPORTATION   Carbon monoxide, nitrogen oxides, and hydrocarbons   The major air pollutants from transportation are carbon monoxide (77.2%), nitrogen oxides (7.7%), and hydrocarbons (13.7%). This is understandable because gasoline, the main fuel involved in transportation, is a complex mixture of hydrocarbons. The incomplete combustion of the gasoline hydrocarbons due to insufficient O_2 or poor mixing of fuel and air in transportation devices, chiefly automobiles, produces carbon monoxide:

$$2C + O_2 \longrightarrow 2CO$$

Nitrogen oxides, on the other hand, are produced by the side reactions

$$N_2 + O_2 \longrightarrow 2NO$$
$$2NO + O_2 \longrightarrow 2NO_2$$

which take place during gasoline combustion. At temperatures of 1200–1800°C that are reached during combustion, significant amounts of NO are produced. However, the reaction of NO with O_2 to give NO_2 does not account for much NO_2, even in the presence of excess air, due to the particular temperature and concentration dependence of the rate of the NO_2-producing reaction.

Because transportation is the largest single source of carbon monoxide, nitrogen oxides, and hydrocarbons, highly populated urban areas show the highest atmospheric concentrations. Within these communities the concentrations of these three gases follow a regular daily pattern which is clearly related to human activities and traffic volume. Averages of measurements obtained at many air-monitoring stations throughout the United States indicate the following maximum CO concentrations: 115 ppm inside vehicles in downtown traffic, 75 ppm inside vehicles on expressways, 50 ppm in commercial areas, and 23 ppm in residential areas. The expected CO background in relatively clean, unpolluted air is 0.025–1.0 ppm. Peak NO concentrations of 1–2 ppm are common, but NO_2 concentrations are usually only about 0.5 ppm. Clearly, these concentrations are many times lower than those for CO. Bear in mind, however, that nitrogen oxides are more toxic than CO at equivalent concentrations (Table 15.4).

Transportation vehicles introduce enough CO into the air to more than double the CO concentration in the worldwide atmosphere every 5 years. However, the observed yearly increase in the global CO concentration has been much less than expected on this basis, indicating the existence of a natural CO removal mechanism. This removal mechanism has been found to involve a natural soil sink that is dependent on particular soil microorganisms for its operation. Although the

United States has a soil sink with a capacity to remove over three times the annual CO discharge into the atmosphere, our largest CO-producing areas often have the least amount of available soil sink. Thus, the nonuniform distribution of both atmospheric CO and the soil sink results in CO pollution at significant concentrations.

Attempts have been made to develop devices that will reduce atmospheric CO pollution to the low levels prescribed in the federal emission standards (see Table 15.9 on p. 416). Clearly, most of this effort is directed toward the automobile exhaust system. Both thermal and catalytic exhaust-system reactors (Fig. 15.9) which are able to oxidize CO to CO_2 have been developed by the nation's automobile industries. The *thermal reactor* consists of a high-temperature chamber in which the hot exhaust gases containing CO are mixed with outside air to provide sufficient oxygen to permit the conversion of CO to CO_2. The *catalytic reactor* consists of a heated chamber in which the exhaust gases containing CO are mixed with air and then passed over a bed of granular catalyst. In this process the oxidation of CO to CO_2 is accomplished at a much lower temperature than is possible in a thermal reactor. However, there are many problems still to be solved before these reactors can be expected to perform efficiently. The main problem with thermal reactors is the lack of an economical heat- and corrosion-resistant material to use in the chamber construction. The greatest problem with catalytic reactors is the lack of a durable catalytic material — a material that will not easily be deactivated or poisoned by the adsorption of materials on their surfaces. One of the most effective catalytic poisons is lead which is present in most gasolines. Lead is also a dangerous pollutant in our environment, thus offering another reason for the development of lead-free gasolines (see p. 409).

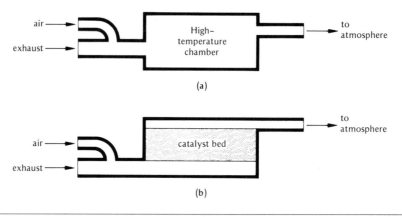

FIGURE 15.9

Exhaust system reactors: (a) thermal exhaust system; (b) catalytic exhaust system.

Natural processes are also active in removing nitrogen oxides from the atmosphere. The end product of nitrogen oxide pollution is nitric acid, which is precipitated as nitrate salts in either rainfall or dust. Both field studies and laboratory investigations have successfully linked these reaction products to the corrosion and weakening of a variety of materials. Because the primary source of nitrogen oxide pollution is the automobile exhaust, the control of nitrogen oxide pollution is being attempted through the use of catalytic reactors, similar to the ones shown in

Fig. 15.9. However, in these reactors a reduction catalyst would convert NO and NO₂ into the relatively inert and nonhazardous N_2.

Hydrocarbon pollutants from automobiles reach the atmosphere through evaporation from fuel tanks, carburetors, and spills. In addition, incomplete fuel combustion produces exhaust gases containing unreacted hydrocarbons. The control of hydrocarbon emissions from automobiles has taken the form of evaporation control and the conversion of unburned hydrocarbons in the exhaust into harmless products. Evaporation control involves the collection and return of vapors from the fuel tank and carburetor to the fuel-induction system for burning in the engine. For a possible solution to the exhaust problem, catalytic exhaust reactors that will oxidize the unreacted hydrocarbons into H_2O and CO_2 are being investigated.

Photochemical oxidants and smog The principal effects attributed to air pollution by hydrocarbons and nitrogen oxides are thought to be caused by the products resulting from atmospheric reactions of these primary pollutants in the presence of sunlight. The *secondary pollutants* thus formed are called photochemical oxidants. The term *photochemical oxidant* is used to describe an atmospheric substance produced by a chemical process that requires light (a photochemical process), which will oxidize materials not readily oxidized by oxygen. Important substances in this category are *ozone* (O_3) and the family of compounds known as

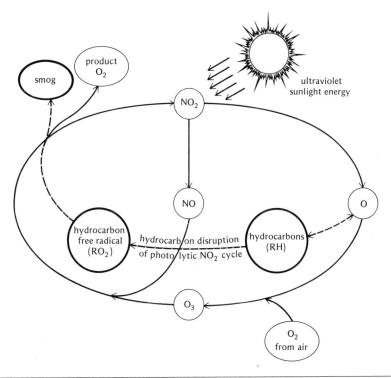

FIGURE 15.10

Photolytic cycle of NO₂ with hydrocarbon disruption. Redrawn, with modifications, from *Air Quality Criteria for Photochemical Oxidants*, National Air Pollution Control Administration Publication No. AP-63, 1970, and from *Air Quality Criteria for Nitrogen Oxides*, National Air Pollution Control Administration Publication No. AP-84, 1971. By permission from the Environmental Protection Agency, U.S. Department of Health, Education, and Welfare, Washington, D.C.

peroxyacylnitrates $\left(R-C \underset{OONO_2}{\overset{\displaystyle \parallel O}{\diagdown}} \right)$, commonly called PAN compounds. These

substances constitute the more harmful components of *smog*.

The production of photochemical oxidants results from hydrocarbon disruption of the naturally occurring NO_2 photolytic cycle—a direct consequence of an interaction between sunlight and NO_2 (Fig. 15.10). The various steps in the NO_2 photolytic cycle are

$$NO_2 + \text{sunlight} \longrightarrow NO + O$$
$$O + O_2 \longrightarrow O_3$$
$$O_3 + NO \longrightarrow NO_2 + O_2$$

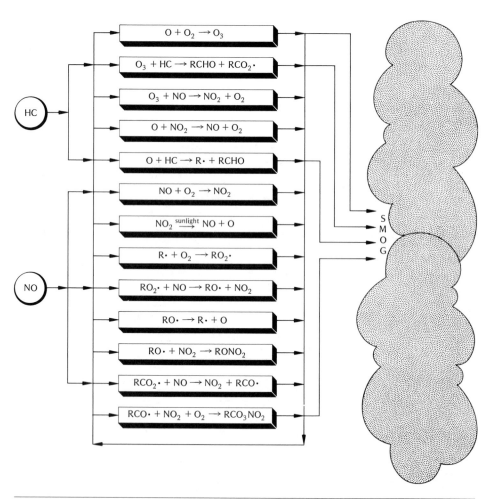

FIGURE 15.11

Reactions involved in the formation of smog from hydrocarbons (HC) and nitric oxide (NO) contaminants in the atmosphere. Reproduced, by permission, from *Chemical & Engineering News*, p. 57 (May 16, 1966). Copyright © 1966 by the American Chemical Society.

Because O_2 and NO are formed and consumed in equal quantities, atmospheric NO and NO_2 concentrations would not change if it were not for competing reactions that involve the hydrocarbons emitted from the same source (transportation) as the nitrogen oxides. The hydrocarbons interact in a way that causes the cycle to become unbalanced, so that NO is converted into NO_2 faster than NO_2 is dissociated into NO and O. For example, hydrocarbons react very rapidly with oxygen atoms to produce a reactive intermediate called a hydrocarbon free radical (RO_2). Free radicals of this type then react with a number of different substances, including NO, NO_2, O_2, O_3, and other hydrocarbons (Fig. 15.11) producing ozone and photo-chemical oxidants belonging to the PAN family.

Many observations suggest that much of the degradation of materials now attributed to weathering is actually the result of attack by photochemical oxidants such as O_3 and PAN. These photochemical oxidants are also known to cause injury to vegetation. They are the active irritants in smog. Control of smog depends on control of these secondary pollutants, which, in turn depends on control of their primary sources, namely, hydrocarbons and nitrogen oxides. Although much is known about the complex mechanisms of smog formation, relatively little progress has been made to remedy the problem. Thus, the smog problem is still with us and more and more people are suffering from respiratory diseases.

In many parts of the world, smog occurs as a result of a *temperature inversion* (Fig. 15.12). A warmer upper-air layer acts as a blanket to prevent the cooler contaminated air near the ground from rising. This concentrates the polluted air at ground level. Many of the most serious air pollution episodes have occurred during temperature inversions.

Airborne lead pollution   Ninety-eight percent of all the lead emissions into the atmosphere are generated by the combustion of gasoline additives, tetraethyl lead

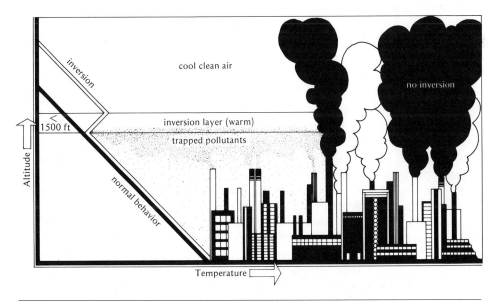

FIGURE 15.12

Temperature inversion. Redrawn, by permission, from Paul R. Ehrlich and Anne H. Ehrlich, *Population, Resources, Environment: Issues in Human Ecology,* 2nd ed., p. 124. Copyright © 1972 by W. H. Freeman and Company.

[Pb(C₂H₅)₄] and tetramethyl lead [Pb(CH₃)₄], in automobile engines. The chemical route followed by these additives, which function as antiknock agents* during gasoline combustion, results in the emission of many different lead compounds in automobile exhaust. The two principal lead compounds emitted are $PbBrCl$ and $PbBrCl \cdot 2PbO$. Halogen-containing compounds are formed because of the presence of chemical scavengers such as ethylene dibromide ($C_2H_4Br_2$) and ethylene dichloride ($C_2H_4Cl_2$) which are added to combine chemically with any lead compounds left in the engine as a result of combustion of the antiknock agents.

The amount of lead in the air has increased markedly as the result of human activities, mainly gasoline combustion for automobile transportation. Atmospheric lead concentrations in urban areas (1.11 $\mu g/m^3$) are 5–50 times that of nonurban areas (0.022–0.21 $\mu g/m^3$). However, not all inhaled lead is absorbed or retained by the human body. The toxic action of lead in the body has been traced in part to an enzyme inhibition of the formation of hemoglobin as a result of a strong interaction between Pb^{2+} and the sulfur groups of some amino acids of the enzyme. However, this may not be the only toxic effect. Because the major source of lead pollution in air is automobile exhaust, an obvious solution to this problem is not to use lead additives in gasoline.

MAJOR AIR POLLUTANTS FROM STATIONARY FUEL COMBUSTION

Sulfur oxides and nitrogen oxides   The major air pollutants from fuel combustion (stationary sources) are sulfur oxides (55.1%) and nitrogen oxides (22.6%). The role of nitrogen oxides in the atmosphere has already been discussed. The combustion of any sulfur-containing material produces two oxides of sulfur:

$$S + O_2 \longrightarrow SO_2$$
$$2SO_2 + O_2 \longrightarrow 2SO_3$$

Even in the presence of excess air, mostly SO_2 is formed, and SO_3 accounts for only 1–10 percent of the total sulfur oxides produced. However, gaseous SO_3 can exist in the air only if the moisture content of the air is very low. When sufficient moisture is present, the SO_3 dissolves in the water to form droplets of H_2SO_4:

$$SO_3 + H_2O \longrightarrow H_2SO_4$$

Normally, sufficient moisture is present in the air so that H_2SO_4 mist rather than SO_3 is found in the atmosphere. Because the amount of atmospheric H_2SO_4 is greater than can be justified on the basis of primary SO_3 emissions alone, it has been suggested that some of the atmospheric SO_2 is converted to SO_3 (and then to H_2SO_4) by catalytic oxidation in the presence of sunlight. The catalytic effects of nitrogen oxides, pollutants also found in the atmosphere, are well known and were used for many years in the industrial production of H_2SO_4 by the lead-chamber process.

Maximum sulfur oxide concentrations in the vicinity of 1 ppm are quite common and may reach 2 ppm occasionally in urban atmospheres. These levels are below those needed to produce detectable effects in animals and healthy humans but well above the 0.5 ppm level that can cause injury to many plants. However, individuals suffering from chronic diseases of the respiratory and cardiovascular systems are highly susceptible to prolonged SO_2 exposure at 0.1–0.5 ppm levels characteristic of air pollution episodes. Also, much damage to building materials is caused by the sulfuric acid mists in polluted environments.

---

* Substances added to gasoline to reduce small explosions that are heard as ''knocks'' in the internal combustion engine.

Because fuel combustion in stationary processes is responsible for 73 percent of all man-made sulfur oxide pollution, much of the effort toward solving sulfur oxide pollution has involved processes for removing sulfur from fuels before combustion. The most commonly used fuel in the United States is coal and it contains an average of about 2.7 percent sulfur. Much research has gone into the process of converting this coal into gas from which the sulfur-containing compounds are removed before being used as fuel. Processes for the removal of sulfur from residual oils are also receiving attention, and some of these processes are being put into operation in the petroleum industry. However, it is the removal of SO_2 from flue gases that is receiving most attention at this time. One process involves the injection of limestone ($CaCO_3$) into the combustion zone of the furnace in power plants and then scrubbing the flue gas by passing it through a slurry of milk of lime whereby at least 90 percent of the SO_2 is converted to calcium sulfate or gypsum ($CaSO_4$). A major disadvantage of the method is the huge quantities of waste gypsum produced — approximately 1 ton of gypsum is produced for each 3 tons of coal burned. Other processes include (a) scrubbing by NH_3, $NaOH$, Na_2SO_3, or seawater to convert the SO_2 to sulfites (SO_3^{2-}) and (b) treating with H_2S or CO to convert the SO_2 to sulfur. In some cases, the by-products have high purity and are marketable (for example, sulfur) or can be converted into marketable forms [such as H_2SO_4 and $(NH_4)_2SO_4$].

Airborne mercury pollution In addition to mercury pollution of the water and soil environment resulting from direct usage of the element and its compounds (see Chapter 14), mercury vapor is released into the atmosphere during the combustion of fossil fuels (coal and oil). Estimates of the average Hg content in fossil fuels vary from about 0.1 to 3.3 ppm. This amount seems small, but on a worldwide basis the annual fossil-fuel consumption of about 5 billion tons with, say, an average content of 0.5 ppm Hg could release 2500 tons of mercury. A similar quantity of mercury vapor may escape into the air during the roasting of naturally occurring sulfide ores of mercury. Due to the volatility of mercury, much greater amounts (as great as 100,000 tons) may be released to the atmosphere from earth that is freshly exposed in thousands of earth-moving projects.

Mercury vapor in air exists normally at levels of 10^{-9} g/m^3 or less. However, it has a great capacity for diffusion through the lungs into the blood and then into the brain and as a result may cause serious central nervous system damage.

MAJOR AIR POLLUTANTS FROM INDUSTRIAL PROCESSES **Particulates, sulfur oxides, and carbon monoxide** The major air pollutants from industrial processes are carbon monoxide (30%), sulfur oxides (19%), and particulates (36%). The role of carbon monoxide and sulfur oxides in the atmosphere has already been discussed. Industrial processes that contribute substantially to CO pollution are the iron and steel, petroleum, and paper industries. Industrial processes that contribute to sulfur oxide pollution are the metallurgical and petroleum industries. In Chapter 8, it was shown that many useful elements such as Cu, Zn, Hg, and Pb occur naturally in the form of sulfide ores. During the extraction of the metals, the sulfur is usually converted into oxides, some of which escape into the atmosphere. Recently, however, the metallurgical industry has begun to use catalytic oxidation methods for converting SO_2 gas into sulfuric acid and thus made some progress toward removing sulfur oxide pollution from the atmosphere.

About 41 percent of the total airborne particulates originate from industrial processes as small solid particles and liquid droplets. Fuel combustion in stationary sources also makes a significant contribution. For example, a single source, coal

TABLE 15.6
Fly ash emissions from a variety of coal combustion units[a]

Component (% calculated as)	% of fly ash
carbon	0.37–36.2
iron (as Fe_2O_3 or Fe_3O_4)	2.0 –26.8
magnesium (as MgO)	0.06– 4.77
calcium (as CaO)	0.12–14.73
aluminum (as Al_2O_3)	9.81–58.4
sulfur (as SO_2)	0.12–24.33
titanium (as TiO_2)	0– 2.8
carbonate (as CO_3^{2-})	0– 2.6
silicon (as SiO_2)	17.3 –63.6
phosphorus (as P_2O_5)	0.07–47.2
potassium (as K_2O)	2.8 – 3.0
sodium (as Na_2O)	0.2 – 0.9
undetermined	0.08–18.9

[a] Data from U.S. Department of Health, Education, and Welfare, *Air Quality Criteria of Particulate Matter,* 1969, p. 27. (NAPCA Publication No. AP-49).

TABLE 15.7
Trace metals that may pose health hazards in the environment[a]

Element	Sources	Health effects
antimony	industry	shortened life-span in rats
arsenic	coal, petroleum, detergents, pesticides, mine tailings	may cause cancer
beryllium	coal, industry, rocket fuels, nuclear power reactors	acute and chronic system poison; cancer
cadmium	coal, zinc mining, water mains and pipes, tobacco smoke	cardiovascular disease and hypertension in humans suspected; interferes with zinc and copper metabolism
germanium	coal	little innate toxicity
lead	auto exhaust, paints	brain damage, convulsions, behavioral disorders, death
mercury	coal, electrical batteries, industries	nerve damage and death
nickel	diesel oil, residual oil, coal, tobacco smoke, chemicals and catalysts, steel and nonferrous alloys	lung cancer
selenium	coal, sulfur	may cause dental caries; carcinogenic in rats; essential to mammals in low doses
vanadium	petroleum, chemicals, catalysts, steel and nonferrous alloys	probably no hazard at current levels
yttrium	coal, petroleum	carcinogenic in mice over long-term exposure

[a] Data from *Chemical and Engineering News,* **49,** 30 (July 19, 1971).

combustion, produces nearly 9 million tons of a variety (Table 15.6) of particulates each year in our country. Because many different polluting substances are found in the form of particulates, only their physical properties (and not chemical properties) can be correlated. The most important physical property is the particle size. Relationships exist between the source and size of airborne particulates: Particulates with diameters between 0.1 and 1μ ($1\ \mu = 10^{-4}$ cm) are primarily combustion products; those with diameters between 1 and 10 μ are process dusts, soil, and combustion products from local industries; and those larger than 10 μ in diameter are produced by mechanical processes such as wind erosion, grinding, spraying, and pulverizing of materials. Figure 15.13 shows the range of sizes characteristic of some selected airborne particulates. Note that the word *aerosol* is commonly used to refer to any small particle in the air. All these particles have a lifetime in the suspended state of between a few seconds and several months. This lifetime depends on the settling rate which, in turn, depends on the size and density of the particles and the turbulence of the air.

Particulates in the atmosphere may exert a definite influence on the amount and type of sunlight that reaches the earth. Further, they may act as condensation sites and influence the formation of clouds, rain, and snow. Airborne particles, including soot, dust, fumes, and mist, can cause a wide range of damage to materials and plants. Particulate matter that enters the human respiratory system can exert a toxic effect due to the toxic nature of the particles themselves by interfering with the removal of other more harmful material from the respiratory tract. However, intrinsically toxic particulates are found in the atmosphere only in trace amounts. The

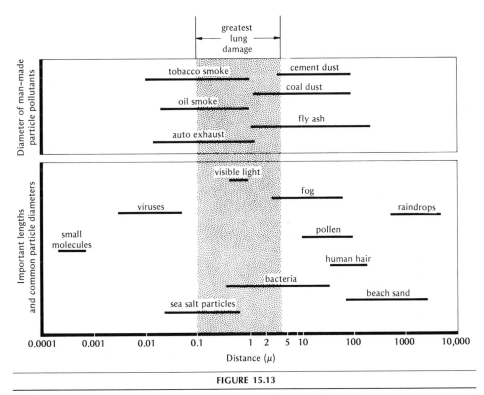

FIGURE 15.13

Typical sizes of airborne particulates. From J. Calvin Giddings, *Chemistry, Man and Environmental Change*, p. 244. Copyright © 1973 by J. Calvin Giddings. Reprinted with permission from Harper & Row, Inc.

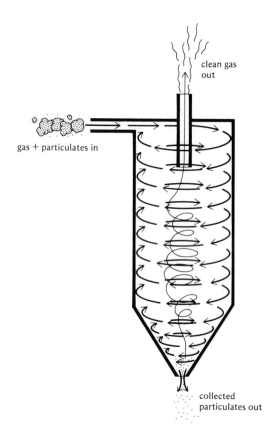

clean gas
out

gas + particulates in

collected
particulates out

FIGURE 15.14

A typical cyclone collector.

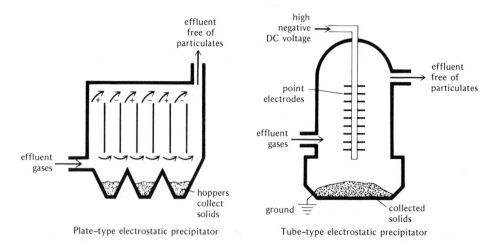

effluent
free of
particulates

effluent
gases

hoppers
collect
solids

Plate–type electrostatic precipitator

high
negative
DC voltage

point
electrodes

effluent
gases

effluent
free of
particulates

ground

collected
solids

Tube–type electrostatic precipitator

FIGURE 15.15

The two basic types of electrostatic precipitator. From Spencer L. Seager and H. Stephen Stoker, *Environmental Chemistry: Air and Water Pollution*, p. 81. Copyright © 1972 by Scott, Foresman and Company. Redrawn by permission of the publisher.

study of the effects of metallic particulates is an area of considerable interest and activity at this time (Table 15.7). Airborne lead and mercury pollutions have been discussed earlier.

Techniques presently used for controlling particulate emissions are dependent on particle size. Particles with a diameter greater than 50 μ are generally removed using gravity settling tanks or porous media (filters) that allow the gas to flow through. Smaller particles are removed using cyclone collectors or electrostatic precipitators. The *cyclone collector* (Fig. 15.14) makes use of the fact that the air flowing in a tight circular spiral produces a centrifugal force on suspended particles, forcing them to move outward through the air stream to a wall where they are collected. The *electrostatic precipitator* (Fig. 15.15) is based on the fact that particles may also be electrically charged and that a collecting surface that bears an electric charge of the opposite sign, therefore, attracts them. Two basic design types, the plate and the tube, are presently used. The plate type consists of a series of parallel metal plates with large electric potentials that are of opposite sign on every other plate. In the tube type, a negatively charged wire runs down the center of a tube that is grounded or positively charged (Fig. 15.15). Such units have a 95 percent removal efficiency for particles in the diameter range of 5–20 μ and are used on a very large scale, notably for reducing smoke from power plants that burn fossil fuels.

EFFECTS OF AIR POLLUTION; PROBLEMS AND ISSUES IN THE CONTROL OF AIR POLLUTION Although direct evidence of adverse effects from polluted air began to accumulate around the beginning of the fourteenth century when coal

TABLE 15.8
Air pollutants, their receptors, and effects[a]

Air pollutants	Receptors	Effects
particulates, sulfur oxides, fluorides, polycyclic organic matter, Pb, Be, As, Cd, Zn, Hg, Se, Cl₂, pesticides, radioactive substances	animals	central nervous system effects, respiratory problems.
particulates, sulfur oxides, oxidants, CO, hydrocarbons, nitrogen oxides, fluorides, Pb, polycyclic organic matter, odors (incl. H₂S), asbestos, Be, HCl, Cl₂, As, Cd, Va, Ni, Mn, Zn, Cu, Ba, B, Hg, Se, Cr, pesticides, radioactive substances, aeroallergens.	humans	eye irritation, respiratory system irritation and damage, oxygen deficiency in blood, esthetic effects such as low visibility and unpleasant odors
particulates, sulfur oxides, oxidants, nitrogen oxides, hydrogen sulfide, HCl, Cl₂, Hg, Cr.	materials	corrosion, loss of strength, loss of color, weathering, soiling of clothes, homes, public buildings and monuments.
particulates, sulfur oxides, oxidants, hydrocarbons, nitrogen oxides, fluorides, Pb, Be, HCl, Cl₂, As, Hg, pesticides	vegetation	leaf spotting, leaf dropping, tissue death and decay, decreased rate of photosynthesis and decreased crop yields

[a] Data from the *Second Annual Report of the Council on Environmental Quality*, Washington, D. C., 1971, p. 105.

was first used extensively, only in the past few decades has man begun to realize the extent and complexity of air pollution effects. For a number of reasons, the determination of pollution effects is difficult and imprecise even today. For example, pollutants are numerous, varied, and in many cases difficult to detect. In many cases it is nearly impossible to determine the precise degree of exposure of a given receptor (human, plant, etc.) to a specific pollutant. What is worse is that some pollutants do not in themselves cause problems but may be dangerous in combination with other pollutants. Despite these limitations, the effects and the receptors in which they can occur have been identified for a wide variety of air pollutants. This information is summarized in Table 15.8. Notice the great diversity among the types of particles in the air, which include radioactive substances (see Chapter 6), aeroallergens (see Chapter 12), and pesticides (see Chapter 12).

TABLE 15.9
United States air quality standards (to be reached by 1975)[a]

Air pollutant	Primary standard	Secondary standard
carbon monoxide	10 mg/m³ (9 ppm) as a maximum 8-hr concentration not to be exceeded more than once a year. 40 mg/m³ (35 ppm) as a maximum 1-hr concentration not to be exceeded more than once a year	same as primary standard
nitrogen oxides	100 μg/m³ (0.05 ppm) annual arithmetic mean	same as primary standard
hydrocarbons	160 μg/m³ (0.24 ppm) as a maximum 3-hr concentration (6 to 9 A.M.) not to be exceeded more than once a year	same as primary standard
photochemical oxidants	160 μg/m³ (0.08 ppm) as a maximum 1-hr concentration not to be exceeded more than once a year	same as primary standard
sulfur oxides	80 μg/m³ (0.03 ppm) annual arithmetic mean; 365 μg/m³ (0.14 ppm) as a maximum 24-hr concentration not to be exceeded more than once a year	60 μg/m³ (0.02 ppm) annual arithmetic mean; 260 μg/m³ (0.1 ppm) as a maximum 24-hr concentration not to be exceeded more than once a year; 1300 μg/m³ (0.5 ppm) as a maximum 3-hr concentration not to be exceeded more than once a year
particulate matter	75 μg/m³ annual geometric mean; 260 μg/m³ as a maximum 24-hr concentration not to be exceeded more than once a year	60 μg/m³ annual geometric mean; 150 μg/m³ as a maximum 24-hr concentration not to be exceeded more than once a year

[a] Data from *U. S. Environmental Protection Agency Report*, Washington, D. C., October, 1971.

From time immemorial, airborne pollen grains and microorganisms such as bacteria, fungi, molds, or spores, have caused many effects that are detrimental to man, including hay fever, bronchial asthma, fungous infections, and airborne bacterial diseases. Man-made pollutants (Table 15.8) have given added dimensions to these age-old detrimental effects. More and more people are suffering from emphysema, bronchitis, asthma, and lung cancer. General untoward symptoms and irritations, including overall discomfort, nervous impairment, eye irritation, and unpleasant reactions to offensive odors have become the order of the times. What has been done about it? The national strategy at this time is expressed in regulations and timetables for the prevention, abatement, and control of pollutants. Under the Clean Air Act of 1970, the U.S. Environmental Protection Agency has established *national air quality standards* to be reached by 1975. For the hazardous air pollutants, there exist primary standards that are intended to protect health and secondary standards that are intended to protect public welfare (Table 15.9). Clearly, the primary impact of control regulations, especially emission limits, will be on the readily identifiable sources of pollution.

Should air pollution alone be controlled or should the sources of air pollution be eliminated and/or replaced? It is up to you to decide and then urge the passage and enforcement of the best pollution control and abatement legislation. At this time one thing is certain: The control and abatement of air pollution must be based on the reduction of the emission of air pollutants at their source by conversion to cleaner fuels or combustion processes. We have already seen that control technology exists for many pollutants, particularly at the present level of air pollution. Unfortunately, the problem goes beyond the question of what control method to apply at the source. The problem involves social and economic factors as well. Are we willing to accept the complex and serious social consequences from the cultural disruptions which may be the price for improving our air environment?

QUESTIONS

15.21 What is meant by air pollution? How does the composition of clean dry air differ from that of polluted air?

15.22 Distinguish between natural and man-made air pollution.

15.23 Why is the United States concerned with air pollution control much more than the other countries of the world?

15.24 What are primary pollutants? List their important sources.

15.25 Discuss the significance of using toxicities as a basis for evaluating pollutants.

15.26 List the major air pollutants arising from the use of automobile transportation. What approach is being investigated to control these pollutants?

15.27 Name two important constituents of photochemical smog. What single term is used to describe both of these substances? What are the effects caused by these substances on the environment? What has been done toward minimizing these effects?

15.28 What is meant by a temperature inversion? Describe the processes that lead to it.

15.29 Give an example of the reaction by which an excessively high level of atmospheric NO_2 might inhibit photochemical smog formation.

15.30 What single commonly used substance is the source of the widest variety of possibly hazardous trace metal pollutants?

15.31 List the major air pollutants from stationary fuel combustion and industrial processes. What are their effects on the environment?

15.32 Why is NO_2 a more significant species than SO_2 insofar as taking part in atmospheric chemical reactions?

15.33 What would be the most likely fate of ammonia introduced into an atmosphere also receiving a high level of SO_2 pollution?

15.34 How do the proposed control methods for sulfur oxide pollution differ from those projected for carbon monoxide and nitrogen oxides?

15.35 Why is hydrogen sulfide generally a less serious air pollutant than sulfur dioxide despite the fact that more hydrogen sulfide enters the global atmosphere each year?

15.36 "Particulate air pollutants are varied and cannot easily be classified." Discuss this statement.

15.37 What is the principle of the electrostatic precipitator?

15.38 What effects do particulate pollutants exert on human health?

15.39 Suggest a method by which you can determine the extent of particulate pollution in your locality.

15.40 In what receptors are air pollutants generally known to cause effects? Give an example of a typical problem in each receptor.

15.41 Analyze the information given in Table 15.8 and relate each effect to specific air pollutants in as many cases as possible.

15.42 Discuss the advantages and disadvantages of controlling air pollution rather than eliminating and/or replacing the sources of air pollution.

SUGGESTED PROJECT 15.1

Automobile emissions (or adjustin' combustion)

Examine your car to determine the sources of automobile pollutant emissions. Then examine the antipollution devices. Suggest methods for further reduction of emissions.

SUGGESTED PROJECT 15.2

The contaminated atmosphere (or Who cares for air?)

The types of pollutants that foul the air over a particular locality are determined by the prevailing industrial and domestic activities. Prepare a list of the sources of air pollutants in your community. If information is locally available, find the composition and amount of air pollutants from each source. Formulate a program of air quality control for your area, keeping in mind the interstate transport of air pollutants.

16

Energy and our environment

A limited view of history has hypnotized them [energy critics] into seeing energy only in terms of a means of ruthlessly extracting resources from nature, using them foolishly (and often unjustly) and then dumping them back into nature in amounts and places where she cannot handle them. The immediate reaction to all this is simply —stop it! . . . Offering that approach as a panacea is unrealistic and unimaginative.

GLENN T. SEABORG (1912–)
American chemist

In Chapter 3 we learned that energy is one of the basic qualities of our environment and that it can be stored and transformed from one form to another, but it cannot be created or destroyed. If energy is neither created nor destroyed, why do we need fear energy shortages? Simply because when energy is transformed from one form to another, it is transformed from a more useful form to a less useful form. Consider, for example, the familiar automobile engine in which the chemical energy of gasoline is transformed first into heat energy and then into mechanical energy. We all know that only a part of the heat energy is converted to the energy of motion of the automobile and that the rest is wasted. This is true of other energy transformations as well (Fig. 16.1). In general, whenever energy is changed in form, some energy always ends up as heat that passes on to (cooler) objects at some lower temperature. This low-temperature heat cannot in practice run machines, generate electricity, or perform other useful work. Thus, it can be said that energy runs downhill from higher to lower grades. In consequence, the crisis, if any, is one not of energy, but of *useful energy* — energy in a form that can be utilized by humans.

It is estimated that an average person today consumes approximately 200,000 cal of energy per day compared to the 2000 cal that our prehistoric ancestors used to keep alive. This increase in energy

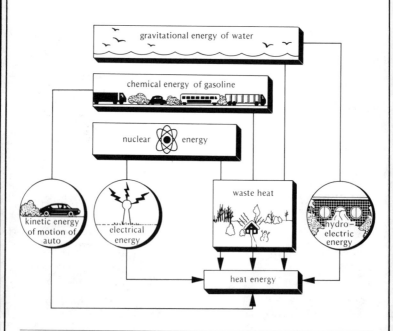

FIGURE 16.1

Some energy pathways.

consumption is clearly a consequence of a technological age in which we use myriads of energy-dependent processes for our comforts. Whatever it is, an electrical generator or an automobile engine, it produces irretrievable heat because of the friction of its moving parts. None of these engines runs without using energy derived from our natural resources. As our population has grown over the years, so have our energy demands, energy-producing devices, and useless heat. And, it is impossible to build perpetual motion machines that will run forever producing useful energy from its own resources alone. The principle of energy conservation says that no machine can give out energy forever unless it is taking it in somewhere.

This chapter is devoted to a critical study of our energy conversion devices, energy demand, and energy resources. We shall also examine the environmental problems caused by the production, processing, and utilization of our energy resources.

SOME ENERGY-CONVERSION DEVICES

At the present time the common, large-scale, energy-conversion devices are the steam engine, the steam turbine operating in conjunction with an electrical generator, and the internal combustion engine. All these devices use fuels such as coal, natural gas, and petroleum for the thermal energy that they convert into mechanical energy. However, they are different devices in that they derive their operating powers differently.

In the conventional *steam engine,* steam is generated in a water boiler by burning coal or oil beneath it and led through an insulated pipe to push a sliding piston that is able to move back and forth inside of a chamber (Fig. 16.2). The piston is connected to a wheel that, on revolving, provides mechanical power and causes the piston to move back and forth in a reciprocating fashion. The *steam turbine* (Fig. 16.3) operates by means of high-pressure superheated (500–600°C) steam impinging on blades mounted around the edges of wheels fixed on an axis about which they can rotate. These wheels, called *rotors,* turn at very high speeds and are joined to electrical generators for power production. In an actual commercial steam turbine, there are many rotors on a single shaft, and these increase in diameter from the steam inlet to the exhaust. The actual efficiencies of steam turbines are relatively high (30–40%) compared to steam engines (5–10%).

Through the use of electrical generators, the motive power (mechanical energy) of the steam turbine is transformed into electric power. *Electrical generators* are based on the principle that a magnetic field can set electrons in motion. Because the flow of electrons in a wire constitutes an electric current, it is clear that a wire that is part of a closed circuit and that is moving through a magnetic field will produce an electric current. The large-scale generation of electricity is based on the use of steam turbines to rotate the copper coils in electrical generators. The overall efficiency with which mechanical energy is thus converted to electrical energy is about 30 percent.

Unlike the steam engine or turbine, the *internal combustion engine* derives its operating power directly from the combustion of fuel inside it. These engines include the gasoline engine, the diesel engine, the jet engine, and various types of rockets. *Gasoline engines* operate as the result of forces generated by explosions of gasoline–air mixtures inside of cylinders with movable pistons. Figure 16.4 illustrates the most common type of gasoline engine. It operates in a four-stroke cycle: During the first stroke a mixture of gasoline vapor (chiefly C_8H_{18}) and air is drawn into the cylinder; during the second stroke the mixture is compressed; during the third stroke the mixture is ignited by a spark (from a spark plug) and the resulting

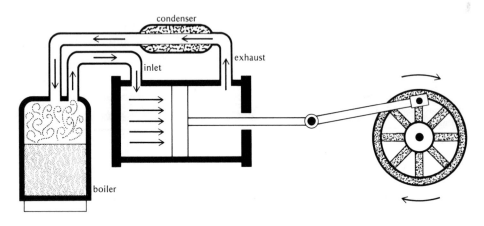

FIGURE 16.2

Schematic representation of a conventional steam engine. From Robert W. Medeiros, *Chemistry: An Interdisciplinary Approach*, p. 456. Copyright © 1971 by Litton Educational Publishing, Inc. Redrawn with permission of Van Nostrand Reinhold Company.

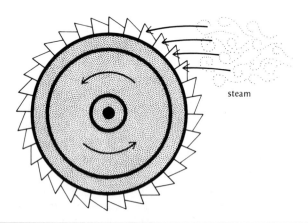

FIGURE 16.3

Simplified steam turbine rotor. From Robert W. Medeiros, *Chemistry: An Interdisciplinary Approach*, p. 458. Copyright © 1971 by Litton Educational Publishing, Inc. Redrawn with permission of Van Nostrand Reinhold Company.

explosion drives the piston back; during the fourth stroke the gaseous products of the explosion are vented to the outside and the engine is ready to begin a new cycle. In the gasoline engines used in automobiles, the wheel to which the piston is connected is, in turn, attached to a drive shaft that ultimately transmits motive power to the wheels of the automobile. In the final analysis, the power required for moving an automobile is derived from the pressure exerted by hot, rapidly expanding gases against a movable piston.

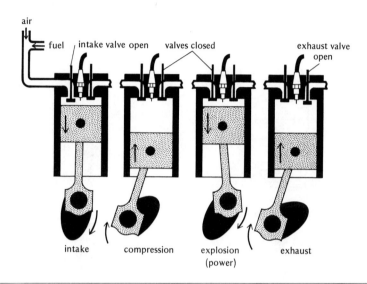

FIGURE 16.4

The four-cycle gasoline engine. From Charles W. Keenan and Jesse H. Wood, *General College Chemistry*, 3rd ed., p. 680. Copyright © 1957 Harper & Row, Inc.; copyright © 1961, 1966 by Charles William Keenan and Jesse Hermon Wood. Redrawn with permission from Harper & Row, Inc.

All *jet engines* and *rockets* operate on the principle that *every action results in an equal and opposite reaction.* In a typical turbojet engine (Fig. 16.5a) that is used in airplanes, air is drawn into the front opening of the engine and is compressed by a screw-type air compressor attached to the same shaft as a turbine in the rear of the engine. The compressed air passes into a combustion chamber into which kerosene (chiefly $C_{12}H_{26}$) is injected. As a result of the combustion, there is an appreciable increase in the rate of gas outflow through the exhaust and in consequence in the thrust of the jet engine. The chief difference between rockets and jet engines is that the latter continuously scoop in their oxidizer (oxygen) from the atmosphere, whereas the former carry a self-contained quantity of oxidizer (liquid oxygen) (Fig. 16.5b). Thus, jet engines must operate within the earth's atmosphere, in contrast to rockets that may continue to perform anywhere in outer space.

Whereas the internal combustion engine supplies the motive power needed for transportation, the steam turbine in conjunction with rotating copper coils in an electrical generator supplies most of the electric power used for a variety of purposes in everyday life. As already pointed out, both these devices use the combustion of hydrocarbon fuels which, unfortunately, contribute heavily to air pollution. If technological breakthroughs are achieved in tapping energy from natural resources without causing environmental pollution, other energy-conversion devices may become important. Specific devices for consideration at this time are the MagnetoHydro-Dynamic (MHD) generators, fuel cells, and solar batteries. Nuclear devices are considered later in the chapter.

Magnetohydrodynamics is a way of making electricity by substituting a hot, flowing ionized gas (called *plasma*) for the rotating copper coils in an electrical

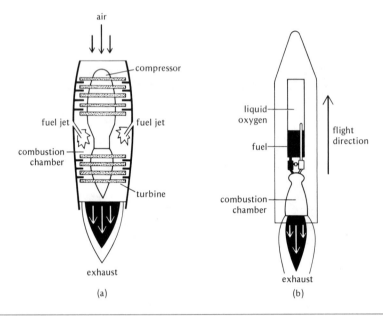

FIGURE 16.5

Cross-sectional diagrams of (a) turbojet engine and (b) liquid fuel rocket. From Robert W. Medeiros, *Chemistry: An Interdisciplinary Approach*, pp. 461 and 462. Copyright © 1971 by Litton Educational Publishing, Inc. Redrawn with permission of Van Nostrand Reinhold Company.

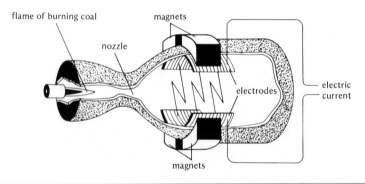

FIGURE 16.6

Cross-sectional diagram of an MHD generator. Adapted from *National Geographic Magazine,* **142** (5), p. 673 (November 1972), with permission of the National Geographic Society.

generator. In such a generator (Fig. 16.6) the hot gases from the combustion of coal are used to generate a metal plasma that jets through a nozzle at supersonic speed (1000 mi/hr) and streams between magnets yielding an electric current. The current can be taken off by electrodes extending into the plasma. The efficiency of the process is about 50 percent, giving a significant saving in fuel over a conventional electrical generator system that has only 30 percent overall efficiency. MHD generators are just beginning to appear in trial industrial service.

Fuel cells are rapidly growing in importance as a means for generating electric power. When the Apollo astronauts went to the moon, their spacecraft used fuel cells to convert hydrogen and oxygen into electricity. The principle behind their operation is the occurrence of a redox reaction that involves a transfer of electrons (see Chapter 5). Consider, for example, a hydrogen–oxygen fuel cell as shown in Fig. 16.7. Here, the two gases, hydrogen and oxygen, are led into the cell where each comes in contact with porous carbon electrodes separated by an electrolyte such as potassium hydroxide. The electrodes contain catalysts embedded in the carbon that cause the following reactions to proceed at convenient rates:

$$2H_2 + 4OH^- \longrightarrow 4H_2O + 2e^- \qquad \text{(anode)}$$
$$O_2 + 2H_2O + 4e^- \longrightarrow 4OH^- \qquad \text{(cathode)}$$

$$2H_2 + O_2 \longrightarrow 2H_2O \qquad \text{(overall)}$$

Note that the hydroxide ion is produced at the cathode and is consumed at the anode. Therefore, the KOH used as the electrolyte is not depleted; however, there

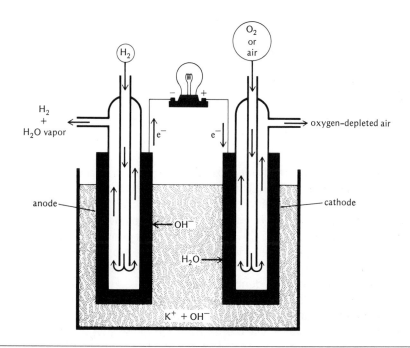

FIGURE 16.7

A hydrogen–oxygen fuel cell. Adapted, by permission, from David O. Johnston, John T. Netterville, James L. Wood, and Mark M. Jones, *Chemistry and the Environment*, p. 115. Copyright © 1973 by W. B. Saunders Company.

must be a high concentration of OH⁻ ions at the anode while the cell is in operation. It is possible to maintain this concentration of ions as the cell produces electricity by diffusion of OH⁻ ions from the cathode. Fuel cells in which acid electrolytes are used produce hydrogen ions at the anode and consume hydrogen ions at the cathode. In addition to hydrogen, which may be used in cells with either alkaline or acid electrolytes, other fuels may also be used, for example ammonia with alkaline electrolytes and hydrocarbons with acid electrolytes. In fact, fuel cells that utilize a great number of different redox reactions have been constructed. The most outstanding advantage is their high (sometimes up to 80%) efficiency of conversion of chemical energy to electrical energy. A fuel cell system based on the hydrogen from natural gas and the oxygen from air can produce water as well as electric power with little or no environmental pollution.

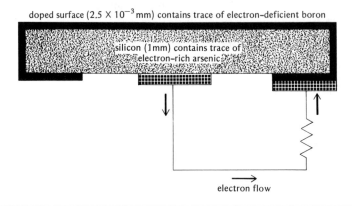

doped surface (2.5 X 10⁻³ mm) contains trace of electron–deficient boron

silicon (1mm) contains trace of electron–rich arsenic

electron flow

FIGURE 16.8

The Bell solar battery (silicon photodiode). Adapted, by permission, from David O. Johnston, John T. Netterville, James L. Wood, and Mark M. Jones, *Chemistry and the Environment*, p. 129. Copyright © 1973 by W. B. Saunders Company.

Solar cells are becoming increasingly important in space exploration and satellite communication systems. They convert sunlight directly to electric power. The principle behind their operation is the photoconductivity of certain substances. A typical solar cell built by Bell Telephone scientists consists of a wafer of two layers of almost pure silicon (Fig. 16.8). The lower thicker layer contains a trace of electron-rich arsenic and the upper thinner layer a trace of electron-deficient boron. Even without sunlight, the "extra" electrons of arsenic diffuse into the electron-deficient boron, driven by the strong tendency to pair electrons. However, when sunlight strikes the boron layer, electrons in the boron layer are unpaired and the freed electrons are repelled to the arsenic layer and into the external circuit. This pulls electrons into the boron layer via the external circuit so that a balance between electron pairing and electron repulsion is maintained. About 10 percent of the energy received on the cell surface can thus be converted into electrical energy. This is not much, but the advantage is that such a cell will keep on generating electricity indefinitely merely by lying in sunlight.

QUESTIONS

16.1 List three different energy transformations and illustrate the fact that all energy tends to become environmental heat.

16.2 Justify the statement that "energy runs downhill."

16.3 Name two important energy-conversion devices that are in use where you live and discuss the mechanisms of their operation.

16.4 What are the basic differences between the jet engines used for our airplanes and the rockets for our Apollo spacecrafts?

16.5 What is the principle of the MHD generator? Is it preferable to a conventional electrical generator?

16.6 Distinguish between fuel cells and solar cells.

16.7 Describe a fuel cell and explain how it functions.

16.8 Explain the operating principle of a solar battery.

ENERGY DEMAND AND RESOURCES

The sun is our natural source of energy. We all know that every living organism on the earth owes its ability to function and, indeed, its very existence to energy that is directly or indirectly derived from the sun. How solar energy is produced is an old question. Unlike our earth, the sun is a sphere of incandescent gas (sometimes called a plasma) nearly 1 million mi in diameter. The massive energy output of the sun is thought to be the result of a self-sustaining thermonuclear reaction in which hydrogen nuclei are fused to form helium nuclei (see Chapter 6):

$$4{}_1^1H \longrightarrow {}_2^4He + 2{}_1^0e + energy$$

It is estimated that the sun loses 6 million tons of mass each second in liberating radiant energy. Yet its total mass is so great that it can be expected to continue at this rate for billions of years.

Our earth, which is located nearly 100 million mi from the sun, intercepts an incredibly tiny fraction (about 3 ten-millionths) of this radiant energy. This minute fraction in itself is about 100,000 times greater than the entire world's presently installed electric power-generating capacity. A still smaller fraction of this solar radiation is captured by the green leaves of plants and converted by *photosynthesis* into the energy in wood, corn, wheat, sugar cane, and other natural products. This chemically stored solar energy becomes the essential biological energy source, supplied in the form of *food*, for humans as well as animals (see Chapter 11).

On a geological time scale, we have the fossil fuels that are the subterranean remains of green plants and animals that once grew and then were buried in sedimentary sands, muds, and limes, under conditions of incomplete oxidation. Our present supply of fossil fuels includes coal, oil, natural gas, oil shale, and tar sands. Unlike the renewable food energy from the sun, via the intermediary of the green plants, the earth's deposits of fossil fuels are finite in amount and nonrenewable during time periods of less than millions of years.

To say that our consumption of energy resources is growing exponentially is to state the obvious (Fig. 16.9). The world's present energy requirements per year are about 10^{18} kcal. The United States with 6 percent of the world's population uses 35 percent of this energy. About 25 percent of the energy resources used each year in the United States goes for production of electricity; about 30 percent for mining, smelting, manufacturing, and other industrial processes; about 25 percent for fueling our transport vehicles; and, about 20 percent for heating and cooking purposes. Today, coal, oil, and natural gas supply 90 percent of our total energy needs and 80 percent of all our electricity (Table 16.1). Only 15 percent of the electricity comes from water power and 3–4 percent from nuclear power. Over the last two decades the rate of depletion of our fossil energy resources has more than doubled due to continuous everincreasing withdrawals. Thus, our immediate concern at this time is with the nonrenewable energy resources such as fossil fuels. Once the fossil fuel picture is clear, we shall examine some of the other possible energy resources for the future.

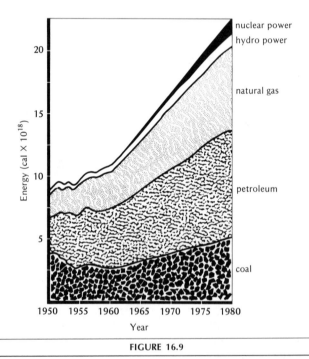

FIGURE 16.9

The increasing consumption of major energy resources in the United States (projections beyond 1970). Adapted, by permission, from David O. Johnston, John T. Netterville, James L. Wood, and Mark M. Jones, *Chemistry and the Environment*, p. 111. Copyright © 1973 by W. B. Saunders Company.

FOSSIL ENERGY Of the fossil fuels, *coal* is in greatest supply. However, most of it is "unclean" and its combustion leads to air pollution, especially from sulfur oxides (see Chapter 15). Improvement in coal preparation involves the removal of sulfur (present mostly as iron sulfides) either from coal before combustion or from the process gases after combustion. Various methods that involve conversion or regeneration are available for stack-gas sulfur removal. Characteristic of the conversion-type process is the wet carbonate scrubbing that produces calcium or magnesium sulfates and oxidation that produces sulfuric acid. Regenerative processes involve a solid or liquid that reacts chemically with and removes sulfur oxides, the sorbent being regenerated in a separate step, usually with the production of sulfur. Sorbents at present in use in coal-combustion plants include alkalized alumina, magnesia, potassium bisulfite, and potassium formate.

As for *petroleum,* not only is there less oil than coal, it is also being consumed at a faster rate. Here again, with new regulations and a ten-fold increased demand since the beginning of the century, it has become necessary to desulfurize petroleum effectively. Originally, the petroleum industry used its hydrodesulfurization program only for the lighter petroleum fractions that constituted about 30 percent of all oil produced. At present, the heavier petroleum fractions, such as residual oil, are also being desulfurized.

Natural gas, obtained from petroleum production, is the fossil fuel in shortest supply and greatest demand. Cheapest, most convenient, and cleanest of the fossil fuels, it provides a third of the total energy and a fifth of the electricity the United States needs at present. However, we are using up our gas reserves faster than we are discovering new ones. To compensate for this, plans are under way to gasify coal, producing the same methane that is the basis of natural gas.

The supply of coal and petroleum is sufficient, in terms of present consumption levels, to meet immediate future needs (Table 16.2). If these substances continue to be used principally for their energy contents, and if they continue to supply the bulk of the global energy requirements, the time required to exhaust the world's

TABLE 16.1
United States energy sources in 1970

Source	Total energy (%)	Electric energy (%)
coal	20.1	55.0
oil	43.0	7.3
natural gas	32.8	19.3
water	3.8	15.1
nuclear	0.3	3.3

TABLE 16.2
Estimated fossil fuel consumption, demand, and recoverable resources[a]

Fuel	Coal (10^9 tons)	Petroleum (10^9 barrels)	Natural gas (10^{12} ft^3)
present yearly consumption	0.5	4.9	19.0
anticipated consumption in year 2000	0.9–3.0	7.3–16.4	34.8–55.7
anticipated cumulative demand for the next 30 years	27–42	195–308	855–1130
estimated recoverable resources	780	532 (crude oil) 600 (shale oil)	2400

[a] Data from G. A. Mills, H. R. Johnson, and H. Perry, *Environmental Science and Technology,* **5,** 30 (1971).

petroleum and coal resources would be about 100 to 200 years, perhaps 400 years at the most. However, the situation concerning natural gas is different in that the projected rates of development will not fulfill the projected need.

Two additional fossil fuel sources are the tar sands (heavy oil sands) and oil shales. The so-called *tar* or *heavy oil sands* are impregnated with what is essentially a heavy crude oil that is too viscous to permit recovery by natural flow into wells. The best known of such deposits and possibly the world's largest are in Alberta, where since 1967 a large-scale plant extracts 45,000 barrels of tar-sand oil daily. Other tar-sand resources are as yet almost unexploited. Because of the chemical similarity to crude oil, the oil from tar-sand is processed by the same refineries that process crude oil. *Oil shales* differ from tar or heavy oil sands in that their hydrocarbon contents are in a solid rather than a viscous-liquid form. The best-known oil shale deposits are in Colorado, Utah, and Wyoming. From these deposits a solid fuel called *kerogen* is obtained. Unlike tar-sand oil, it differs considerably from crude oil in chemical content, includes objectionable nitrogen and other impurities, and thus poses special problems in refining. Hence, the oil shales appear to be more promising as a resource of raw materials for the chemical industry than as a major source of energy. However, the disposal of tons of extracted shale rock could be a serious environmental problem because it would damage wildlife habitats.

Synthetic fuels from fossil energy Several processes for the preparation of synthetic natural gas by gasifying coal have been investigated. In one such process developed by the U.S. Bureau of Mines, coal is reacted with steam and oxygen in a fluidized bed at about 600–1000 psi, to produce a mixture of CH_4, H_2, CO, H_2S, and CO_2. After the CO_2 and H_2S are removed, the CO and H_2 are reacted to form additional methane. The use of air in place of oxygen to achieve technological and economical improvement in coal gasification is also under study. Underground gasification of coal and gasification of oil shale offer additional possibilities for gas supply if new technical advances can be made. It should, however, be emphasized that the price of synthetic pipeline gas will be too high compared to natural gas. Only if the price of natural gas increases enough or that of coal gasification decreases sufficiently can synthetic gas from coal become a commercial reality.

It is also possible to convert coal to liquid fuels, including high-quality gasoline. In fact, during World War II the German chemists developed two methods for the production of synthetic, liquid hydrocarbon fuels. One of these, the *Fischer-Tropsch process,* uses the catalytic reaction of carbon monoxide with hydrogen:

$$2CO + H_2 \longrightarrow (CH_2)_n + CO_2$$

The carbon monoxide is obtained by the incomplete combustion of coal in air. When the catalyst used is iron containing small amounts of copper and potassium the products obtained are similar to gasoline. Here copper and potassium increase the effectiveness of the iron catalyst and are called *catalyst promoters*. By altering the amounts of promoters, temperature, pressure, and so on, it is possible to obtain a wide distribution of products, ranging from low molecular mass compounds, such as methane, to high molecular mass compounds, such as waxes. With the other process, known as the *Bergius process,* a liquid fuel is obtained from the reaction of hydrogen with coal at elevated temperatures and high pressures over an iron and molybdenum oxide catalyst:

$$nC + (n + 1)H_2 \longrightarrow C_nH_{2n+2}$$

These two processes are suitable for the production of gasoline in the event of a serious depletion of our crude oil reserves or a war in which our normal supplies of crude oil are cut off. At the present time neither of these processes can compete with the more economically feasible use of the crude petroleum as a fuel source. However, the cost of refining gasoline from petroleum is spiralling upward, and, probably within the next two or three decades, it will be economically feasible to make gasoline synthetically. As a matter of fact, by that time it will as well be necessary to produce synthetic gasoline because of depletion of our petroleum resources.

QUESTIONS

16.9 What fossil fuels are available in your locality?

16.10 Is the electrical energy where you live produced by burning fossil fuels? If so, what are the pollution problems associated with the use of fossil fuels?

16.11 Which is a more efficient use of energy: burning coal to produce electricity that, in turn, is used to heat your home or burning gas to heat the home directly?

16.12 Suppose the coal available for power production in your town contains 2 percent sulfur. If the power plant burns 500 tons of coal a day, what is the maximum weight of
a. sulfur
b. sulfur dioxide
that could be released into your air environment each day? Can you suggest a method whereby the sulfur can be removed?

16.13 Describe two different methods by which synthetic fuels can be prepared from fossil energy.

16.14 What is your attitude toward using up the fossil fuels hastily?

SUGGESTED PROJECT 16.1

Energy consumption and your household (or cooking with gas)

Natural gas, oil, and electricity are the chief sources of energy used for heating, cooking, and lighting purposes in every household. Find out the monthly consumption of natural gas [in therms (Btu); 1 therm = 252 cal], of oil (in gallons), and of electricity (in kilowatt hours; 1 kWh = 860 kcal) in your home and then project the annual energy requirements for your home. Based on this estimate, determine the annual requirement of each of these energy sources for (1) your town, (2) your state, and (3) the whole country.

Gasoline provides the motive power for our automobiles. Find out the weekly consumption of gasoline (in gallons) by the automobile(s) used by you and your family and then project the annual gasoline requirements for your household. Based on this estimate determine the annual requirement of gasoline for (1) your town, (2) your state, and (3) the whole country.

SUGGESTED PROJECT 16.2

How to exploit a fuel source (or fueling around)

Suppose there is in your vicinity a fuel source that is a mixture of sand, water, and oil. Devise a scheme by which the oil can be separated from the mixture, for eventual use as a fuel. Following approval by your instructor test your scheme using artificially prepared mixtures. If you are living near an oil spill area, you may test your scheme using natural samples.

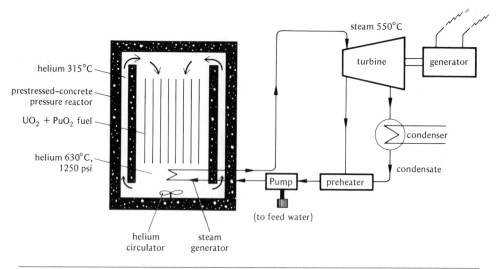

FIGURE 16.10

Schematic diagram of a gas-cooled fast breeder reactor. From the Proceedings of the National Academy of Sciences, Symposium on "Energy for the Future," 108th Annual Meeting in April 1971. (First published in August 1971.) Adapted with permission of Professor Manson Benedict and the National Academy of Sciences, Washington, D.C.

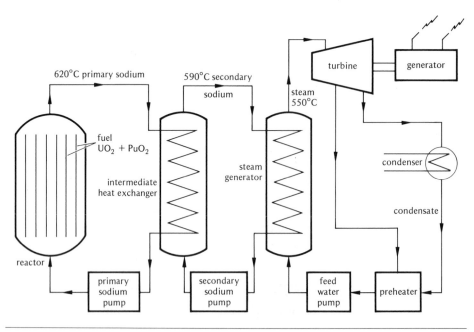

FIGURE 16.11

Schematic diagram of a liquid-metal-cooled, fast breeder reactor. From the Proceedings of the National Academy of Sciences Symposium on "Energy for the Future," 108th Annual Meeting in April 1971. (First published in August 1971.) Adapted with permission of Professor Manson Benedict and the National Academy of Sciences, Washington, D.C.

NUCLEAR ENERGY With depleting fossil energy resources, many of our energy specialists believe that sources of power for the future are to be sought among the mineral fuels and, above all, in nuclear energy. In Chapter 6 we learned of the release of energy from two contrasting nuclear processes, namely, *fission* and *fusion*. We also learned that the fission device, called the *nuclear reactor,* taps the large energy released when uranium atoms fission or split. The type of nuclear reactor used throughout the world today for electric power generation obtains most of its energy from slow neutron fission of the scarce uranium-235. In these reactors (see Fig. 6.12), ordinary water, under pressure and at temperatures up to 300°C, is used both as coolant to transport the heat released in fission and as moderator to slow down the fast neutrons initially produced in fission. The heat carried away by the circulating water is used to produce steam for a turbine generator, just as in a coal- or oil-fired plant. Although these water-cooled reactors barely tap 1 percent of the available fission energy, there are about 50 commercial nuclear power plants already operating in the United States and several are under construction.

However, our uranium-235 supply is limited. It is estimated that this supply will be exhausted by the end of the century if we continue to build only the types of nuclear power plants operating today. To obtain the full potential of nuclear energy we have to develop effective means for utilizing uranium-238 which constitutes 99.3 percent of natural uranium. The most promising type of reactor for this purpose is the *fast breeder reactor* in which uranium-238 is converted to plutonium-239, which then undergoes fission with fast neutrons:

$$^{238}_{92}U + ^{1}_{0}n \longrightarrow ^{239}_{94}Pu + 2^{0}_{-1}e$$
$$^{239}_{94}Pu + ^{1}_{0}n \longrightarrow ^{143}_{58}Ce + ^{94}_{36}Kr + 3^{1}_{0}n + energy$$

The breeder is not really new, for it was the world's first nuclear reactor to generate electric power as early as 1951. It demonstrated that nuclear reactors could produce more fuel than they consumed. At the present time two types of breeder reactors appear to be promising. These are the *gas-cooled fast breeder reactor* (Fig. 16.10) and the *liquid metal-cooled fast breeder reactor* (Fig. 16.11). Each type would use a mixture of plutonium oxide and natural uranium oxide as fuel, would have a thermal efficiency close to 40 percent, and would be capable of producing close to 1.5 g of new plutonium for every gram of plutonium consumed. This high breeding ratio gives these reactor types a tremendous advantage over other types of breeders. The gas-cooled breeder uses helium at 50–75 atm as coolant, whereas the liquid metal breeder uses molten sodium at low pressure.

It is estimated that the breeder reactor will provide the world with electrical energy for thousands of years, far beyond the capability of all fossil fuels now known or likely to be discovered. The main advantage of the breeder over the conventional water-cooled reactor is its greater efficiency. It converts more of the nuclear heat to electricity, at the same time producing more fuel than it consumes. Thus, there is less heat loss (thermal pollution) and less radioactive waste (radiation hazard). Further, it operates at much lower pressure, thus decreasing the chance of leakage of radioactive gases. For these reasons, the breeder appears to be, as President Nixon stated "our best hope today for meeting the nation's growing demand for economical clean energy."

However, there are some technical problems with breeders. For example, the metals (mostly steel) used in the reactor assembly swell and become brittle under bombardment by the fast neutrons necessary for the process. The relatively poor heat removal characteristic of helium gas, compared with liquid sodium, is a major draw-

(a)

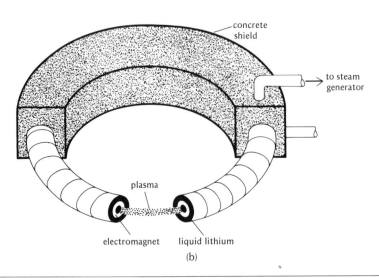

concrete
shield

to steam
generator

plasma

electromagnet   liquid lithium

(b)

FIGURE 16.12

(a) Scyllac. Photograph courtesy of the Los Alamos Scientific Laboratory, Los Alamos, New Mexico.
(b) Drawing of the torus (quartz doughnut) of the Scyllac, which eliminates open ends and loss of plasma.
The eventual working reactor for the fusion of atoms to produce energy will form a complete circle.
Adapted from the *National Geographic Magazine,* **142**(5), p. 671 (November 1972), with permission of
the National Geographic Society.

back of the gas-cooled breeder. The problems of handling sodium complicate the liquid metal breeder. Our efforts to cope with these difficulties require prolonged experimentation with ever-larger demonstration breeders before their commercial production is attempted. At the present time, the Atomic Energy Commission and the power industry, not only in the United States, but also in the Soviet Union, France, Germany, Japan, and Britain are attempting to develop the breeder for commercial use within this decade.

Instead of heavyweight nuclei releasing energy by splitting as in the fission of uranium-235 or plutonium-239, lightweight nuclei can combine or fuse together at extremely high temperature to release energy by *thermonuclear fusion* as in the sun and the stars. Unlike fission energy, the fusion process creates no residual radio-active by-products that will require processing and storing for thousands of years in order to preserve our environment even as we use its resources. However, thus far, controlled thermonuclear fusion is not technically feasible. To make controlled fusion work, one must heat a very tenuous ionized, or electrified, gas plasma to temperatures on the order of a 100 million degrees, hotter than the interior of the sun; contain the gas so that it cannot touch the walls of the vessel; and hold it in this condition long enough for a fusion reaction to take place — a few tenths or even hundredths of a second.

The most hopeful approaches to the achievement of controlled fusion are those that involve the fusion of two deuterium atoms,

$$\mathstrut_1^2D + \mathstrut_1^2D \longrightarrow \mathstrut_2^3He + \mathstrut_0^1n + 3.2 \text{ Mev}$$
$$\mathstrut_1^2D + \mathstrut_1^2D \longrightarrow \mathstrut_1^3T + \mathstrut_1^1H + 4.0 \text{ Mev}$$

or of one deuterium and one tritium atom,

$$\mathstrut_1^2D + \mathstrut_1^3T \longrightarrow \mathstrut_2^4He + \mathstrut_0^1n + 17.6 \text{ Mev}$$

(Here Mev denotes million electron volts; 1 Mev $\simeq 3.789 \times 10^{-14}$ cal.) The tritium can be "bred" in the reactor by neutrons produced in the fusion process:

$$\mathstrut_3^6Li + \mathstrut_0^1n \longrightarrow \mathstrut_2^4He + \mathstrut_1^3T + 4.8 \text{ Mev}$$

However, the scarcity of lithium renders this reaction a much less promising source of energy than the deuterium–deuterium reaction for which deuterium is abundant in our ocean waters. It is estimated that 1 m³ of water contains about 1.028×10^{25} atoms of deuterium having a mass of 34.4 g and a potential fusion energy equivalent to that generated by the combustion of 269 metric tons of coal or 1360 barrels of crude oil. Thus the energy released by the controlled thermonuclear fusion of the deuterium atoms present in the oceans would be about 50 million times that of the world's original supply of fossil fuels. Clearly, we may run out of certain types of fuels but we would not run out of energy if efforts to control fusion were to succeed before the turn of the century.

Where does development of the fusion reactor stand today? Obviously, if the atoms of deuterium and tritium were to be striking the container walls they would soon impart enough energy to the walls to melt and even vaporize them. What, for example, will contain the fuel gases at the fantastically high temperatures required? Scientists have been studying the use of magnetic walls to contain the fusing gases (see Chapter 6). Among the world's largest fusion experiments is the Scyllac at the Los Alamos Scientific Laboratory in New Mexico. It consists of a quartz doughnut called a *torus*, about 50 ft around and fully operational at the moment (Fig. 16.12). The deuterium plasma is formed inside the torus which is surrounded by heavy magnets. To these magnets, a maze of electrical cables carries discharges of 60,000 V from capacitor banks that store enormous electric charges. As the current charges the magnets, intense magnetic fields several thousand times that of earth's are generated. Within the electromagnet, electrical pulses heat and magnetically squeeze or compress the plasma, so that nuclei collide and fuse, releasing heat energy. However, the heating hitherto achieved in the Scyllac is only up to 10 million °K, well below the 100 million °K required for fusion. Moreover, even at 10 million °K, it has been possible to contain the plasma only for five-millionths of a second, again well below the two-hundredths of a second estimated for the occurrence of fusion reactions. However, in the Soviet Union, hopeful results have been reported with devices called Tokamak that can contain deuterium plasma one-millionth as dense as air at a 100 million °K for one-hundredth of a second.

Besides the magnetic walls, a special kind of light device, the *laser*, holds promise of harnessing useful fusion power by providing scientists with plasmas at higher temperatures and longer confinement times than ever achieved before. In a possible fusion power plant operation, the released heat energy will be absorbed by molten lithium and carried off to produce steam for generating electricity. Controlled fusion producing more energy than it consumes will probably be achieved before the end of this decade. Even then the technology for practical fusion power plants will probably require 20 or 30 more years of development. Hopefully, fusion may replace nuclear fission power long before the depletion of fissioning energy sources, because it does not produce radioactive wastes, the disposal of which may soon become our number one garbage problem. Can you imagine future generations carrying Geiger-Müller counters (see Chapter 6) to monitor and stand guard over some of our nuclear excreta for thousands of years?

QUESTIONS

16.15 Distinguish between fission and fusion. Which theoretically gives greater energy?

16.16 What are fast breeder reactors? How are they different from the conventional reactors at present used in nuclear power plants?

16.17 What are the major problems associated with achieving a controlled fusion reaction?

16.18 Justify the statement that "we may run out of certain types of fuels but we will not run out of energy."

16.19 Discuss the advantages and disadvantages of fusion power in relation to our present energy resources and environmental problems.

OTHER SOURCES OF ENERGY Hydroelectric energy Water reservoir energy represents the largest concentration of solar energy that is produced by any natural process. The sun's energy evaporates water from the ocean's surface, changing salty water to freshwater in the form of clouds, snow, and rain. The snow melts and rain falls, and as the freshwater runs back in the form of a river to the sea, we can harness its energy to turn a turbine to produce electricity. It is estimated that the total potential hydroelectric power of the world is about four times as large as total installed electric power capacity. If the world's hydroelectric power were fully developed, it would be of a magnitude comparable to the world's total present rate of energy consumption. However, with prime sites already harnessed and contributing 15 percent of our present electrical output, the United States nears full utilization of this clean energy resource. Important as they could be in presently underdeveloped parts of the world, conventional sources of hydroelectric power are erratically distributed, and over a period of time reservoirs clog up with silt and sediment carried by the dammed streams.

Tidal energy   The source of tidal energy is the gravitational force of the moon and the sun acting on our oceans. Tidal power is similar in all essential respects to hydroelectric power except that, whereas hydroelectric power is obtained from the energy of unidirectional stream flow, tidal electric power is obtained from the oscillatory flow of water in the filling and emptying of partially enclosed coastal basins during the semidiurnal rise and fall of the oceanic tides. This energy is now partially converted into tidal electric power in France by enclosing the Rance river estuary in Brittany with dams to create a difference in water level between the ocean and the estuary basin, and then using the waterflow, while the basin is either filling or emptying, to drive submarine-like, hydraulic turbine generators with reversible propellers. Besides this large installation, which produces 240,000 kW (1 kW = 8.6 $\times$ 10^5 cal) of power, there has been little exploitation of the world's tidal sites for power production, perhaps because surveys by energy experts show that all the exploitable tidal power of the world together approximate less than 1 percent of the world's exploitable, conventional hydroelectric power. Thus, although tidal energy may provide appreciable amounts of electricity with rather small environmental intrusion at specific locations, it will never contribute an important fraction of our future energy demands. However, like solar energy, this source will continue essentially forever.

Wind energy   The sun's energy is also responsible for some of our atmospheric phenomena, such as causing winds to blow. We have all heard about the historic role of windmills and Yankee clippers in man's use of wind energy. However, there are few places in the world where the wind is strong enough and steady enough to warrant its harnessing for the large-scale production of power. Nevertheless, it has been suggested that large windmill generators with rotors of glass-reinforced plastic be floated on the open seas where they would catch the strong prevailing winds and transform their energy to electric power.

Direct solar energy   Solar energy is daily renewable and enormous in amount. Other than the natural processes of photosynthesis and maintenance of the atmospheric, hydrologic, and oceanic circulations, at present the principal uses of solar energy appear to be small-scale and specialized in nature. These include water and house heating, air conditioning, distillation, solar furnaces, solar cookery, and numerous thermoelectric, photoelectric, and other means of electrical conversion or storage of solar energy. The usual method exploits the fact that a black surface absorbs almost all the solar energy falling on it. Usually water or a bed of rocks serves as a storage medium for the heat until needed, and glass is used as a shield to reduce heat loss by radiation and convection.

Larger amounts of energy have been obtained by building collectors that follow the sun in its path across the sky and higher temperatures by using lenses and mirrors to concentrate the sun's energy on a small area. For example, near Odeillo in the French Pyrenees, 63 mirror panels, each 18 ft in height, slowly swivel with the moving sun and reflect its rays onto a 148-ft high, parabolic mirror that focuses the light on the world's largest solar furnace creating temperatures of 3500°C. New coatings that are superior to glass in increasing the greenhouse effect (solar energy gets in but heat does not escape) may put an end to conventional assumptions about the efficiency of solar collectors. However, solar energy is variable in that on cloudy days not much gets through, none at night, and in winter less is available than in summer. To overcome the loss due to reflection and absorption by atmosphere and clouds, it has been suggested by the American engineer Peter Glaser that large solar collectors in earth orbit be used. Collectors covered with solar cells, such as those already used by many spacecraft, would convert sunlight to electric power and then beam to earth on microwaves.* Certainly, the project will depend on development of a reliable space-shuttle system to ferry the solar panels into orbit and on much cheaper solar cells.

Equally exciting is the recent proposal of the American energy specialists Meinel and Meinel. They envisage a solar farm (Fig. 16.13) in which ridged lenses tilted toward the sun train rays down through a slot into a glass pipe, the mirrored inside surface of which reflects the heat onto a coated inner steel tube that circulates nitrogen gas. Reaching temperatures of 550°C, the gas flows to tanks of molten salts capable of storing the heat. Steam heated by the salts can be used to drive conventional turbine generator to make electricity. It is estimated that the above system will have a solar-to-electrical conversion efficiency of at least 25 percent, making the generation of electricity at even large central power plants a possibility.

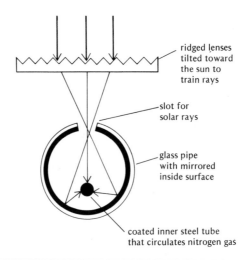

FIGURE 16.13

A proposed solar farm. Adapted from the *National Geographic Magazine,* **142** (5), p. 677 (November 1972), with permission of the National Geographic Society.

If solar energy is, indeed, harnessed through large electric power plants, there will only be some thermal pollution, mostly aquatic due to discharging waste heat to local water bodies. However, if the energy is collected in one geographic area and then transmitted elsewhere, the natural energy balance may be sufficiently upset to influence circulation patterns. Further, the use of solar-collecting satellites could cause climatic perturbations because large amounts of energy that would

---

* Microwaves are electromagnetic radiations having a wavelength in the approximate range from 1×10^{-1} to 10^2 cm, the region between infrared and shortwave radio wavelengths (see Fig. 3.16).

have been reflected, or would not have struck the earth at all, would thus be collected and eventually dissipated on earth as heat. Nevertheless, some energy specialists and environmentalists as well believe that the thermal impact of solar energy will be much smaller than that of any other alternative to meet our future energy demand.

Geothermal energy Derived from the Greek *geo* and *therme*, meaning "earth-heat," geothermal refers to the heat from the earth's interior. One of the energy inputs into the earth's surface environment consists of the heat conducted from the earth's interior as a result of the increasing temperature with depth. Another energy input consists of the heat convected to the earth's surface by volcanoes and hot springs. In special geological situations in volcanic areas, underground water is trapped in porous or fractured rocks and becomes superheated from volcanic heat. Wells drilled into such reservoirs of superheated water (1000–1100°C) permit the steam to be conducted to the surface where it can be used as an energy source for a conventional, steam–electric power plant.

The earliest use of geothermal energy for power was in Italy in 1904. Large geothermal electric power plants have only recently been built and others are in process of construction. In the United States we have a large geothermal facility at the geysers in northern California, which produces about 200 megawatts (abbreviated MW; 1MW = 8.6×10^8 cal) of power compared to a world production (chiefly in Italy, New Zealand, and Japan) of 1200 MW at the present time. It is estimated that the ultimate magnitude of geothermal power production will probably be in the tens of thousands of megawatts. However, this projected amount represents only a small fraction of the world's total energy requirements and can last for only a limited period of time. Nevertheless, geothermal energy is capable of sustaining a large number of small power plants in a limited number of localities for a limited amount of time, provided new reservoirs of geothermal steam can soon be located.

Geothermal power is relatively cheap, clean, and almost pollution free. The principal environmental problem with geothermal energy may be the possibility of surface subsidence if water is not pumped back underground after being passed through the turbines. Additionally, the water from underground reservoirs is laden with mineral salts that could constitute a nuisance if not returned underground. However, the possibility exists that the waste heat from the generation process at seaside locations can be used to desalt seawater.

QUESTIONS

16.20 Why do you think solar energy might be unpromising for large-scale electric power generation?

16.21 What proposals have been made for the direct utilization of solar energy on a large scale?

16.22 List some of the principal uses of solar energy at the present time.

16.23 How is the sun's energy responsible for
   a. water reservoir energy?
   b. wind energy?
   c. tidal energy?

16.24 In what respects does tidal power differ from hydroelectric power?

16.25 What is meant by geothermal energy? How is this source of energy tapped for power production?

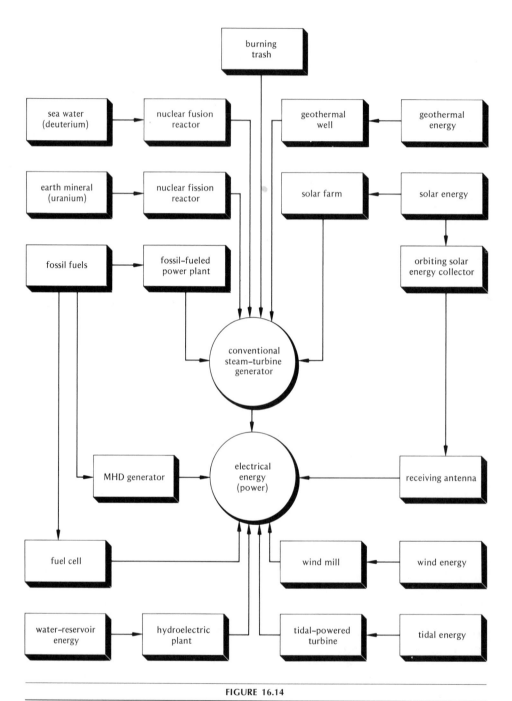

FIGURE 16.14

Nature's energy sources and man's conversion methods. Adapted from the *National Geographic Magazine,* **142** (5), p. 666 (November 1972), with permission of the National Geographic Society.

ENVIRONMENTAL PROBLEMS

Nature's energy sources and man's conversion methods are summarized in Fig. 16.14. The production, processing, and utilization of fossil fuels cause the most, if not all, of our environmental problems. The most significant of these pollution problems, and the impact on land, water, and air are summarized in Table 16.3. In the United States, about 2 million acres of land are disturbed each year by surface mining of coal. When proper land reclamation is not practiced, water pollution from acid mine drainage and silt damage occur. Underground coal mining can cause subsidence unless the mining systems are designed to prevent deterioration and failure of abandoned mine pillars. By the washing of coal only to improve its quality, about 90 million tons of waste are produced annually. At times, these wastes ignite creating air pollution; also, rainwater leaches salts and acid from the wastes to contaminate nearby streams. Utilization of coal also produces solid waste in the form of ash and slag. About 30 million tons of these materials are collected each year; an estimated 8 million tons are discharged into the atmosphere.

Nearly 80 percent of all air pollution in the United States is caused by fossil fuel combustion. About 95 percent of all sulfur oxides, 85 percent of all nitrogen oxides, and over half of the carbon monoxide, hydrocarbons, and particulate matter are produced by fossil fuel use. The production of oxides of nitrogen or carbon monoxide does not vary greatly for any of the fossil fuels in use. Natural gas produces no sulfur oxide emissions, but the use of coal and residual oil produces 74 percent of all the oxides of sulfur emitted into the air. Because of the relatively high sulfur content of coal and the fact that seven times as much coal as oil is used in electricity-generating plants, coal accounts for nearly two-thirds of the sulfur oxides emitted into the atmosphere. Residual oil is high in sulfur and difficult to burn efficiently. Industrial use of oil tends to contribute larger amounts of carbon monoxide, hydro-

	TABLE 16.3		
	Environmental impacts of fossil energy sources		
Fossil energy source	Impacts on resources		
	Land	Water	Air
coal	disturbed land from production; solid wastes from processing; ash and slag disposal problems in utilization	acid mine drainage from production; increased water temperatures by utilization	sulfur oxides, nitrogen oxides, particulate matter, etc., during utilization
oil		oil spills, transfer, and brines during production; increased water temperatures from utilization	carbon monoxide, nitrogen oxides, hydrocarbons, etc., during utilization
natural gas		increased water temperatures	

carbons, and oxides of nitrogen than do household and commercial uses, which consume about 25 percent of the fuel oil. However, the largest use of oil is in the transportation industry, to which approximately one-half of all the air pollution in our country (see Chapter 15) can be attributed. Clearly, the management of fossil fuels is critical not only for future energy resources but also for the minimization of environmental problems.

Water problems that arise from increasing demand for energy are twofold: (1) *water quality,* which relates to individual energy sources, and (2) *water temperature,* which is common to use of all fuel commodities. Poor water quality, whether from chemical pollution or sedimentation, is a debasement of our environment resulting from both surface and underground mining for fuels. However, by far the most important water problem resulting from fuel use is thermal pollution. Over 80 percent of all thermal pollution arises from the generation of electricity. The amount of heat rejected to cooling water represents 45 percent of the heating power of the fuel used in the most efficient fuel plants, and 55 percent in nuclear plants. If the projected electricity demands of the future are accurate the demand for fresh cooling water will be larger than its supply. Not only will the river or ocean water used for cooling become warmer, but so will the air.

Wastes resulting from nuclear generation of electricity involve only small tonnages of materials but have a very great potential for environmental damage for long periods because of their radioactivity. Most radioactive waste is generated in the fuel-processing plants and is in liquid or slurry form. It is stored in steel tanks for a year or more to dissipate thermal energy before attempting more permanent storage or disposal. The principal solid wastes are radioactive trash and reactor and machinery parts that have acquired an induced radioactivity from neutron bombardment. The present practices for the ultimate storage of high-level liquid or slurry wastes are considered satisfactory. However, the disposal of solid wastes by burial in trenches 10–15 ft deep is considered unreliable by some environmentalists because these trenches are within the domain of circulating soil moisture, some of which returns to the surface by evaporation and plant transpiration, and some of which descends to the water table. Thus, these wastes are not being isolated from the biological environment at present, and it is questionable to what extent the same practices can be continued as nuclear plants become more numerous. Clearly, an essential requirement for a nuclear power industry, based on fissionable fuel, is a system for the safe management and disposal of radioactive waste that is hazardous to the health and genetic security of the nation.

Energy from controlled incineration of nonradioactive trash may help alleviate part of our future energy problems and environmental problems as well. In the past the so-called urban and agricultural refuse, which is mostly organic matter, has been used as landfill or incinerated causing significant air pollution. Techniques have now been developed to recover energy by controlled incineration, by pyrolysis to make gas and oil, or by hydrogenation to produce a low-sulfur oil. Recent experiments by the U.S. Bureau of Mines have shown that heating a ton of garbage to 380°C for 20 min in the presence of carbon monoxide and water under high pressure produced over two barrels of low-sulfur oil. Another experiment uses the hot combustion gases from incineration to drive a gas turbine that turns the generator. The only effluents from such a plant are warm, filtered combustion gases and a solid residue much smaller than the original waste. One problem with installations such as this is that the combustion products of the wide variety of plastics in municipal garbage, some of which may not be removed by the usual filtering, may have serious effects on human health (see Chapter 10). In the United States alone, it is estimated that nearly 3 billion tons of garbage are produced each year. If this garbage is burned in power plants it would produce more than half the electricity we are now generating.

The notion of turning our wastes into fuel is not a new one. Many countries burn their wastes for cooking and heat. For many years the better part of the garbage from the city of Paris has been burned to produce steam for heating and for electric power plants. In the United States, only recently the city of St. Louis began burning shredded trash with pulverized coal to make electricity. It is estimated that a fifth of the municipal waste in St. Louis is being converted each day into 300 tons of odorless, clean burning fuel which has a fuel value of 150 tons of high-grade coal. If all of America's municipal wastes were used in this way, each year we could recycle a considerable part of our wasted energy resources.

In any conversion of energy from one form to another, there is a loss of energy as heat. This heat will have to be retrieved and used as a resource if we are to control and abate thermal pollution in the future. Hopefully, we can find ways for huge amounts of heat to be used in systems not even thought of at this time. Otherwise, we may not survive the heat that is released in producing and using ever-increasing quantities of energy.

QUESTIONS

16.26 What are the environmental problems caused by the use of fossil fuels?
16.27 In what environmental aspects is natural gas superior to coal and oil? If you have adequate supplies of all these fossil fuels, which would you prefer as the energy source?
16.28 Discuss air pollution versus fossil fuel combustion.
16.29 What do you suggest that we do in the future with our
   a. municipal wastes?
   b. agricultural wastes?
   c. industrial wastes?
   d. radioactive wastes?
16.30 Suggest possible ways and means for retrieval and use of waste heat so that thermal pollution may be abated.

SUGGESTED PROJECT 16.3

Conservation of energy (or resourcefulness for fun and survival)

Present and future demand for energy is massive. Every nation must create a political and economic environment that will encourage energy conservation. Based on the available energy resources, the United States has the following options:

1. Use more coal, both directly and as a source of synthetic oil and gas.
2. Advance the development of shale oil as a raw material for the chemical industries.
3. Speed the pace of construction of fission power plants based on present nuclear technology.
4. Speed the pace of nuclear fusion program and develop new nuclear technology.

5. Develop new resources such as water energy (see Chapter 5), tidal energy, wind energy, solar energy, and geothermal energy.
6. Implement fuel economy programs for automobile and airplane transportation.
7. Reduce the amount of energy used for heating, air conditioning, and lighting residential and commercial buildings.
8. Rely more on the animal frame than on machines for performing work.

Examine your locality for its energy needs and recommend options that will control energy consumption and environmental pollution at the same time. Support your recommendation with statistical data and information. Make suggestions for resolving conflicts, if any, between environmental goals and energy resource development.

APPENDIXES

Appendix A

Table of atomic masses (based on carbon-12)[a]

Symbol	Element	Atomic number	Atomic mass	Symbol	Element	Atomic number	Atomic mass
Ac	actinium	89		Hg	mercury	80	200.59
Al	aluminum	13	26.9815	Mo	molybdenum	42	95.94
Am	americium	95		Nd	neodymium	60	144.24
Sb	antimony	51	121.75	Ne	neon	10	20.179
Ar	argon	18	39.948	Np	neptunium	93	237.0482
As	arsenic	33	74.9216	Ni	nickel	28	58.71
At	astatine	85		Nb	niobium	41	92.9064
Ba	barium	56	137.34	N	nitrogen	7	14.0067
Bk	berkelium	97		No	nobelium	102	
Be	beryllium	4	9.01218	Os	osmium	76	190.2
Bi	bismuth	83	208.9806	O	oxygen	8	15.9994
B	boron	5	10.81	Pd	palladium	46	106.4
Br	bromine	35	79.904	P	phosphorus	15	30.9738
Cd	cadmium	48	112.40	Pt	platinum	78	195.09
Ca	calcium	20	40.08	Pu	plutonium	94	
Cf	californium	98		Po	polonium	84	
C	carbon	6	12.011	K	potassium	19	39.102
Ce	cerium	58	140.12	Pr	praseodymium	59	140.0977
Cs	cesium	55	132.9055	Pm	promethium	61	
Cl	chlorine	17	35.453	Pa	protactinium	91	231.0359
Cr	chromium	24	51.996	Ra	radium	88	226.0254
Co	cobalt	27	58.9332	Rn	radon	86	
Cu	copper	29	63.546	Re	rhenium	75	186.2
Cm	curium	96		Rh	rhodium	45	102.9055
Dy	dysprosium	66	162.50	Rb	rubidium	37	85.4678
Es	einsteinium	99		Ru	ruthenium	44	101.07
Er	erbium	68	167.26	Sm	samarium	62	150.4
Eu	europium	63	151.96	Sc	scandium	21	44.9559
Fm	fermium	100		Se	selenium	34	78.96
F	fluorine	9	18.9984	Si	silicon	14	28.086
Fr	francium	87		Ag	silver	47	107.868
Gd	gadolinium	64	157.25	Na	sodium	11	22.9898
Ga	gallium	31	69.72	Sr	strontium	38	87.62
Ge	germanium	32	72.59	S	sulfur	16	32.06
Au	gold	79	196.9665	Ta	tantalum	73	180.9479
Hf	hafnium	72	178.49	Tc	technetium	43	98.9062
He	helium	2	4.00260	Te	tellurium	52	127.60
Ho	holmium	67	164.9303	Tb	terbium	65	158.9254
H	hydrogen	1	1.0080	Tl	thallium	81	204.37
In	indium	49	114.82	Th	thorium	90	232.0381
I	iodine	53	126.9045	Tm	thulium	69	168.9342
Ir	iridium	77	192.22	Sn	tin	50	118.69
Fe	iron	26	55.847	Ti	titanium	22	47.90
Kr	krypton	36	83.80	W	tungsten	74	183.85
La	lanthanum	57	138.9055	U	uranium	92	238.029
Lr	lawrencium	103		V	vanadium	23	50.9414
Pb	lead	82	207.2	Xe	xenon	54	131.30
Li	lithium	3	6.941	Yb	ytterbium	70	173.04
Lu	lutetium	71	174.97	Y	yttrium	39	88.9059
Mg	magnesium	12	24.305	Zn	zinc	30	65.37
Mn	manganese	25	54.9380	Zr	zirconium	40	91.22
Md	mendelevium	101					

[a] Adopted by the International Union of Pure and Applied Chemistry, June 1969.

Appendix B

Electronic configurations of the elements

Atomic number	Element symbol	1s	2s	2p	3s	3p	3d	4s	4p	4d	4f	5s	5p	5d	5f
1	H	1													
2	He	2													
3	Li	2	1												
4	Be	2	2												
5	B	2	2	1											
6	C	2	2	2											
7	N	2	2	3											
8	O	2	2	4											
9	F	2	2	5											
10	Ne	2	2	6											
11	Na	2	2	6	1										
12	Mg	2	2	6	2										
13	Al	2	2	6	2	1									
14	Si	2	2	6	2	2									
15	P	2	2	6	2	3									
16	S	2	2	6	2	4									
17	Cl	2	2	6	2	5									
18	Ar	2	2	6	2	6									
19	K	2	2	6	2	6		1							
20	Ca	2	2	6	2	6		2							
21	Sc	2	2	6	2	6	1	2							
22	Ti	2	2	6	2	6	2	2							
23	V	2	2	6	2	6	3	2							
24	Cr	2	2	6	2	6	5	1							
25	Mn	2	2	6	2	6	5	2							
26	Fe	2	2	6	2	6	6	2							
27	Co	2	2	6	2	6	7	2							
28	Ni	2	2	6	2	6	8	2							
29	Cu	2	2	6	2	6	10	1							
30	Zn	2	2	6	2	6	10	2							
31	Ga	2	2	6	2	6	10	2	1						
32	Ge	2	2	6	2	6	10	2	2						
33	As	2	2	6	2	6	10	2	3						
34	Se	2	2	6	2	6	10	2	4						
35	Br	2	2	6	2	6	10	2	5						
36	Kr	2	2	6	2	6	10	2	6						
37	Rb	2	2	6	2	6	10	2	6			1			
38	Sr	2	2	6	2	6	10	2	6			2			
39	Y	2	2	6	2	6	10	2	6	1		2			
40	Zr	2	2	6	2	6	10	2	6	2		2			
41	Nb	2	2	6	2	6	10	2	6	4		1			
42	Mo	2	2	6	2	6	10	2	6	5		1			
43	Tc	2	2	6	2	6	10	2	6	6		1			
44	Ru	2	2	6	2	6	10	2	6	7		1			
45	Rh	2	2	6	2	6	10	2	6	8		1			
46	Pd	2	2	6	2	6	10	2	6	10					
47	Ag	2	2	6	2	6	10	2	6	10		1			
48	Cd	2	2	6	2	6	10	2	6	10		2			
49	In	2	2	6	2	6	10	2	6	10		2	1		
50	Sn	2	2	6	2	6	10	2	6	10		2	2		

Atomic number	Element symbol	1	2	3	4s	4p	4d	4f	5s	5p	5d	5f	6s	6p	6d	6f	7s
51	Sb	2	8	18	2	6	10		2	3							
52	Te	2	8	18	2	6	10		2	4							
53	I	2	8	18	2	6	10		2	5							
54	Xe	2	8	18	2	6	10		2	6							
55	Cs	2	8	18	2	6	10		2	6			1				
56	Ba	2	8	18	2	6	10		2	6			2				
57	La	2	8	18	2	6	10		2	6	1		2				
58	Ce	2	8	18	2	6	10	2	2	6			2				
59	Pr	2	8	18	2	6	10	3	2	6			2				
60	Nd	2	8	18	2	6	10	4	2	6			2				
61	Pm	2	8	18	2	6	10	5	2	6			2				
62	Sm	2	8	18	2	6	10	6	2	6			2				
63	Eu	2	8	18	2	6	10	7	2	6			2				
64	Gd	2	8	18	2	6	10	7	2	6	1		2				
65	Tb	2	8	18	2	6	10	9	2	6			2				
66	Dy	2	8	18	2	6	10	10	2	6			2				
67	Ho	2	8	18	2	6	10	11	2	6			2				
68	Er	2	8	18	2	6	10	12	2	6			2				
69	Tm	2	8	18	2	6	10	13	2	6			2				
70	Yb	2	8	18	2	6	10	14	2	6			2				
71	Lu	2	8	18	2	6	10	14	2	6	1		2				
72	Hf	2	8	18	2	6	10	14	2	6	2		2				
73	Ta	2	8	18	2	6	10	14	2	6	3		2				
74	W	2	8	18	2	6	10	14	2	6	4		2				
75	Re	2	8	18	2	6	10	14	2	6	5		2				
76	Os	2	8	18	2	6	10	14	2	6	6		2				
77	Ir	2	8	18	2	6	10	14	2	6	7		2				
78	Pt	2	8	18	2	6	10	14	2	6	9		1				
79	Au	2	8	18	2	6	10	14	2	6	10		1				
80	Hg	2	8	18	2	6	10	14	2	6	10		2				
81	Tl	2	8	18	2	6	10	14	2	6	10		2	1			
82	Pb	2	8	18	2	6	10	14	2	6	10		2	2			
83	Bi	2	8	18	2	6	10	14	2	6	10		2	3			
84	Po	2	8	18	2	6	10	14	2	6	10		2	4			
85	At	2	8	18	2	6	10	14	2	6	10		2	5			
86	Rn	2	8	18	2	6	10	14	2	6	10		2	6			
87	Fr	2	8	18	2	6	10	14	2	6	10		2	6			1
88	Ra	2	8	18	2	6	10	14	2	6	10		2	6			2
89	Ac	2	8	18	2	6	10	14	2	6	10		2	6	1		2
90	Th	2	8	18	2	6	10	14	2	6	10		2	6	2		2
91	Pa	2	8	18	2	6	10	14	2	6	10	2	2	6	1		2
92	U	2	8	18	2	6	10	14	2	6	10	3	2	6	1		2
93	Np	2	8	18	2	6	10	14	2	6	10	5	2	6			2
94	Pu	2	8	18	2	6	10	14	2	6	10	6	2	6			2
95	Am	2	8	18	2	6	10	14	2	6	10	7	2	6			2
96	Cm	2	8	18	2	6	10	14	2	6	10	7	2	6	1		2
97	Bk	2	8	18	2	6	10	14	2	6	10	8	2	6	1		2
98	Cf	2	8	18	2	6	10	14	2	6	10	10	2	6			2
99	Es	2	8	18	2	6	10	14	2	6	10	11	2	6			2
100	Fm	2	8	18	2	6	10	14	2	6	10	12	2	6			2
101	Md	2	8	18	2	6	10	14	2	6	10	13	2	6			2
102	No	2	8	18	2	6	10	14	2	6	10	14	2	6			2
103	Lr	2	8	18	2	6	10	14	2	6	10	14	2	6	1		2

Appendix C

CHAPTER 2

2.12: 50 mi
2.13: 55 yd
2.14: 758 ml
2.15: 30¢
2.16: $500,000
2.17: 3 ppm
2.18: 1 ppm

CHAPTER 3

3.17: 12.01 amu
     14.01 amu
     16.00 amu
3.18: 28.0 g/mole N_2
      16.04 g/mole CH_4
      58.4 g/mole NaCl
      78.1 g/mole C_6H_6
      342 g/mole $C_{12}H_{22}O_{11}$
3.26: a. $-182.97°C$
      b. $90.18°K$
3.27: 1945.4°F
3.31: 100 cal
3.32: 0.2 cal/g °C
3.41: 2.5×10^{14} cal

CHAPTER 7

7.5: C, 75%
     H, 25%
7.6: CH_4
7.7: HO
7.8: H, 3.1%
     O, 65.3%
     P, 31.6%
7.9: $C_6H_{12}O_6$
7.22: 47 g

CHAPTER 10

10.18: 0.36 mole
10.21: 100

CHAPTER 15

15.6: N_2, 80.9 mole%
      O_2, 19.06 mole%

CHAPTER 16

16.12: a. 10 tons
       b. 20 tons

Index

451